Machine Learning for Subsurface Characterization

Machine Learning for Subsurface Characterization

Siddharth Misra

Hao Li

Jiabo He

Gulf Professional Publishing is an imprint of Elsevier
50 Hampshire Street, 5th Floor, Cambridge, MA 02139, United States
The Boulevard, Langford Lane, Kidlington, Oxford, OX5 1GB, United Kingdom

Library of Congress Cataloging-in-Publication Data
A catalog record for this book is available from the Library of Congress

British Library Cataloguing-in-Publication Data
A catalogue record for this book is available from the British Library

ISBN: 978-0-12-817736-5

For information on all Gulf Professional publications
visit our website at https://www.elsevier.com/books-and-journals

Publisher: Brian Romer
Acquisition Editor: Katie Hammon
Editorial Project Manager: Aleksandra Packowska
Production Project Manager: Selvaraj Raviraj
Cover Designer: Mark Rogers

Typeset by SPi Global, India

Dedication

This book is dedicated to our parents: Rabindra, Surekha, Tao, Jianying, Weidong, and Beihong.

We are grateful to our parents for their unconditional love, sacrifices, and encouragement. They were always there to hold our hands and support us through the most difficult times.

We thank our parents for making our lives so beautiful!

Contents

Contributors

Numbers in parenthesis indicate the pages on which the authors' contributions begin.

Pritesh Bhoumick (39), PricewaterhouseCoopers (PwC), Houston, TX, United States

Aditya Chakravarty (39), Harold Vance Department of Petroleum Engineering, Texas A&M University, College Station, TX, United States

Eliza Ganguly (315), Harold Vance Department of Petroleum Engineering, Texas A&M University, College Station, TX, United States

Yifu Han (339,369), Schlumberger Technology Company, Beijing, China

Jiabo He (65,103,129), School of Computing and Information Systems, University of Melbourne, Parkville, VIC, Australia

Hao Li (129,157,183,219,243), The University of Oklahoma, Norman, OK, United States

Siddharth Misra (1,39,65,103,129,157,183,219,243,289,315,339,369), Harold Vance Department of Petroleum Engineering, Texas A&M University, College Station, TX, United States

Oghenekaro Osogba (1), Texas A&M University, College Station, TX, United States

Mark Powers (1), The University of Oklahoma, Norman, OK, United States

Chandra S. Rai (39), The University of Oklahoma, Norman, OK, United States

Pratiksha Tathed (339), BP, Houston, TX, United States

Yaokun Wu (289,315), Texas A&M University, College Station, TX, United States

Preface

In Gartner's list of top 10 strategic technology trends for 2019, "artificial intelligence (AI)–driven development" is placed at the very top. In MIT Technology Review's 10 breakthrough technologies for 2019, curated by Bill Gates, AI-driven automation and AI assistants are mentioned as revolutionary innovations. Terms like artificial intelligence (AI), machine learning (ML), deep learning (DL), and big data have been used interchangeably—they are related but not the same.

Artificial intelligence (AI) is a branch of computer science focused on developing algorithms inspired by certain aspects of natural intelligence to perform tasks requiring human intelligence, such as visual perception, speech recognition, problem-solving, and reasoning [1]. AI is the grand goal, and machine learning and deep learning are some of the many techniques to achieve AI.

Machine learning (ML) is a subset of methods to achieve AI, wherein the focus is to develop algorithms that learns from large datasets, also referred as the "big data." ML algorithms build a data-driven model based on input data (combination of features and targets) with an objective of using the data-driven model to make predictions, recommendations, decisions, and various other intelligence-related tasks, without needing any explicitly programmed instructions/rules. Deep learning (DL) is a subset of machine learning methods that processes "big data" to construct numerous layers of abstraction that builds functional mapping of the features/attributes to the targets. The feature-target mappings learned by the DL algorithms can be used to make predictions, recommendations, decisions, and various other intelligence-related tasks. Machine learning (including deep learning methods) builds data-driven models that improve over time as the model is fed more and more data. Big data is an important component of machine learning and deep learning. Big data is defined as extremely large datasets that cannot be analyzed, searched, or interpreted using traditional data mining/processing methods.

In engineering and most other domains, when people say "AI," they really mean machine learning (includes deep learning) [2]. ML works by recognizing patterns using complex mathematical/statistical techniques and algorithms and many a times brute-force computing. Deep learning is subtly different from simple/traditional machine learning. DL methods do not require the manual step of extracting/engineering features to accomplish the learning task. DL instead requires us to feed large amounts of data to get reliable results. In addition, DL model development requires a high-performance computing to process huge data volumes at a reasonable speed. DL is a powerful pattern recognition method, but the DL approach severely limits explainability of its outcomes and interpretability of the DL models. DL heavily relies on data annotation/labeling quality. ML/DL method tends to be impressive when considering its statistical performance over many samples, but they can be highly erroneous in individual cases.

Over the last 7 years, incredible advances in machine learning have been made with the advent of deep neural networks that are trained on "'Big Data" using very fast GPUs. These advances have benefited from the accumulation of digitized data and ubiquitous deployment of robust sensor systems. In addition, there is a wealth of openly available technologies that make it simpler and cheaper to build and run machine learning algorithms. Many of the tools are easily accessible and inexpensive, for example, public clouds like Microsoft Azure and Amazon Web Services, allowing massive data crunching exercises without the need to buy tons of hardware [3]. These advances have led to the state of the art in computer vision and speech recognition, such that machines have now exceeded the powers of human sensory perception in certain areas [4].

Machine learning has ushered a whole new way of doing business by propelling progress in automation, sensor-based industrial monitoring, and algorithmic analysis of business processes. Now, computers can learn the tasks to assist humans rather than merely doing as they're told. AI as a research area has been around in computer science since the 1950s (including its subfields such as machine learning and deep learning). Recent boom in AI implementations and its popularity has been due to better algorithms leading to improved accuracy, faster GPUs providing large compute power, large datasets for training the ML and DL algorithms, easily accessible ML platforms for developing data-driven models, and cloud services providing easier access to computational resources [5].

Due to the proliferation of data and rapid advances in the predictive analytics, machine learning is attracting large financial investments. Venture capitalists funded 1028 AI-related startups last year. Technical conferences and workshops promising to explain AI and demonstrate the power of AI have become a common and widespread trend. The annual meeting of the World Economic Forum in Davos this year included close to 10 panels related to AI, for example, "Designing Your AI Strategy" and "Setting Rules for the AI Race" [6]. Any technology advancing at a fast pace and with such breathless enthusiasm should be brought under a thorough reality check.

Here are few tasks related to O&G upstream exploration and production that are suitable for ML/DL implementations:

- Detecting minute changes, variations, and patterns in high-dimensional datasets
- Finding similarity and dissimilarity among systems/processes at a granular level

- Fast decision-making by processing high-speed data flow generated from multiple sources/channels
- Develop data-driven models that improve over time to better represent the physical processes, systems, and phenomena
- Facilitate precision engineering and characterization, especially for diagnosis and insight generation
- Intelligently automate mundane, repetitive, low-risk tasks

Machine learning–driven success stories

With widespread adoption of sensing and larger quantities of digital data within reach, there are numerous potential applications for machine learning. When humans feed well-structured data, the ML algorithms can extract patterns, trends, and relationship to recommend the most appropriate procedure to accomplish a task. Companies like Amazon and Netflix invest heavily to train machine learning models to build robust recommendation engines for making relevant and accurate predictions that match the user's tastes/preferences. Alphabet's life sciences division Verily has been successful with assessment of specific disease risk and its progression to facilitate preventive measures. IBM Watson Health is discovering off-label uses of existing drugs, improving chronic disease management, and providing drug safety evaluation. ML improves speed, efficiency, and effectivity of discovery, as corroborated by many drug discovery projects. AlphaGo is one of the feats of machine learning. AlphaGo uses deep learning techniques that combine "supervised learning from human expert games and reinforcement learning from games of self-play."

Narrowly applied AI will be crucial in automation, preventive maintenance, and rapid decision-making. Even if AI gets the right answer only 90% of the time, the benefits of being able to instantly react to incoming data streams with certain accuracy will be extremely valuable [4]. Human beings have limited memory and cognitive resources for solving very complex problems. Additional help from ML models in terms of handling complexity will improve human productivity and efficiency. ML-based decision-making is not affected by human emotional and physical states. Compared with human specialist, ML can process volumes of information from varied fields in a short amount of time to help generate valuable insights and reliable forecast. In addition to this, machine learning models after robust training are immediately scalable, with the potential for simultaneous use by any number of users. When given continuous supply of data and sufficient time, ML methods can concurrently improve in both breadth and depth, unlike a human, for example, a ML interface can be an expert in geology, geophysics, and geological engineering, simultaneously [7]. ML can analyze large populations to identify large-scale patterns for developing holistic approaches. At the same time, ML methods can be also designed to aggregate data for precision/personalized applications. ML will enable automation of analytical activities such as segmentation, optimization, and predictive modeling. ML will help us reduce our efforts in repetitive, cumbersome, and unexciting tasks, availing us to take on higher-value tasks. ML is a tool for enhancing human capabilities instead of replacing humans. An important thing to remember is that all this requires human involvement, assessment, and feedback, without which it is difficult to develop reliable, consistent, and robust ML/DL models.

Challenges and precautions when using machine learning

ML algorithms are only as good as the data that go into them. A trained model fails on a data that is dissimilar to the training data. ML models are not suitable for edge/rare cases because the model cannot learn enough statistical information about such cases; as a result, the models produce unreliable results with high uncertainty for such cases. Poor data hygiene leads to "garbage in, garbage out" scenario. Consequently, a lot of effort is required to transform the messy, unstructured data into clean structured data suitable for being consumed by the ML model [8]. ML rely heavily on manual services for creating labels/targets/annotations and for data cleaning/preprocessing, following that additional services are required for structuring the data; all this makes the ML workflow slow, tedious, and time consuming.

ML systems are yet to figure out ways to accomplish unsupervised learning, to learn from very limited amount of data under limited computational resources, or to train without a lot of human intervention. DL workflow requires a huge amount of information and large computational resources to succeed at even basic tasks. ML tends to perform poorly in learning new concepts and extending that learning to new contexts. A major concern is the so-called "curse of dimensionality," where having too many features/attributes (high-dimensional data) and not enough observations/samples (small dataset) hinders the model development and performance. Also, several ML proponents ignore the complex challenges faced in the real world when taking a ML model from a research paper or a controlled study to an engineered product for real-world deployment [9]. ML practitioners have noticed that these methods generally pick up patterns and relationships that are inconsistent and do not honor logical reasoning. Such unexpected relationships and trends learned by these methods will ultimately invalidate the results during real-world deployment. Moreover, it has been demonstrated that it is easy to trick ML models to learn inconsistent patterns or to generate unreliable results. A branch of DL referred as adversarial attacks is dedicated to developing new ways of fooling deep learning techniques. ML-driven automated systems can be severely affected by such adversarial attacks. ML models are not suitable for edge/rare cases because the model cannot learn enough statistical information about such cases; as a result, the models produce unreliable results with high uncertainty for such cases. None of the recent ML research has shown a lot of progress in these areas.

A lot of the AI/ML hype originates from the extrapolation of current trends and recent successes. When vendors describe the built-in machine learning capabilities of their products, there are three common customer responses: (1) product capabilities are overhyped, (2) vendors do not know what they are talking about, and (3) customer is exhausted with yet another machine learning product. There is a perception that, when vendors are promoting their "AI platforms," these are repackaged versions of traditional business intelligence or analytical tools. The big data hype a decade back was very similarto the current AI/ML wave. That time, big data was the new business intelligence, and big data was projected to solve everything [2].

Another big challenge to the adoption of ML is whether the technical domain experts will trust and adopt the ML models. Domain experts have a limited understanding of the reasons due to which an ML system makes a certain recommendation, in other words explain how an answer or insight was produced by the ML system. From a domain expert's point of view, the ML predictions/results are too generic and nonspecific—lacking deep insights. Without the interpretability of the model and explainability of the results, there is some level of faith that must be put in ML-based strategies. The biggest barrier to machine learning in several industries is the culture that values a domain expert's intuition over data-driven solutions. Quite often, the problem lies in how AI, ML, and DL are portrayed, for example, when it is said "neural networks are inspired by neurons in the brain" or "convolutional neural networks are inspired by human visual processing system." It is not clear to anyone outside of the inner circle how to start applying the AI technologies. They should be made aware that the barriers to entry are quite low. AI, machine learning, and deep learning are not hard-to-grasp, science-fiction concepts but are based on mathematical, statistical, and computational foundations. ML and DL may be hard to understand but are indeed simple to implement (caution: implementation is easy, but the evaluation is very challenging).

ML methods work well when a complex task requiring human intelligence is broken into simpler less-intelligent, pattern recognition–type problems. For example, machine learning can fill missing words in a sentence and translate the sentences to different languages; however, these methods are far from deriving the concept or intent of a sentence. ML methods tend to be effective for narrowly focused tasks. For example, ML-assisted conversational chatbots can now perform goal-oriented conversations: setting up an appointment between two people, wherein the goal is limited to coordinating the calendars of two people.

Irrespective of successes and failures, efforts to infuse ML into organizations are spreading like wildfire and are a reality. ML implementations, design, and approaches are evolving too fast; the ML practitioners are having trouble staying abreast of leading ML practices. In a haste primarily driven by the fear of missing out, organizations are entering the ML race without sufficient planning. A premature adoption of ML adversely affects the outcomes, lending a bad name to machine learning. For avoiding such scenarios, Harvard Business Review (HBR) suggests taking a portfolio approach to truly harness machine learning. HBR recommends having a mix of projects, ones that have the potential to generate quick wins and long-term projects for end-to-end transformation of business processes. ML models perform the best after being exposed to large historical/real-time datasets and stringent evaluation for a certain duration of time, which ensures that the ML models are robust to edge cases and that they do not pick up inconsistent patterns.

Recommendations for harnessing the power of machine learning

1. When you master machine learning techniques, you can truly benefit from the ever-growing vast datasets available to you. ML is a great tool to have at your disposal, like computers, word processors, and mobile phones. As computing speeds are expected to double five times over the next 10 years, machine learning tools will serve as inexpensive tool to extract information and insights from the enormous troves of data.
2. When you plan on using ML techniques, ensure you have a large, high-quality dataset both to build the data-driven models and to test them. Also, you need to ensure that the dataset you are using for building the models should be available in the real world for ensuring a robust deployment of the ML models. It could be argued that the data is more important than the ML algorithms because ML algorithms are only as good as the data that go into them. For example, Google, Facebook, Netflix, and Amazon are leaders in ML applications not only because of their intelligent algorithms and skilled data scientists but also because of the high-quality digital data they have about people and products.
3. A vendor's demo of ML workflow may work well on the vendor's data; this does not mean that the vendor's ML workflow will work equally well when applied to your data. Even when you see great results with your data, the real-world deployment of the ML workflow will unearth severe limitations in the ML implementations. Nonetheless, your efforts to fix these challenges will make your ML implementations more robust.
4. Domain knowledge is a very important ingredient in building effective ML models. ML users should know the limitations of ML methods and when these methods can go wrong, or else, ML methods will learn relationships that are totally spurious or tend to get overtrained without us knowing about such gross errors. ML users should be aware that ML methods can pick up patterns and relationships that are inconsistent without any physical basis.
5. ML tools are very good at learning clearly defined tasks, like identifying people in photographs or accurately transcribing speech. ML tools currently cannot understand human motivations or draw nuanced conclusions. For now, ML methods work well when a complex task requiring human intelligence is broken into simpler less-intelligent, pattern recognition–type problems.

6. Human beings have limited memory, cannot visualize data in high dimensions, and have restricted cognitive resources for solving very complex problems. ML tools help us handle complexity leading to improved human productivity and efficiency. Notably, ML tools if designed properly can remove human bias from decision-making.
7. Large firms are using ML tools to solve large-scale, high-visibility business and engineering problems. Smaller firms should try to identify the neglected, mundane tasks and deploy ML to solve them. The hype around ML has made large firms to invest their energy on eye-catching, news-worthy, marketable tasks. Moreover, there have been massive extrapolations of current ML trends and successes toward many exciting yet superficial future scenarios. More useful applications of ML can only emerge when we try to solve mundane and "boring" applications, which may never get the limelight.

Concluding remarks

At the start, the field of genetics didn't have any understanding or even theory of DNA. Genetics in early days tried to answer simple, narrow tasks, such as "Why some people have black hair?". In the course of few decades, with the advancements in biology, chemistry, microscopy, and computations, now, we can sequence the whole human genome and understand physical basis of diseases and traits. In the same vein, the field of AI is slowly marching toward the grand vision of general intelligence, and ML/DL tools are few techniques helping us progress the field of AI by harnessing the power of big data [10]. Once 3d printing and virtual reality were in their hype phase. Both the technologies are now coming out of the Trough of Disillusionment (Gartner Hype Cycle) with real and useful applications. At the peak of hype, these technologies were touted to accomplish grand tasks, for which they were not ready. AI/ML technologies are in the hype cycle but will soon come out of the hype much stronger and more productive. While AI and its subsets are powerful tools capable of shaping a wide range of industries and the way we live, they are not the ultimate solution to the problems faced by us and our planet.

Siddharth Misra
Harold Vance Department of Petroleum Engineering, Texas A&M University, College Station, TX, United States

References

[1] https://www.wired.com/insights/2015/02/myth-busting-artificial-intelligence/.
[2] https://builttoadapt.io/why-the-ai-hype-train-is-already-off-the-rails-and-why-im-over-ai-already-e7314e972ef4.
[3] https://www.fico.com/blogs/analytics-optimization/hype-and-reality-in-machine-learning-artificial-intelligence/.

[4] https://www.datanami.com/2018/08/29/sorting-ai-hype-from-reality/.
[5] https://becominghuman.ai/hype-reality-of-artificial-intelligence-in-healthcare-a2be4f3452cd.
[6] https://www.trustradius.com/buyer-blog/ai-machine-learning-buzzwords.
[7] https://www.information-age.com/ai-hype-reality-healthcare-123468637/.
[8] https://www.statnews.com/2018/12/06/artificial-intelligence-pharma-health-care/.
[9] https://www.forbes.com/sites/valleyvoices/2016/11/16/cutting-through-the-machine-learning-hype/#691b61254653.
[10] https://readwrite.com/2017/06/05/understanding-reality-artificial-intelligence-dl1/.

Further reading

[11] https://towardsdatascience.com/beyond-the-hype-the-value-of-machine-learning-and-ai-artificial-intelligence-for-businesses-892128f12dd7.

Acknowledgment

Writing a book is a long journey, especially when writing the first one. My wife, Swati, was instrumental in getting this book to the finish line. She provided constant support, encouragement, tasty snacks, and energy drinks while I was absorbed in writing the book. I thank my loving family members, Mom, Dad, Samarth, Dr. Ganesh, and Ms. Swagatika, who gave me good company and checked on the status of the book as they cheered from the sidelines. Friends in Texas, Oklahoma, Colorado, California, Washington, and India took away the stress of writing the book through their humor, love, and all the exciting activities we did together over the last 1 year. Special thanks to the professors in Texas A&M, University of Oklahoma, and University of Texas, who put aside their Football rivalries to provide me much-needed guidance to create a great book that can stand the test of time.

I would like to express my heartfelt gratitude to Dr. Carlos Torres-Verdin, my Ph.D. advisor, who introduced me to the world of research and taught me to strive for perfection. I am also grateful to Dr. Chandra Rai and Dr. Carl Sondergeld, who are legends in petrophysics and rock physics. When I embarked on a career in academia as an assistant professor, Dr. Rai and Dr. Sondergeld supported me to grow as an independent researcher and demystified industry-academia collaboration.

I thank Katie Hammon from Elsevier for trusting a new author like me and for giving me the opportunity to write my first book. Katie ensured that the book that I had in my mind got all the support from Elsevier so that it can be translated to reality. A big thanks to Selvaraj Raviraj and Aleksandra Packowska from Elsevier, who ensured a timely completion at the highest quality.

A special thanks to all the students and researchers that I have worked with so far over the last 4 years. All of you motivate me to be better at what I do and to create new knowledge that can be published as books and journal papers.

Siddharth Misra

Chapter 1

Unsupervised outlier detection techniques for well logs and geophysical data

Siddharth Misra*, Oghenekaro Osogba[†,a] and Mark Powers[‡]
*Harold Vance Department of Petroleum Engineering, Texas A&M University, College Station, TX, United States, †Texas A&M University, College Station, TX, United States, ‡The University of Oklahoma, Norman, OK, United States

Chapter outline

---

[a] Formerly at the University of Oklahoma, Norman, OK, United States

Machine Learning for Subsurface Characterization. https://doi.org/10.1016/B978-0-12-817736-5.00001-6

1 Introduction

From a statistical standpoint, outliers are data points (samples) that are significantly different from the general trend of the dataset. From a conceptual standpoint, a sample is considered as an outlier when it does not represent the behavior of the phenomenon/process as represented by most of the samples in a dataset. Outliers are indicative of issues in data collection/measurement procedure or unexpected events in the operation/process that generated the data. Detection and removal of outliers is an important step prior to building a robust data-driven (DD) and machine learning-based (ML) model. Outliers skew the descriptive statistics used by data analysis, data-driven and machine learning methods to build the data-driven model. A model developed on data containing outliers will not accurately represent the normal behavior of data because the model picks the unrepresentative patterns exhibited by the outliers. As a result, there will be nonuniqueness in the model predictions. Data-driven models affected by outliers have lower predictive accuracy and generalization capability.

Outlier handling refers to all the steps taken to negate the adverse effects of outliers in a dataset. After detecting the outliers in a dataset, how they are handled depends on the immediate use of the dataset. Outliers can be removed, replaced, or transformed depending on the type of dataset and its use. Outlier handling is particularly important as outliers could enhance or mask relevant statistical characteristics of the dataset. For instance, outliers in weather data could be early signs of a weather disaster; ignoring this could have catastrophic consequences. However, before considering outlier handling, we must first detect them.

Outliers in well logs and other borehole-based subsurface measurements occur due to wellbore conditions, logging tool deployment, and physical characteristics of the geological formations. For example, washed out zones in the wellbore and borehole rugosity significantly affects the readings of shallow-sensing logs, such as density, sonic, and photoelectric factor (PEF) logs, resulting in outlier response. Along with wellbore conditions, uncommon beds and sudden change in physical/chemical properties at a certain depth in a formation also result in outlier behavior of the subsurface measurements. In this chapter, we perform a comparative study of the performances of four unsupervised outlier detection techniques (ODTs) on various original and synthetic well-log datasets.

1.1 Basic terminologies in machine learning and data-driven models

Before discussing more about outliers, the authors would like to clearly distinguish the following terms: dataset, sample, feature, and target. Data-driven (DD) and machine learning-based (ML) methods find statistical/probabilistic functions by processing a relevant dataset to either relate features to targets (referred as supervised learning) or appropriately transform features and/or samples (referred as unsupervised learning). Various types of information (i.e., values of features and targets) about several samples constitute a dataset. A dataset is a collection of values corresponding to features and/or targets for several samples. Features are physical properties or attributes that can be measured or computed for each sample in the dataset. Targets are the observable/measurable outcomes, and the target values for a sample are consequences of certain combinations of features for that sample. For purposes of unsupervised learning, a relevant dataset is collection of only the features for all the available samples, whereas a dataset is collection of features and corresponding targets for all the available samples for purposes of supervised learning. A dataset comprises of one or many targets and several features for several samples. An increase in the number of samples increases the size of the dataset, whereas an increase in the number of features increases the dimensionality of dataset. A DD/ML model becomes more robust with the increase in the size of the dataset. However, with increase in dimension of the dataset, a model tends to overfit and becomes less generalizable, unless the increase in dimension is due to the addition of informative, relevant, uncorrelated features. Prior to building the DD/ML model using supervised learning, a dataset is split into training and testing datasets to ensure the model does not overfit the training dataset and generalizes well to the testing dataset. Further, the training dataset is divided into certain number of splits to perform cross validation that ensures the model learns from and is evaluated on all the statistical distributions present in the training dataset. For evaluating the model on the testing dataset, it is of utmost importance to avoid any form of mixing (leakage) between the training and testing datasets. Also, when evaluating the model on the testing dataset, one should select relevant evaluation metrics out of the several available metrics with various assumptions and limitations.

1.2 Types of machine learning techniques

Machine learning (ML) models can be broadly categorized into three techniques: supervised learning, unsupervised learning, and reinforcement learning. In supervised learning (e.g., regression and classification), a data-driven model is developed by first training the model on samples with known features/attributes and corresponding targets/outcomes from the training dataset; following that, the trained model is evaluated on the testing dataset; and finally, the data-driven model is used to predict targets/outcomes based on the features/attributes of new, unseen samples during the model deployment. In unsupervised learning (e.g., clustering and transformation), a data-driven model learns to generate an outcome based on the features/attributes of samples without any prior information about the outcomes. In reinforcement learning (which tends to be very challenging), a data-driven model learns to perform a specific task by interacting with an environment to receive a reward based on the actions performed by the model toward accomplishing the task. In reinforcement learning, the model learns the policy for a specific task by optimizing the cumulative reward obtained from the environment. These three learning techniques have several day-to-day applications; for instance, supervised learning is commonly used in spam detection. The spam detection model is trained on different mails labeled as spam or not spam; after gaining the knowledge from the training dataset and subsequent evaluation on the testing dataset, the trained spam detection model can detect if a new mail is spam or not. Unsupervised learning is used in marketing where customers are categorized/segmented based on the similarity/dissimilarity of their purchasing trends as compared with other customers; for instance, Netflix's computational engine uses the similarity/dissimilarity between what other users have watched when recommending the movies. Reinforcement learning was used to train DeepMind's AlphaGo to beat world champions at the game of Go. Reinforcement learning was also used to train the chess playing engine, where the model was penalized for making moves that led to losing a piece and rewarded for moves that led to a checkmate.

A machine learning method first processes the training dataset to build a data-driven model; following that, the performance of the newly developed model is evaluated against the testing dataset. After confirming the accuracy and precision of the data-driven model on the testing dataset, these methods are deployed on the new dataset. These three types of dataset, namely, training, testing, and new dataset, comprise measurements of certain specific features for numerous samples. The training and testing datasets, when used in supervised learning, contain additional measurements of the targets/outcomes. A supervised learning technique tries to functionally relate the features to the targets for all the samples in the dataset. On the contrary, for unsupervised learning, the data-driven model development takes place without the targets; in other words, there are no targets to be considered during the training and testing stages of unsupervised learning. Obviously, information about the targets is never available in the new dataset because the trained models are deployed on the new dataset to compute the desired targets or certain outcomes.

1.3 Types of outliers

In the context of this work, outliers can be broadly categorized into three types: point/global, contextual, and collective outliers [1]. Point/global outliers refer to individual data points or samples that significantly deviate from the overall distribution of the entire dataset or from the distribution of certain combination of features. These outliers exist at the tail end of a distribution and largely vary from the mean of the distribution, generally lying beyond 2 standard deviations away from the mean; for example, subsurface depths where porosity is >40 porosity units or permeability is >5 Darcy should be considered as point/global outliers. From an event perspective, a house getting hit by a meteorite is an example of point outlier. The second category of outliers is the contextual/conditional outliers, which deviate significantly from the data points within a specific context, for example, a large gamma ray reading in sandstone due to an increase in potassium-rich minerals (feldspar). Snow in summer is an example of contextual outlier. Points labeled as contextual outliers are valid outliers only for a specific context; a change in the context will result in a similar point to be considered as an inlier. Collective outliers are a small cluster of data that as a whole deviate significantly from the entire dataset, for example, log measurements from regions affected by borehole washout. For example, it is not rare that people move from one residence to the next; however, when an entire neighborhood relocates at the same time, it will be considered as collective outlier. Contextual and collective outliers need a domain expert to guide the outlier detection.

2 Outlier detection techniques

An outlier detection technique (ODT) is used to detect anomalous observations/samples that do not fit the typical/normal statistical distribution of a dataset. Simple methods for outlier detection use statistical tools, such as boxplot and Z-score, on each individual feature of the dataset. A boxplot is a standardized way of representing the distributions of samples corresponding to various features using boxes and whiskers. The boxes represent the interquartile range of the data, and the whiskers represent a multiple of the first and third quartiles of the variable; any data point/sample outside these limits is considered as an outlier. The next simple statistical tool for feature-specific outlier detection is the Z-score, which indicates how far the value of the data point/sample is from its mean for a specific feature. A Z-score of 1 means the sample point is 1 standard deviation away from its mean. Typically, Z-score values greater than or less than +3 or −3, respectively, are considered outliers. Z-score is expressed as

$$Z - \text{score} = \frac{x_i - \overline{x}}{\sigma} \tag{1.1}$$

where σ and $\overline{x}$ is the standard deviation and mean of the distribution of feature x, respectively, and x_i is the value of the feature x for the ith sample.

Outlier detection based on simple statistical tools generally assume that the features have normal distributions while neglecting the correlation between features in a multivariate dataset. Advanced outlier detection method based on machine learning (ML) can handle correlated multivariate dataset, detect abnormalities within them, and do not assume a normal distributions of the features. Well logs and subsurface measurements are sensing heterogeneous geological mixtures with a lot of complexity in terms of the distributions of minerals and fluids; consequently, these measurements generally do not necessarily exhibit Gaussian distribution and generally exhibit considerable correlations within the features. Data-driven outlier detection techniques built using machine learning are more robust in detecting outliers as compared with simple statistical tools.

Outliers in dataset can be detected using either supervised or unsupervised ML technique. In supervised ODT, outlier detection is treated as a classification problem. The outlier-detection model is trained on dataset with samples prelabeled as either normal data (inliers) or outliers. The trained model then assigns labels to the samples in a new, unseen, unlabeled dataset as either inliers or outliers based on what was learned from the training dataset. Supervised ODT is robust when the model is exposed to a large, statistically diverse training set (i.e., dataset that contains every possible instance of normal/inlier and outlier samples), whose samples are accurately labeled as normal/inlier or outlier. Unfortunately, this is difficult, time-consuming, and sometimes impossible to obtain because it requires significant human expertise in labeling and expensive data acquisition to obtain a large dataset. In contrary, unsupervised ODT overcomes the requirement of labeled dataset. Unsupervised ODTs generally assume the following: (1) The number of outliers is much smaller than the normal samples, and (2) outliers do not follow the overall "trend" in the dataset. A list of popular outlier detection techniques is listed in Appendix A.

Both supervised and unsupervised ODTs are used in various industries. For instance, in credit fraud detection, neural networks are trained on all known fraudulent and legitimate transactions, and every new transaction is assigned a fraudulent or legitimate label by the model based on the information from the training dataset. It could also be trained in an unsupervised manner by flagging transactions that are dissimilar from what is normally encountered. In medical diagnosis, ODTs are used in early detection and diagnosis of certain diseases by analyzing the patient data (e.g., blood pressure, heart rate, and insulin level) to find patients for whom the measurements deviate significantly from the normal conditions. Zengyou et al. [2] used a cluster-based local outlier factor algorithm to detect malignant breast cancer by training their model on features related to breast cancer. ODTs are also used in detecting irregularities in the heart functioning by analyzing the measurements from an echocardiogram (ECG) for purposes of early diagnosis of certain heart diseases. In the oil and gas industry, Chaudhary et al. [3] was able to improve the performance of the stretched exponential production decline (SEPD) model by detecting and removing outliers from production data by using the local outlier factor method. In another oil and gas application, Luis et al. [4] used one-class support vector machine (OCSVM) to detect possible operational issues in offshore turbomachinery, such as pumps and compressors, by detecting anomalous signals from their sensors. When implementing an unsupervised ODT, a prior knowledge of the expected fraction of outliers improves the accuracy of outlier detection. In many real-world applications, these values are known. For example, in the medical field, there is a good estimate of the fraction of people who contract a certain rare disease, or in a factory assembly line, there is a good estimate of the fraction of defective mechanical parts. Unfortunately, when working with well log and other geophysical dataset, the expected fraction of outliers is not necessarily known a priori because this fraction depends on several factors (operating conditions during logging, type of formation, sensor physics, etc.). This is a significant challenge in applying unsupervised ODTs on well-log data and other geophysical data.

Under unsupervised conditions, accuracy and robustness of the ODT rely on the values of hyperparameters. Hyperparameters are user-defined parameters specified prior to applying a data-driven method on a dataset. Hyperparameters control the learning of the data-driven method and determine the final functional form of the data-driven model. Hyperparameters govern the learning process, whereas parameters (weights) are consequence of the learning process. Choice of hyperparameters can make one unsupervised outlier-detection model to perform poorly as compared to other outlier-detection models on the same dataset. Unfortunately, when using unsupervised ODTs on well logs and subsurface data, there is no prior information about the hyperparameters. Generally, an unsupervised ODT needs to be applied on the well-log and geophysical dataset without any hyperparameter tuning and without any prior information of the hyperparameters. The primary motivation of our study is to identify the best-performing unsupervised ODT method that needs minimal hyperparameter tuning and manual interventions.

3 Unsupervised outlier detection techniques

In this article, we apply four unsupervised ODTs on well logs to identify the formation depths that exhibit anomalous or outlier log responses. The ODTs were used in an unsupervised manner without much hyperparameter tuning. Each formation depth can be considered as a sample, and the various logs acquired at a specific depth can be considered as features. Being unsupervised approach, there is no target or desired outcome for a given set of feature values (feature vector) of a sample. An unsupervised ODT processes the feature vectors corresponding to the available samples that contain both normal (inlier) and anomalous (outlier) behavior to identify the depths that exhibit outlier behavior. Unsupervised ODT are based on distance, density, decision boundary, or affinity, which are used to quantify the relationships among the features governing the inlier and outlier behavior of samples. In this section, we will introduce four unsupervised ODTs, namely, isolation forest (IF), one-class SVM (OCSVM), local outlier factor (LOF), and density-based spatial clustering of applications with noise (DBSCAN). In this study, all methods are implemented from the scikit-learn package.

3.1 Isolation forest

Isolation forest (IF) assumes that the outliers tend to lie in sparse regions of the feature space and have more empty space around them than the densely clustered normal/inlier data [5]. Since outliers are in less populated regions of the dataset, it generally takes fewer random partitions to isolate them in a segment/partition. In other words, since outliers are few and different, they are more susceptible to isolation [6]. IF is an unsupervised ODT that uses a forest of randomly partitioned trees to isolate outlier samples in terminating nodes. IF performs recursive random partitioning/splitting of the feature space by randomly subsampling features and corresponding threshold values of the features. This generates treelike structure, where the number of splittings required to isolate an sample in a terminating node is equivalent to the path length from the root node to the terminating node. This path length, averaged over a forest of such random trees, is a measure of normality of a sample, such that anomalies/outliers have noticeably shorter path lengths; in other words, it is easy to partition the outliers with a few number of partitionings of the feature space. A decision function categorizes each observation as an inlier or outlier based on the path length of the observation compared with the average path length of all observations. Unlike most other unsupervised ODTs that use distance and density as measures for outlier detection, IF uses isolation as a measure.

IF has low computational requirements, is fast to deploy, has low computational time complexity, and can be parallelized for faster computation. IF does not require feature scaling and dimensionality reduction. Like other tree-based methods, IF does not need much tuning because hundreds of different trees (with different subsamples of features and feature thresholds) are parallelly trained on the dataset. Nonetheless, users who intend to control the performance of outlier detection need to tune the following hyperparameters: amount of contamination in the dataset, number of trees/estimators, maximum number of samples to be used in each tree, and maximum number of subsampled features used in each tree. Hyperparameters govern the learning process. Fig. 1.1A illustrates the outlier detection by the isolation forest when applied to a simple two-dimensional dataset containing 25 samples having two features/attributes (represented by the x- and y-axes). *Red samples (gray in the print version)* are outliers, and the shade of *blue (light gray in the print version)* in the background is indicative of degree of normality of samples lying in the shaded region, where *darker blue shades (dark gray in the print version)* correspond to outliers that are easy to partition. Isolation forest is effective in detecting point and tends to fail in detecting collective outliers and contextual outliers.

3.2 One-class SVM

One-class support vector machine (OCSVM) is a parametric unsupervised ODT suitable when the data points (i.e., samples) are mostly "normal" data with very

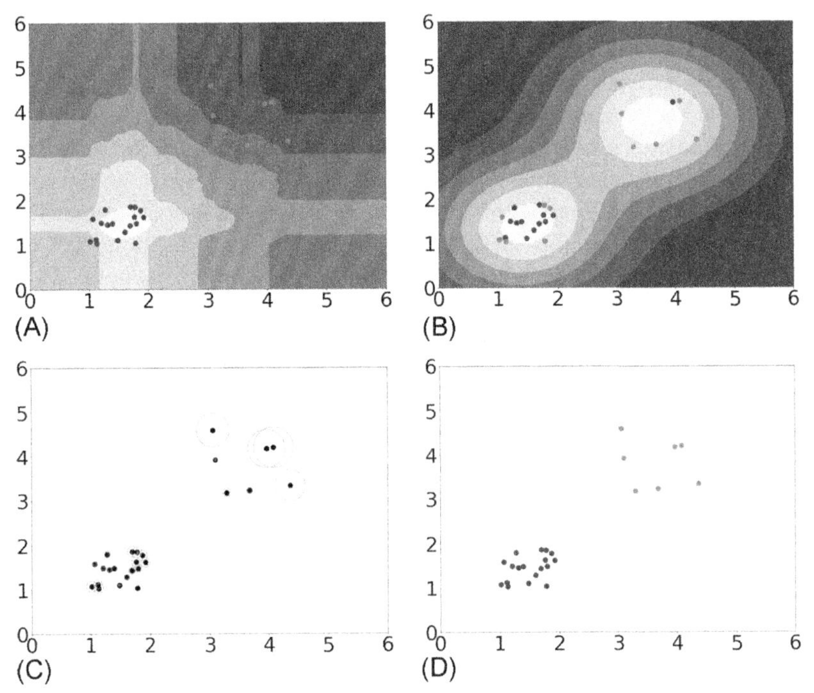

FIG. 1.1 Performances of the four unsupervised outlier detection techniques, namely, (A) isolation forest with hyperparameters: max_samples = 10, n_estimators = 100, max_features = 2, and contamination = "auto"; (B) one-class SVM with hyperparameters: nu = 0.5 and gamma = 0.04; (C) local outlier factor with hyperparameters: n_neighbors = 20, metric = "minkowiski," and $p = 2$; and (D) DBSCAN with hyperparameters: eps = 0.5, min_samples = 5, metric = "minkowiski," and $p = 2$, on the synthetic two-dimensional dataset containing 25 samples. *Red samples (light gray in the print version)* indicate outliers, and *blue samples (dark gray in the print version)* indicate inliers. All other hyperparameters except those mentioned earlier have default values.

few outliers (minimally contaminated). OCSVM builds a representational model of normality (inliers) by processing the dataset, wherein most of the samples are considered as inliers. OCSVM is based on the support vector machine that finds the support vectors and then separates the data into separate classes using hyperplanes/hyperspheres. OCSVM finds a minimal hypersphere in the kernel space (transformed feature space) that circumscribes maximum inliers (normal samples). The hypersphere determines the normality in the dataset. OCSVM nonlinearly projects the data into a high-dimensional kernel space and then maximally separates the data from the origin of the kernel space by finding an optimal hypersphere. As a result, OCSVM may be viewed as a regular two-class SVM where most of the training data (i.e., inliers) lie in the first class, and the origin is taken as a dominant member of the second class containing the outliers. Nonetheless, there is a trade-off between maximizing the distance of the hyperplane from the origin and the number of training data points contained in the hypersphere (region separated from the origin by the hyperplane). An optimization routine is used to process the available data to select certain samples as support vectors that parameterize the decision boundary defining the hypersphere to be used for outlier detection [7].

OCSVM implementation is challenging for high-dimensional data, is slower to train and deploy, tends to overfit, is suitable when fraction of outlier is small, and needs careful tuning of the hyperparameters. OCSVM requires feature scaling and dimensionality reduction for fast training. Important hyperparameters of OCSVM are the gamma and outlier fraction. The gamma influences the radius of the Gaussian hypersphere that separates the inliers from outliers; large values of gamma will result in smaller hypersphere and "stricter" model that finds more outliers. It acts as the cutoff parameter for the Gaussian hypersphere that governs the separating boundary between inliers and outliers [8]. Outlier fraction defines the percentage of the dataset that is outlier. Outlier fraction helps in creating tighter decision boundary to improve outlier detection. Similar to Fig. 1.1A, Fig. 1.1B illustrates the working of the one-class SVM where the interfaces of two different shades are few possible decision functions that can be used for outlier detection. Fig. 1.1B illustrates the outlier detection by the OCSVM when applied to a simple two-dimensional dataset containing 25 samples having two features/attributes. *Red samples (gray in the print version)* are outliers, and the shade of *blue (light gray in the print version)* in the background is indicative of degree of normality of samples lying in the shaded region, where *darker blue shades (dark gray in the print version)* correspond to outliers that are easy to partition. OCSVM is effective in detecting both point and collective outliers when tuned properly. The ability of OCSVM to detect contextual outlier depends on appropriate feature selection, which can be time-consuming.

3.3 DBSCAN

Density-based spectral clustering of applications with noise (DBSCAN) is a density-based clustering algorithm that can be used as an unsupervised ODT. The density of a region depends on the number of samples in that region and the proximity of the samples to each other. DBSCAN seeks to find regions of high density separated by low-density regions in a dataset. Samples in the high-density regions are labeled as inliers, whereas those in low-density regions are labeled as outliers. The key idea is that for each sample in the inlier cluster, the neighborhood region of certain user-defined size (referred as bandwidth) must contain at least a minimum number of samples, that is, the density in the neighborhood must exceed a user-defined threshold [9]. Samples that do not meet the density threshold are labeled as outliers. DBSCAN requires the tuning of the following hyperparameters that control the outlier detection process: minimum number of samples required to form the inlier cluster; maximum distance between any two samples in an inlier cluster; and parameter p that determines the distance measure in the form of the Minkowski distance, such that Minkowski distance transforms into Euclidean distance for $p=2$.

The DBSCAN model is very effective in detecting point outliers. It can detect collective outliers if they occur as low-density regions. It is not reliable for detecting contextual outliers. DBSCAN is not suitable when inliers are distributed as low-density regions and requires a lot of expertise in selecting optimal hyperparameters that controls the outlier detection. For Fig. 1.1D, DBSCAN exhibits a reliable outlier-detection performance unlike OCSVM.

3.4 Local outlier factor

Local outlier factor (LOF) is an unsupervised ODT based on relative density of region. Simple density-based ODT methods are not as reliable for outlier detection when the clusters are of varying densities; for example, inliers can be distributed as high-density and low-density regions, and outliers can be distributed as high-density region. Local outlier factor mitigates the challenges with DBCAN by using relative density as the measure to assign an outlier score to each sample. LOF compares the local density of a sample with the local densities of its k-nearest neighbors to identify outliers, which are in the regions that have a substantially lower density than their k-nearest neighbors. LOF assigns a score to each sample by computing relative density of each sample as a ratio of the average local reachability density of neighbors to the local reachability density of the sample and flags the points with low scores as outliers [10]. A sample with LOF score of 3 means the average density of this point's neighbors is about three times more than its local density, that is, the sample is not like its neighbors. LOF score of a sample smaller than 1 indicates the sample has higher density than neighbors. The number of neighbors (K) sets how many neighbors are considered when computing the LOF score for a sample.

In Fig. 1.1C, the LOF is applied to the previously mentioned two-dimensional dataset containing 25 samples. Samples at the upper right corner of Fig. 1.1C are outliers, and the radius of the circle encompassing a sample is directly proportional to the LOF score of the sample. LOF score for three of those six points at the upper right corner is low; this is odd considering that from visual inspection, it is obvious those points are outliers as well. We notice that exception because those points are closer to the high-density region, and we set the number of neighbors to be considered when calculating relative density at $K = 20$. The density of the samples in the dense region reduces the LOF sample score of the three points in the upper right-hand corner. For the dataset considered in Fig. 1.1, LOF performance can be improved by reducing the value of the hyperparameter K. Like DBSCAN for unsupervised ODT, LOF is severely affected by the curse of dimensionality and is computationally intensive when there are a large number of samples [11]. Moreover, LOF can be biased because a user selects a cutoff for the LOF scores to label the outliers. Selecting the LOF score threshold can be inconsistent and biased. For a certain dataset, a score greater than 1.2 could represent an outlier, while in another case, the limit could be 1.8. LOF needs attentive tuning of the hyperparameters. Due to the local approach and relative density calculation, LOF is able to identify outliers in a dataset that would not be outliers in another area of the dataset. The major hyperparameters for tuning are the number of neighbors K to consider for each sample and metric p for measuring the distance, similar to DBSCAN, where the general form of Minkowski distance transforms into Euclidean distance for $p = 2$.

3.5 Influence of hyperparameters on the unsupervised ODTs

Hyperparameters are the parameters of the model that are defined by the user prior to training/applying the model on the data. For example, the number of layers and number of neurons in a layer are the hyperparameters of neural network, and the number of trees and the maximum depth of a tree are the hyperparameters of random forest. Hyperparameters control the learning process; for unsupervised ODTs, hyperparameters determine the decision boundaries (e.g., OCSVM), partitions (e.g., IF), similarity/dissimilarity labels (e.g., DBSCAN), and scores (e.g., LOF) that differentiate the inliers from outliers. By changing the hyperparameters, we can effectively alter the performance of an unsupervised ODT. The effects of hyperparameters on the unsupervised ODT are evident when comparing Fig. 1.1 with Fig. 1.2.

When the IF hyperparameter "contamination" is changed from "auto" (i.e., the model labels outliers based on the default threshold of the model) to 0.1, the isolation forest model is constrained to label only 10% of the dataset (~3 samples) as outliers (Fig. 1.2A), thereby worsening the outlier detection for the synthetic 25-sample dataset. IF model therefore labels the "top 3" outliers based on the learnt decision functions, as shown in Fig. 1.2A. When the OCSVM hyperparameter "*nu*" (maximum fraction of possible outliers in the dataset) is changed from 0.5 (Fig. 1.1B) to 0.2 (Fig. 1.2B), OCSVM model becomes more conservative in outlier detection and detects few outliers; thereby improving the outlier detection for the synthetic 25-sample dataset (Fig. 1.1B vs Fig. 1.2B). When the LOF hyperparameter "number of neighbors" is reduced, the effect of the high-density regions on the scores assigned to samples in low-density regions is also reduced, thereby improving outlier detection for the synthetic 25-sample dataset (Fig. 1.1C vs Fig. 1.2C). Finally, when the value of epsilon (bandwidth) is increased from 0.5 (in Fig. 1.1D) to 1 (in Fig. 1.2D), all the data points are considered as inliers (Fig. 2D) because all samples can now meet the density requirement, such that several samples form an independent cluster; thereby worsening the outlier detection for the synthetic 25-sample dataset. Increasing bandwidth means that a sample belongs to an inlier cluster even when it is in a low-density region (i.e., bandwidth defines the minimum number of neighboring samples within a maximum distance around a sample above which the sample can be considered as an inlier cluster).

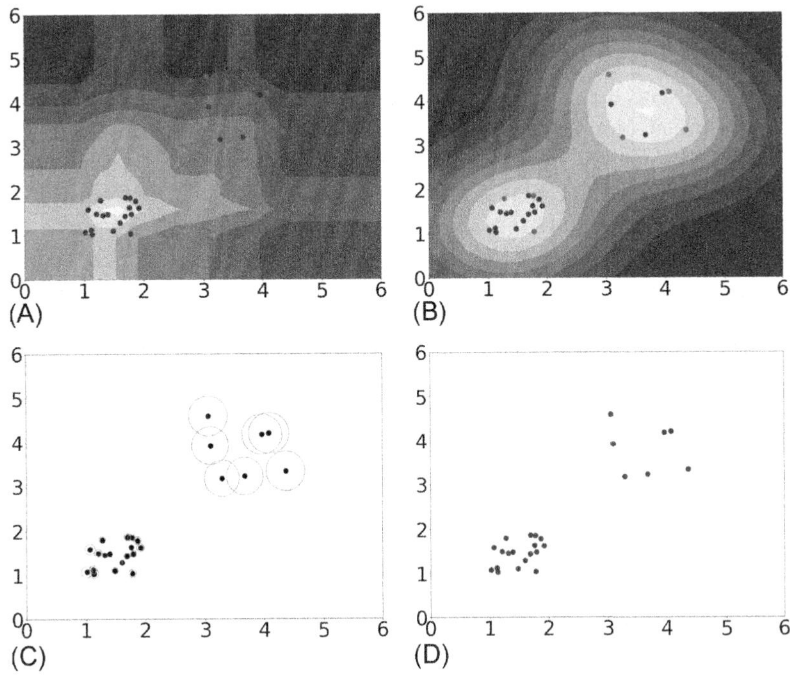

FIG. 1.2 Performances of the four unsupervised outlier detection techniques, namely, (A) isolation forest with hyperparameters: max_samples = 10, n_estimators = 100, max_features = 2, and contamination = 0.1; (B) one-class SVM with hyperparameters: nu = 0.2 and gamma = 0.04; (C) local outlier factor with hyperparameters: n_neighbors = 10, metric = "minkowiski," and $p = 2$; and (D) DBSCAN with hyperparameters: eps = 1, min_samples = 5, metric = "minkowiski," and $p = 2$, on the synthetic two-dimensional dataset containing 25 samples. *Red samples (light gray in the print version)* indicate outliers, and *blue samples (dark gray in the print version)* indicate inliers. All other hyperparameters except those mentioned earlier have default values.

Comparison of Figs. 1.1 and 1.2 highlights the effects of hyperparameter tuning on outlier detection. Choice of hyperparameters can make one method perform poorly compared with the other method on the same dataset. Unfortunately, when using unsupervised ODTs on well logs and subsurface geophysical data, there is no information about the degree of contamination, outlier fraction, bandwidth, or any other hyperparameter. For high-dimensional subsurface dataset, it is a challenge to visualize the data in entirety and identify hyperparameters suited for a dataset. Generally, an unsupervised ODT needs to be applied on the dataset without any hyperparameter tuning and without any prior information of the hyperparameters. The primary motivation of our study is to identify the best-performing unsupervised ODT method that needs minimal hyperparameter tuning.

4 Comparative study of unsupervised outlier detection methods on well logs

Fig. 1.3A presents the workflow for comparative study of unsupervised ODTs. As earlier stated, unsupervised ODTs are used in different industries, including the oil and gas industry. Osborne et al. [12] and Ferdowsi et al. [13] showed the importance of outlier detection for robust data analysis and data-driven predictive modeling. Orr et al. [14] suggested that outliers need not be removed but analyzed to better understand the dataset. Although there is some disagreement on how outliers should be handled, the consensus is that outlier detection is an important process before any form of data analytics (predictive modeling, clustering, ANOVA, etc.). Performance of unsupervised ODT depends on the choice of hyperparameters and the inherent properties of the dataset, which is governed by the process/phenomenon that generated it. Consequently, there is no universally best-performing unsupervised outlier detection technique.

Taking this into consideration, we conducted a comparative study of the performances of four popular unsupervised ODTs on well-log data by following

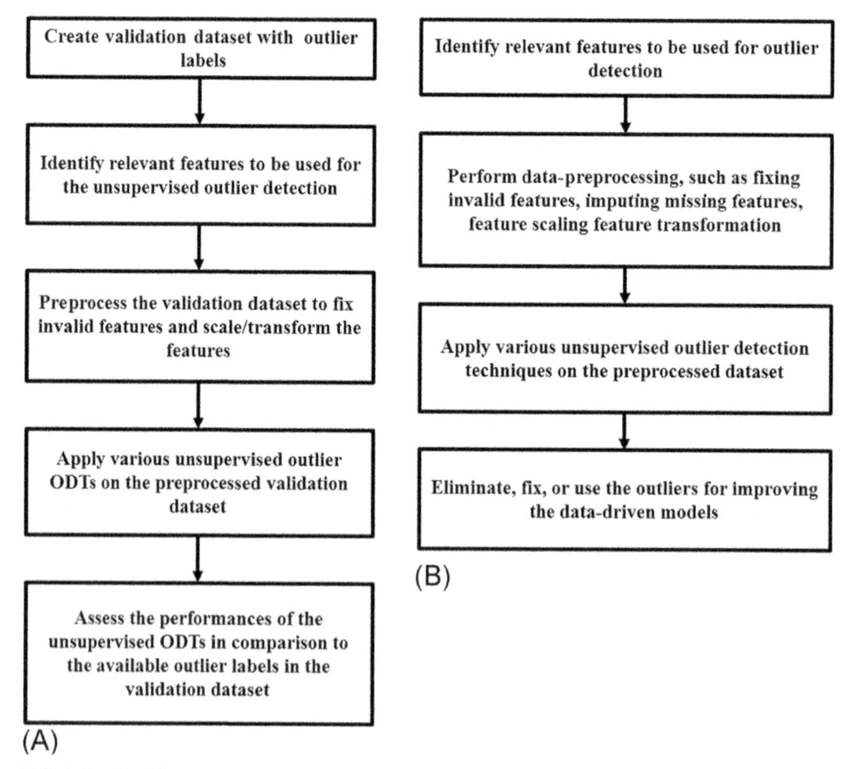

FIG. 1.3 Workflows for (A) comparative study and (B) actual deployment of unsupervised outlier detection methods on well logs and subsurface measurements.

the steps laid out in Fig 1.3A. To ensure a controlled environment for our investigation, we created four distinct validation datasets comprising outlier/inlier labels, which were assigned by a human expert. The ability of the unsupervised ODTs to accurately detect the outliers and inliers is analyzed using various evaluation metrics. It is to be noted that real-world implementations of unsupervised ODTs are generally done without any prior knowledge of outliers and inliers by following the steps laid out in Fig. 1.3B; consequently, there is no way to evaluate the unsupervised ODTs during real-world implementations and choose the best one. Nonetheless, our comparative study will help identify the unsupervised ODT that performs the best on various types of well-log dataset with minimal hyperparameter tuning.

4.1 Description of the dataset used for the comparative study of unsupervised ODTs

Log data used for this work were obtained from two wells in different reservoirs. Gamma ray (GR), density (RHOB), neutron porosity (NPHI), compressional sonic travel time (DTC), and deep and shallow resistivity logs (RT and RXO) from Well 1 are available within the depth interval of 580–5186 ft comprising 5617 depth samples; herein, this dataset will be referred as the onshore dataset. The onshore dataset contains log responses from different lithologies of limestone, sandstone, dolomite, and shale. Gamma ray (GR), density (DEN), neutron porosity (NEU), compressional sonic transit time (AC), deep and medium resistivities (RDEP and RMED), and photoelectric factor (PEF) logs from Well 2 are available within the depth interval of 8333–13327 ft comprising 9986 depth samples; herein, this dataset will be referred as the offshore dataset. The offshore dataset contains log responses from different lithologies of limestones, sandstone, dolomite, shale, and anhydrites.

4.2 Data preprocessing

Data preprocessing refers to the transformations applied to data before feeding them to the machine learning algorithm [15]. Primary use of data preprocessing is to convert the raw data into a clean dataset that the machine learning workflow can process. A few data preprocessing tasks include fixing null/NaN values, imputing missing values, scaling the features, normalizing samples, removing anomalies, encoding the qualitative/nominal categorical features, and data reformatting. Data preprocessing is an important step because a data-driven model built using machine learning is as good as the quality of data processed by the model.

4.2.1 Feature transformation: Convert R to log(R)

Machine learning models tend to be more efficient when the features/attributes are not skewed and have relatively similar distribution and variance. Resistivity measurements range from 10^{-2} ohm-m (brine-filled formation) to 10^3 ohm-m (low-porosity formation) and tend to exhibit log-normal distribution. To reduce the right skewness (i.e., mean $\gg$ mode) and large variance observed in the resistivity data relative to other logs (features), we transformed resistivity to its logarithm. This reduces it skewness and variability and improves the model's predictive performance, as demonstrated in subsequent sections.

4.2.2 Feature scaling: Use of robust scaler

A dataset generally contains features that significantly differ from each other in terms of magnitude, unit, and range. This tends to bias the machine learning methods based on distance, volume, density, and gradients [16]. Without feature scaling, a few features will dominate during the model development. For instance, the features with high magnitudes will weigh in a lot more in the distance calculations than features with low magnitudes, which, for example, will adversely affect *k*-nearest neighbor classification/regression and principal component analysis. Without feature scaling, samples will exhibit high density in few feature dimensions and low density in other feature dimensions. Feature scaling is an important aspect of data preprocessing that improves the performance of the data-driven models. For methods based on distance, volume, and density, it is essential to ensure that the features have similar or near similar scales for improved performance. For methods based on gradient-descent optimization and other forms of gradients, it is recommended to use scaled features for fast convergence. For neural networks, it is crucial to have the features scaled between minimum and maximum values, preferably 0 to 1 or −1 to 1, for fast and robust convergence. Notably, tree-based methods such as random forest regression/classification, AdaBoost, and isolation forest do not require feature scaling. Data from different logs usually range between different scales. For example, the RHOB (1.95–2.95 g/cc), porosity (0.0–0.2 fraction) and GR (50–250 gAPI) logs have vastly different scales.

For purposes of feature scaling, we used the robust scaling method, which can be expressed mathematically as

$$x_{is} = \frac{x_i - Q_1(x)}{Q_3(x) - Q_1(x)} \tag{1.2}$$

where x_{is} is the scaled feature x for the ith sample, x_i is the unscaled feature x for the ith sample, $Q_1(x)$ is the first quartile of the distribution of feature x, and $Q_3(x)$ is the third quartile of the distribution of feature x. The first and third quartiles represent the medians of the lower half and upper half of the data, respectively, not influenced by outliers. We perform robust scaling on the features (logs) because it overcomes the limitations of other scaling methods, like the Standard scaler that assumes the data are normally distributed and the MinMax scaler that assumes that the feature cannot exceed certain bounds/limits due to the physics or measurement governing the feature. For example, MinMax scaler is suitable for pixel intensity in an image, which is bound within a range due to the data acquisition requirements, and is suitable for amplitude of a speech signal, which is bound within a range. The presence of outliers adversely affects the Standard scaler and severely affects the MinMax scaler. Robust scaler overcomes the limitations of MinMax scaler and Standard scaler by using the first and third quartiles for scaling the features instead of the minimum, mean, and maximum values. The use of quartiles ensures that the robust scaler is not sensitive to outliers, whereas the minimum and maximum values used in the MinMax scaler could be the outliers and the mean and standard deviation values used in the Standard scaler are influenced by outliers.

4.3 Validation dataset

We created four distinct validation datasets containing known real/synthetic outliers to assess and compare the performances of the four unsupervised ODTs studied in this chapter. Being unsupervised methods, there is no direct way of quantifying the performances of isolation forest, local outlier factor, DBSCAN, and one-class SVM. Therefore, we implement expert knowledge, physically consistent thresholds, and various synthetic data creation methods to assign an outlier or inlier label to each sample in the validation dataset. Several samples in the labeled validation dataset are synthetic samples generated using physically consistent formulations. Each of the four validation datasets is processed by the each of the four unsupervised ODTs; following that, the inliers and outliers detected by the unsupervised ODT are compared with the prespecified outlier/inlier labels assigned to the samples of the validation dataset by the human expert.

4.3.1 Dataset #1: Containing noisy measurements

Dataset #1 was constructed from the previously mentioned onshore dataset to compare the performance of the four unsupervised outlier detection techniques in detecting depths where the log responses are adversely affected by noise. Noise in well-log dataset can adversely affect its geological/geophysical interpretation as it masks the formation properties. The onshore dataset contains log responses measured at 5617 depths in Well 1 drilled with a bit of size 7.875″. Dataset #1 comprise gamma ray (GR), bulk density (RHOB), compressional sonic travel time (DTC), and deep resistivity (RT) logs from the onshore dataset for the depths, where the borehole diameter is between 7.8″ and 8.2″. This led to 4037 inliers in Dataset #1. Following that, synthetic noisy log responses for 200 additional depths were randomly introduced into the Dataset #1. The noise samples were created such that they belong to the same distribution as the inlier data but are two standard deviations away from the mean of each feature in Dataset #1. Consequently, Dataset #1 contains in total 4237 samples, out of which 200 are point outliers. Comparative study on Dataset #1 involved experiments with

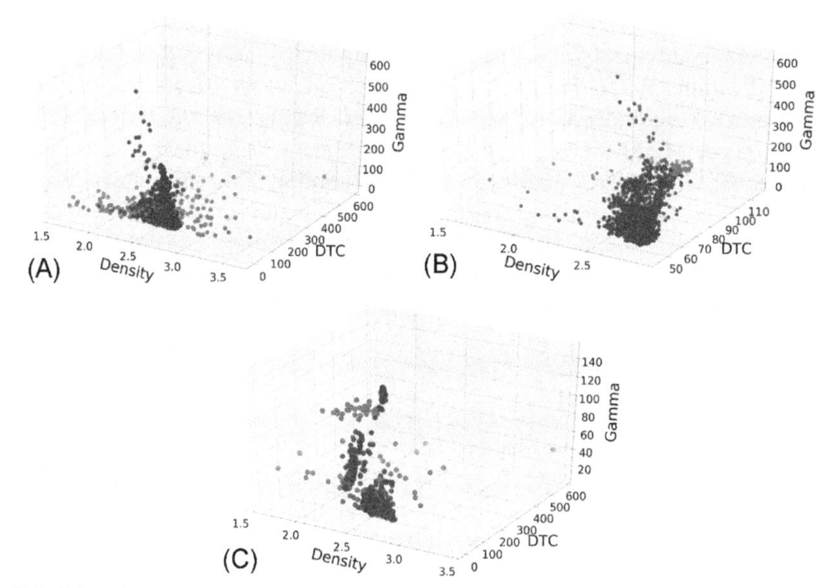

FIG. 1.4 3D scatterplots of (A) Dataset #1, (B) Dataset #2, and (C) Dataset #3.

three distinct feature subsets sampled from the available features, namely GR, RHOB, DTC, and RT logs. The three feature subsets are referred to as FS1, FS2, and FS2*, where FS1 contains GR, RHOB, and DTC; FS2 contains GR, RHOB, and RT; and FS2* contains GR, RHOB, and RT. Fig. 1.4A is a 3D scatterplot of Dataset #1 for the subset FS1, such that the *blue points (dark gray in the print version)* are the labeled known inliers and the *red points (light gray in the print version)* are the labeled known outliers. The outliers are spread evenly around the dataset similar to point outliers. Inlier samples form a cluster with some sparse points spread around the cluster; the outliers are spread randomly and evenly around the inlier cluster (Fig. 1.4A).

4.3.2 Dataset #2: Containing bad holes

Dataset #2 was constructed from the onshore dataset to compare the performances of the four unsupervised outlier detection techniques in detecting depths where the log responses are adversely affected by the large borehole sizes, also referred as bad holes. Dataset #2 comprise GR, RHOB, DTC, deep resistivity (RT), shallow resistivity (RXO), and neutron porosity (NPHI) logs from the onshore dataset for depths, where the borehole diameter is between 7.8″ and 8.2″. Following that, the depths in the onshore dataset where borehole diameter is greater than 12″ were added to Dataset #2 as point and collective outliers. Consequently, Dataset #2 contains in total 4128 samples, out of which 91 are outliers and 4037 are inliers. Inliers in Dataset #2 are the same as those in Dataset #1. Comparative study on Dataset #2 involved experiments with five distinct feature subsets sampled from the available features GR, RHOB, DTC, RT, RXO, and NPHI logs. The five feature subsets are referred to as FS1, FS2, FS2**, FS3, and FS4, where FS1 contains GR, RHOB, and DTC; FS2 contains GR, RHOB, and RT; FS2** contains GR, RHOB, and RXO; FS3 contains GR, RHOB, DTC, and RT; and FS4 contains GR, RHOB, DTC, and NPHI. The five feature subsets were used to analyze the effects of features on the performances of the four unsupervised ODTs. Fig. 1.4B is a 3D scatterplot of Dataset #2 for the subset FS1, where *blue points (dark gray in the print version)* are the known inliers and the *red points (light gray in the print version)* are the known outliers. The outlier points in this dataset are mostly concentrated on one end of the plot, which indicates that the bad hole resulted in pushing the log responses toward a region, similar to collective outliers. The inlier points in Fig. 1.4A and B are the same; the outlier points in this dataset as earlier mentioned are based on hole size; and unsurprisingly, they are mostly located in the shale formation, which are susceptible to washouts and breakouts. Outliers form a cluster at the edge of the inlier dataset with some outlier points randomly spread across the plot.

4.3.3 Dataset #3: Containing shaly layers and bad holes with noisy measurements

Dataset #3 was constructed from the onshore dataset to compare the performances of the four unsupervised outlier detection techniques in detecting thin shale layers/beds in the presence of noisy and bad-hole depths. Dataset #3 comprise GR, RHOB, DTC, RT, and NPHI responses from 201 depth points from a sandstone bed, 201 depth points from a limestone bed, 201 depth points from a dolostone bed, and 101 depth points from a shale bed of the onshore dataset. These 704 depths constitute inliers. Thirty bad-hole depths with borehole diameter $>12''$ and 40 synthetic noisy log responses are the outliers that are combined with the 704 inliers to form the Dataset #3. Consequently, Dataset #3 contains in total 774 samples, out of which 70 are outliers. Comparative study on Dataset #3 involved experiments with four distinct feature subsets sampled from the available features GR, RHOB, DTC, RT, and NPHI logs, namely, FS1, FS2, FS3, and FS4, like that performed on Dataset #2. FS1 contains GR, RHOB, and DTC; FS2 contains GR, RHOB, and RT; FS3 contains GR, RHOB, DTC, and RT; and FS4 contains GR, RHOB, DTC, and NPHI. Fig. 1.4C is a 3D scatterplot of Dataset #3 for the subset FS1, where *blue points (dark gray in the print version)* are the known inliers and the *red points (light gray in the print version)* are the known outliers. Three separate inlier clusters (dolomite and limestone, sandstone, and shale) and two clusters of outliers, noise and bad-hole points, are observed in Fig. 1.4C. Noise data are randomly spread around the inlier cluster, while the bad-hole data form a cluster close to but distinct from the shale inlier cluster.

4.3.4 Dataset #4: Containing manually labeled outliers

The offshore dataset acquired in Well 2 contains seven log responses from different lithologies of limestones, sandstone, dolomite, shale, and anhydrites. The seven logs are gamma ray (GR), density (DEN), neutron porosity (NEU), compressional sonic transit time (AC), deep and medium resistivity (RDEP and RMED), and photoelectric factor (PEF) logs. The offshore dataset was labeled using manual inspection, feature thresholding, and DBSCAN followed by manual verification of the labels (outliers vs inlier) to create the Dataset #4 for the purposes of validation of the four unsupervised ODTs. Construction of Dataset #4 required an expert to closely examine the log responses along the entire length of Well 2 to manually assign outlier labels to certain depths exhibiting anomalous log responses. Manual labels (outlier vs inlier) were assigned after analyzing variance of each log and three-dimensional distributions of logs acquired in Well 2. Few outliers were identified by first viewing the histogram and boxplot for each feature (i.e., log) and then defining the thresholds for each feature. The thresholds were determined based on common industry standards for determining when the logs are outside their normal ranges. We implemented following feature thresholds to determine outliers based on the one-dimensional distribution of a log: (1) density correction (DENC) log >0.12 g/cc, (2) photoelectric factor (PEF) log >8 B/E, and (3) gamma ray (GR) log >350 gAPI.

The seven logs from the offshore dataset were also analyzed using three-dimensional scatter plots to detect outliers based on the three-dimensional distribution of each combination of three logs, one at a time. Seven available logs in the offshore dataset will have 35 ($^7C_3$) possible combinations of three logs. Out of the 35 combinations, 7 combinations were analyzed to manually label the outliers. DBSCAN was used sequentially on each combination of three logs, one combination at a time, to identify the isolated points and clusters that do not belong to the dense cluster of normal data. DBSCAN was used as a clustering technique to identify noise points and clusters that were labeled as outliers because of their location in the low-density region of the feature space. When creating the Dataset #4, DBSCAN was used as a clustering technique and not as a unsupervised ODT. Dataset #4 was designed as validation set to compare the performance of three out of the four unsupervised ODTs, namely isolation forest, OCSVM, and LOF. The seventh subsequent combination of 3 logs provided minimal additional outliers to the dataset indicating that most outliers in three-dimensional space had already been identified using the 6 prior combinations out of the total 35 possible combinations.

DBSCAN is suited when the dimensionality of the dataset is low. DBSCAN has two primary hyperparameters, namely, min_samples and eps, that control the detection of outliers. The two hyperparameters of DBSCAN clustering were tuned for each combination of three logs through visual analysis of the outliers being detected on the scatter plot. This process was continued until the normal cluster of data was identified as inliers and all other points were identified as outliers. Three of the seven scatterplots used are shown in Fig. 1.5. The *blue*

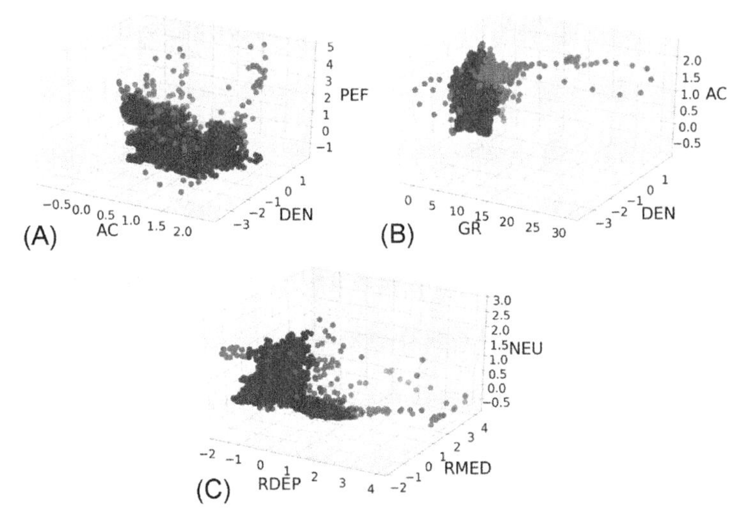

FIG. 1.5 Three of the seven 3D scatterplots containing various combinations of three (A–C) out of the seven available logs in the offshore dataset. DBSCAN clustering was optimized and sequentially applied on each of the combination of 3 logs out of the total 35 possible combinations ($^7C_3$) to identify outliers in the offshore dataset.

points (dark gray in the print version) show the known inliers, and the *red points (light gray in the print version)* show the known outliers, as detected by the DBSCAN model. The model has been tuned to where the main body of normal data is all *blue points (dark gray in the print version)* while all the points that deviate from this main body are *red (light gray in the print version).*

4.4 Metrics/scores for the assessment of the performances of unsupervised ODTs on the conventional logs

In real-world implementations, there is no means and metrics to evaluate the performance of the unsupervised ODTs. However, for purposes of comparative study of the four unsupervised ODTs as planned in this chapter, unsupervised ODTs process the four previously mentioned validation datasets, namely Datasets #1 to #4, to assign a label (either outlier or inlier) to each depth (sample) in the dataset. A label is assigned to each depth based on the log responses (feature vector) at that depth. In real-world applications of unsupervised outlier detection, there is no prior information of outliers, and no such inlier vs. outlier labels are present. For purposes of comparative study of the performances of the unsupervised ODTs, we created the four datasets, namely, Datasets #1, #2, #3, and #4 containing well-defined, manually verified outlier vs. inlier labels. Therefore, for evaluating the unsupervised ODTs, we compare the labels known a priori against the new labels assigned by the unsupervised ODT using metrics/scores popularly used to evaluate the classification methods. In a real-world scenario, these evaluation metrics for classification algorithm cannot be used for unsupervised ODT due to the lack of prior information about the outliers and inliers.

Formulations of evaluation metrics for classification methods are based on true positive, true negative, false positive, and false negative [17]. For purposes of assessing the outlier detection methods, true positive/negative refers to the number of outliers/inliers that are correctly detected as the outlier/inlier by the unsupervised ODT. On those lines, false positive/negative refers to the number of inliers/outliers that are incorrectly detected as the outlier/inlier by the unsupervised ODT. For example, when an actual inlier is detected as an outlier, it is referred as false positive. Following are some simple evaluation metrics that we are using for comparing the performances of unsupervised ODT on the labeled validation dataset. Appendix B presents true positives, true negatives, false positives, and false negatives for certain unsupervised ODTs on certain datasets. It is to be noted that these simple evaluation metrics can be improved by weighting the metrics to address the effects of outlier-inlier imbalance; that is, the number of positives (outliers) tend to be one order of magnitude smaller than the number of negatives (inliers).

4.4.1 Recall

Recall (also referred to as sensitivity) is the ratio of true positives to the sum of true positives and false negatives (i.e. total number of actual/true outliers). It represents the fraction of outliers in dataset correctly detected as outliers. Recall is expressed as

$$\text{Recall} = \frac{TP}{TP + FN} \tag{1.3}$$

Recall is an important metric but should not be used in isolation because it does not take into account the performance of the method in inlier detection. For example, a high recall close to 1 does not mean that the outlier detection method has a great performance because of the possibility of large false positives because actual inliers are being detected as outliers. For example, when an unsupervised ODT detects each data point as an outlier, the recall is 1, but the specificity is 0, indicating that the performance of the unsupervised ODT is unreliable.

4.4.2 Specificity

Specificity is the ratio of true negatives to the sum of true negatives and false positives (i.e., total number of inliers). It represents the fraction of correctly detected inliers by the unsupervised ODT. Specificity is expressed as

$$\text{Specificity} = \frac{TN}{TN + FP} \tag{1.4}$$

It is an important metric in this work as it ensures that inliers are not wrongly labeled as outliers. Specificity should be used together with recall to evaluate the performance of a model. Ideally, we want high recall close to 1 and high specificity close to 1. A high specificity on its own does not indicate a good performance of the unsupervised ODT. For example, if a model detects every data point as an inlier, the specificity is 1, but the recall is 0, indicating that the performance of the unsupervised ODT is bad.

4.4.3 Balanced accuracy score

Balanced accuracy score is the arithmetic mean of the specificity and recall. It is a better metric because it combines both recall and specificity into a single metric for evaluating the performance of the outlier detection model. Balanced accuracy score is expressed as

$$\text{Balanced accuracy score} = \frac{\text{Recall} + \text{Specificty}}{2} \tag{1.5}$$

Its values range from 0 to 1, such that 1 indicates a perfect ODT that correctly detects all the inliers and outliers in the dataset. Consider a dataset containing 1000 inliers and 100 outliers. When an unsupervised ODT model detects each sample as an outlier, the recall is 1, the specificity is 0, and the balanced accuracy score is 0.5. Balanced accuracy score is high when large portion of outliers and inliers in data are accurately detected as outliers and inliers, respectively.

4.4.4 Precision

Precision is a measure of the reliability of outlier label assigned by the unsupervised ODT. It represents the fraction of correctly predicted outlier points among all the predicted outliers. It is expressed mathematically as

$$\text{Precision} = \frac{TP}{TP + FP} \tag{1.6}$$

Similar to recall, precision should not be used in isolation to assess the performance. For a dataset containing 1000 inliers and 100 outliers, when the unsupervised ODT detects only one point as an outlier and it happens to be a true outlier, then the precision of the model will be 1.

4.4.5 F1 score

F1 score is the harmonic mean of the recall and precision, like the balanced accuracy score it combines both metrics to overcome their individual limitations. It is expressed as

$$\text{F1 score} = \frac{2 \times \text{Precison} \times \text{Recall}}{\text{Precision} + \text{Recall}} \tag{1.7}$$

The values range from 0 to 1, such that F1 score of 1 indicates a perfect prediction (provided that there is no imbalance between two labels) and 0 indicates a total failure of the model. Consider the case discussed earlier, wherein the dataset contains 1000 inliers and 100 outliers, and the unsupervised ODT detects only 1 outlier and that outlier is correctly detected by the model. In that case, the precision is 1, the recall is 1/100, specificity is 1, balanced accuracy is close to 0.5, and F1 score is close to 0.02. F1 score and balanced accuracy score help detect the poor performance of the unsupervised ODT. For the purpose of outlier detection, good F1 score indicates good recall and good precision, meaning large portion of outliers in data are accurately detected as outliers and large portion of the detected outliers are originally outliers and not inliers.

4.4.6 Receiver operating characteristics (ROC) curve and ROC-AUC score

Unsupervised ODT generally assigns a score to each sample, such that the score represents the likelihood of the sample to be an outlier. Each unsupervised ODT implements a specific decision threshold to determine whether a sample is outlier, such that all samples with scores greater than the decision threshold are labeled as either inlier or outlier. A robust unsupervised ODT should be insensitive to the variations in the decision threshold, that is, the outliers and inliers detected by the unsupervised ODT should not change a lot with changes in decision threshold. ROC curve is a plot of the true positive rate (TPR; recall) vs the false positive rate (FPR; 1 − specificity) of the unsupervised ODT on a dataset at different decision/probability thresholds. When the threshold of an unsupervised ODT is altered, the performance of the unsupervised ODT changes resulting in the ROC curve. For instance, the isolation forest iteratively partitions the feature space to isolate a sample and assigns an outlier score based on the average path length. Samples with shorter path lengths are given a higher outlier score and are considered more likely to be outliers because it is easy to isolate them. A threshold is set for the isolation forest by defining the score beyond which a sample will be considered an outlier. For the isolation forest, the anomaly scores typically range from −1 to 1 with the threshold set at 0 by default, such that negative values (<0) are labeled outliers and positive value (>0) are labeled inliers.

For reliable and robust outlier detection, an unsupervised ODT should have high recall (high TPR) and high specificity (low FPR) that is relatively insensitive to changes in the decision thresholds. For such scenarios, the ROC curve will shift toward the top left corner of the plot shown in Fig. 1.6A, which indicates a robust performance. As the ROC curve shifts to the left top corner, the area under curve (AUC) tends to 1, which represents a perfect outlier detection for various choices of threshold. At an ROC-AUC of 1, an unsupervised ODT is robust and reliable when the recall and specificity are close to 1 and relatively independent of the choice of thresholds. ROC curve

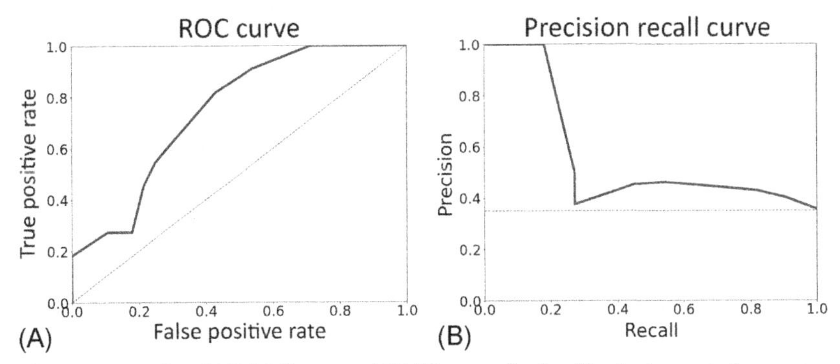

FIG. 1.6 Examples of (A) ROC curve and (B) PR curve of a classifier for demonstration purpose. Area under the blue curves represent the AUC.

exhibiting a gradient close to 1 and ROC-AUC of 0.5 (*red-dotted line (light gray in the print version)* in Fig. 1.6A; referred as the no-skill line) indicates that the unsupervised ODT is performing only as good as randomly selecting certain samples as outliers. A high ROC-AUC score close to 1 indicates that large portion of actual outliers and inliers will be correctly detected without much sensitivity to the decision thresholds of the method. DBSCAN is not designed for supervised tasks; therefore, there is no inbuilt functionality to generate the ROC curve and the ROC-AUC score. ROC curve should be used when numbers of outliers and inliers are nearly equal without any major imbalance in the dataset. ROC curves for various unsupervised methods on Dataset #1 are shown in Figs. 1.D1–1.D3 in Appendix D.

4.4.7 Precision-recall (PR) curve and PR-AUC score

Like the ROC curve, the precision-recall (PR) curve is a plot of the precision vs recall of the unsupervised ODT on a dataset for various decision thresholds. Performance of the unsupervised ODT is considered to be excellent when the detection has high precision and high recall irrespective of the choice of decision threshold, that is, the PR curve shifts toward the top right corner in the plot away from the *red-dotted line (light gray in the print version)* shown in Fig. 1.6B. The *red-dotted line (light gray in the print version)* is referred as the no-skill line, which is defined by the total number of original positives (i.e., true outliers) divided by the total number of original outliers and original inliers (i.e., total number of sample). The performance demonstrated in Fig. 1.6B is a rather poor performance. AUC of the PR curve is used as a measure of the ODT model performance with an AUC of 1 indicating a robust and reliable outlier detection. A good PR curve should be away from the baseline and toward the left top corner in the plot shown in Fig. 1.6B. PR curve should be used when numbers of outliers and inliers are very different resulting in significant imbalance in the dataset. PR curves for various unsupervised methods on Dataset #1 are shown in Figs. 1.D1–1.D3 in Appendix D.

5 Performance of unsupervised ODTs on the four validation datasets

In this section, the performances of the four unsupervised ODTs (IF, OCSVM, LOF, and DBSCAN) are evaluated by comparing the unsupervised detections against the known labels in the validation datasets. Performance of each model is expressed in terms of balanced accuracy score, F1 score, and ROC-AUC score (AUC for the ROC curve). Balanced accuracy score is high when large portion of actual outliers and inliers in data are accurately detected as outliers and inliers, respectively. For purpose of outlier detection, good F1 score indicates good recall and good precision, indicating that a large fraction of actual outliers in data are accurately detected as outliers and a large fraction of the detected outliers are originally outliers and not inliers. A high ROC-AUC score close to 1 indicates that a large portion of actual outliers and inliers are correctly detected without much sensitivity to the decision thresholds of the unsupervised ODT model. These three metrics are simple evaluation metrics. For more robust assessment, these evaluation metrics should be appropriately weighted to address the effects of outlier-inlier imbalance (i.e., the number of positives/outliers is much smaller than the number of negatives/inliers). Appendix B presents true positives, true negatives, false positives, and false negatives for certain unsupervised ODTs on certain datasets in the form of confusion matrix. Appendix C lists the values of hyperparameters of various models for the unsupervised outlier detection. Because our goal is to find the most reliable unsupervised outlier-detection method, these hyperparameters are not tuned/modified and the other hyperparameters (if any) are set at the default values. Values mentioned in Appendix C are kept constant for all the numerical experiments on the four datasets. In a real-world scenario, without any labels to compare and evaluate the outlier/inlier detections, the hyperparameters need to be tuned based on a manual inspection of each of the outliers and inliers.

5.1 Performance on Dataset #1 containing noisy measurements

The unsupervised ODT model performance is evaluated for three feature subsets referred to as FS1, FS2, and FS2*, where FS1 contains GR, RHOB, and DTC; FS2 contains GR, RHOB, and logarithm of RT; and FS2* contains GR, RHOB, and RT. For the subsets FS1 and FS2* of Dataset #1, DBSCAN performs better than the other models, as indicated by the balanced accuracy score. For the subset FS1 of Dataset #1, the DBSCAN correctly labels 176 of the 200 introduced noise samples as outliers and 3962 of the 4037 "normal" data points as inliers; consequently, DBSCAN has a balanced accuracy score and F1 score of 0.93 and 0.78, respectively. For the subset FS2 of Dataset #1, log transform of resistivity negatively impacts the outlier detection performance. Logarithmic transformation of resistivity reduces the variability in the feature. On using deep resistivity (RT) as is (i.e., without logarithmic transformation) in the subset FS2*,

DBSCAN generates similar performance as with the subset FS1 (Table 1.1). All models except isolation forest (IF) are adversely affected by the logarithmic transformation of RT. Visual representation of the performances in terms of balanced accuracy score is shown in Fig. 1.7A.

LOF model does not perform well in detecting noise in a well-log dataset. Based on the ROC-AUC score, LOF performs the worst compared with OCSVM and IF in terms of the sensitivity of the accuracies (precisions) of both inlier and outlier detections to the decision thresholds. Based on F1 score, DBSCAN has the highest reliability and accuracy (precision) in outlier detection; however, hyperparameter tuning should be done to improve the precision of DBSCAN because the current F1 score is not close to 1. One reason for low F1 score is that we have not addressed the inlier-outlier imbalance. All these evaluation metrics used in this study are simple metrics that can be improved by weighting the metrics to address the effects of imbalance (i.e., the number of positives are one order of magnitude smaller than the number of negatives). F1 score of all the methods can be improved by improving the precision. ROC and PR curves for various unsupervised methods on Dataset #1 are shown in Figs. 1.D1–1.D3 in Appendix D.

TABLE 1.1 Performances of the four unsupervised ODTs on Dataset #1

		Dataset #1 results		
		Balanced accuracy score	*F1 score*	*ROC-AUC score*
Isolation forest	FS1	0.84	0.55	0.93
	FS2	0.85	0.37	0.93
	FS2*	0.88	0.63	0.96
One-class SVM	FS1	0.91	0.57	0.95
	FS2	0.81	0.45	0.93
	FS2*	0.92	0.59	0.96
Local outlier factor	FS1	0.73	0.28	0.79
	FS2	0.62	0.18	0.75
	FS2*	0.68	0.24	0.72
DBSCAN	FS1	0.93	0.78	NA
	FS2	0.66	0.42	NA
	FS2*	0.93	0.76	NA

Visual representation of the performances in terms of balanced accuracy score is shown in Fig. 1.7A.
**log(RT) is replaced by RT.*

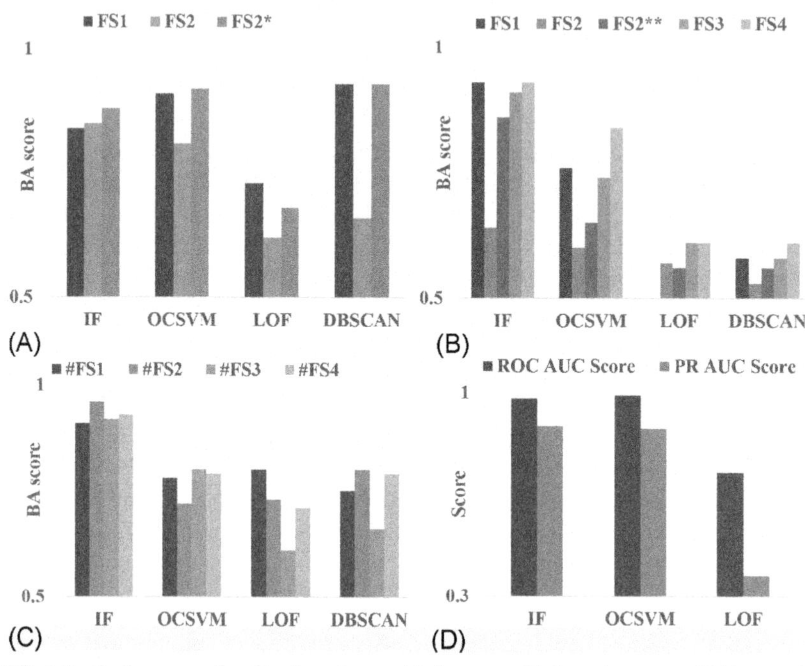

FIG. 1.7 Performance of outlier detection models in terms of balanced accuracy (BA) score for various subsets of (A) Dataset #1, (B) Dataset #2, and (C) Dataset #3, and that in terms of (D) ROC-AUC and PR-AUC scores for Dataset #4.

5.2 Performance on Dataset #2 containing measurements affected by bad holes

In the bad-hole dataset, model performance is evaluated for five feature subsets: FS1, FS2, FS2**, FS3, and FS4. FS1 contains GR, RHOB, and DTC; FS2 contains GR, RHOB, and RT; FS2** contains GR, RHOB, and RXO; FS3 contains GR, RHOB, DTC, and RT; and FS4 contains GR, RHOB, DTC, and NPHI. In each feature set, we have 91 depths (samples) labeled as outliers and 4037 depths labeled as inliers. Isolation forest (IF) performs better than other methods for all the feature sets. DBSCAN and LOF detections are the worst. IF performance for FS2 is worse compared with other feature subsets, because FS2 uses RT, which is a deep-sensing log and is not much affected by the bad holes. Consequently, when RT (deep resistivity) is replaced with RXO (shallow resistivity) in subset FS2**, the IF performance significantly improves indicating the need of shallow-sensing logs for better detection of depths where logs are adversely affected by bad holes. Subset FS3 is created by adding DTC (sonic) to FS2. FS3 has four features, such that DTC is extremely sensitive to the effects of bad holes, whereas RT is not sensitive. In doing so, the performance of IF on FS3 is comparable with that on FS1 and much better than that on FS2. This mandates the use of shallow-sensing logs as features for outlier detection. Visual representation of the performances in terms of balanced accuracy score is shown in Fig. 1.7B.

Outlier detection performance on Dataset #2 clearly shows that when features that are not strongly affected by hole size (e.g., deep resistivity, RT) are used, the model performance drops, as observed in FS2. On the contrary, when shallow-sensing DTC and RXO are used as features, the model performance improves. We conclude that feature selection plays an important role in determining the performance of ODTs, especially in identifying "contextual outliers." IF model is best in detecting contextual outliers, like the group of log responses affected by bad holes. F1 scores are low because the fraction of actual outliers in the dataset is a small fraction (0.022) of the entire dataset, and we do not set contamination level a priori. Being an unsupervised approach, in the absence of constraints such as contamination level, the model is detecting many original inliers as outliers. Therefore, balanced accuracy score and ROC-AUC score are important evaluation metrics (Table 1.2).

5.3 Performance on Dataset #3 containing shaly layers and bad holes with noisy measurements

Performance on Dataset #3 indicates how well a model detects depths where log responses are affected by either noise or bad hole in a heterogenous formation with thin layers of sparsely occurring rock type (i.e., shale). The objective of this evaluation is to test if the models can detect the noise and bad-hole influenced depths (samples) without picking the rare occurrence of shales as outliers. Outlier methods are designed to pick rare occurrences as outliers; however, a good shale zone even if it occurs rarely should not be labeled as outlier by the unsupervised methods.

Comparative study on Dataset #3 involved experiments with four distinct feature subsets sampled from the available features GR, RHOB, DTC, RT, and NPHI logs, namely, FS1, FS2, FS3, and FS4. FS1 contains GR, RHOB, and DTC; FS2 contains GR, RHOB, and RT; FS3 contains GR, RHOB, DTC, and RT; and FS4 contains GR, RHOB, DTC, and NPHI. In all feature sets, 70 points are known outliers, and 704 are known inliers, comprising sandstone, limestone, dolostone, and shales. Isolation forest (IF) model performs better than the rest for all feature sets. Interestingly, with respect to F1 score, IF underperforms on FS2 compared with the rest, due to lower precision and imbalance in dataset. This also suggests that DTC is important for detecting the bad-hole depths, because FS2 does not contain DTC, unlike the rest (Table 1.3). Visual representation of the performances in terms of balanced accuracy score is shown in Fig. 1.7C.

TABLE 1.2 Performances of the four unsupervised ODTs on Dataset #2

Dataset #2 results				
		Balanced accuracy score	*F1 score*	*ROC-AUC score*
Isolation forest	FS1	0.93	0.23	0.97
	FS2	0.64	0.11	0.87
	FS2**	0.86	0.21	0.92
	FS3	0.91	0.22	0.96
	FS4	0.93	0.24	0.99
One-class SVM	FS1	0.76	0.22	0.89
	FS2	0.6	0.11	0.87
	FS2**	0.65	0.14	0.88
	FS3	0.74	0.21	0.88
	FS4	0.84	0.28	0.95
Local outlier factor	FS1	0.38	0.11	0.62
	FS2	0.57	0.07	0.61
	FS2**	0.56	0.08	0.63
	FS3	0.61	0.1	0.62
	FS4	0.61	0.09	0.55
DBSCAN	FS1	0.58	0.18	NA
	FS2	0.53	0.09	NA
	FS2**	0.56	0.17	NA
	FS3	0.58	0.14	NA
	FS4	0.61	0.18	NA

Visual representation of the performances in terms of balanced accuracy score is shown in Fig. 1.7B.
***RT is replaced by RXO.*

5.4 Performance on Dataset #4 containing manually labeled outliers

The offshore dataset contains seven log responses from different lithology, namely, limestones, sandstone, dolomite, shale, and anhydrites. The seven logs are gamma ray (GR), density (DEN), neutron porosity (NEU), compressional sonic transit time (AC), deep and medium resistivities (RDEP and RMED),

TABLE 1.3 Performances of the four unsupervised ODTs on Dataset #3

		Dataset #3 result		
		Balanced accuracy score	*F1 score*	*ROC-AUC score*
Isolation forest	FS1	0.91	0.81	0.97
	FS2	0.96	0.69	0.99
	FS3	0.92	0.84	0.99
	FS4	0.93	0.83	0.99
One-class SVM	FS1	0.78	0.57	0.8
	FS2	0.72	0.47	0.75
	FS3	0.8	0.61	0.81
	FS4	0.79	0.6	0.88
Local outlier factor	FS1	0.8	0.61	0.86
	FS2	0.73	0.24	0.66
	FS3	0.61	0.34	0.79
	FS4	0.71	0.34	0.73
DBSCAN	FS1	0.75	0.95	NA
	FS2	0.8	0.47	NA
	FS3	0.66	0.73	NA
	FS4	0.79	0.73	NA

Visual representation of the performances in terms of balanced accuracy score is shown in Fig. 1.7C.

and photoelectric factor (PEF) logs. Offshore dataset was labeled using manual inspection, feature thresholding, and DBSCAN followed by manual verification of the labels (outliers vs inliers) to create the Dataset #4. Consequently, Dataset #4 contains several manually labeled outliers. This comparative study focuses on IF, OCSVM, and LOF and evaluates their performances using the ROC-AUC score and PR-AUC score. This is a challenging dataset because seven logs from the offshore dataset are being simultaneously processed by the unsupervised methods and then compared with manually verified labels. Increase in the number of features increases the dimensionality of the dataset leading to underperformance of the data-driven methods. IF and OCSVM perform equally well and significantly outperform the LOF method for both the PR-AUC and

TABLE 1.4 Performances of the four unsupervised ODTs on Dataset #4

	Dataset #4 result	
	PR-AUC score	*ROC-AUC score*
Isolation forest	0.89	0.98
One-class SVM	0.88	0.99
Local outlier factor	0.37	0.73

Visual representation of the performances in terms of PR-AUC and ROC-AUC scores is shown in Fig. 1.7D.

ROC-AUC scores. ROC-AUC score for LOF indicates a marginal performance (not a poor performance); however, ROC only considers recall and specificity without accounting for the precision. In this case, PR-AUC score of LOF indicate a very poor performance where precision is considered. Low PR-AUC score of LOF indicates the outlier detection cannot be trusted because of the high fraction of original inliers getting wrongly detected as outliers. PR-AUC and ROC-AUC curves and scores should be analyzed together for the best assessment of unsupervised ODTs. Visual representation of the performances in terms of *PR-AUC and ROC-AUC* score*s is* shown in Fig. 1.7D. ROC curves are appropriate when there is no imbalance, whereas precision-recall curves are suitable for imbalanced datasets (Table 1.4).

6 Conclusions

Four distinct well-log datasets containing outlier/inlier labels were used to perform a comparative study of the performances of four unsupervised outlier detection techniques (ODT), namely, isolation forest (IF), one-class SVM (OCSVM), local outlier factor (LOF), and DBSCAN. Unsupervised ODTs were applied on the dataset without hyperparameter tuning and without any prior information about either the inliers or outliers. Simple evaluation metrics designed for supervised classification methods, such as balanced accuracy score, F1 score, receiver operating characteristics (ROC) curve, precision-recall (PR) curve, and area under curve (AUC), were used to evaluate the performance of the unsupervised ODTs on the labeled validation dataset containing outlier/inlier labels. PR curve, ROC curve, and AUC should be used together to accurately assess the sensitivity of the performance of the unsupervised ODTs to decision thresholds. A robust performance is obtained when AUC is close to 1 indicating that the unsupervised ODT is not sensitive to decision thresholds.

For any specific decision threshold, balanced accuracy score and F1 score should be used together to evaluate the reliability and precision of the unsupervised ODT.

DBSCAN is the most effective in detecting noise in data as outliers, while IF and OCSVM have slightly lower performances in detecting noisy data points as outliers and lower precisions. DBSCAN, IF, and OCSVM are suitable for detecting point outliers, when outliers are scattered around the inlier zone. None of these methods are suitable when outliers occur as dense regions in the feature space as collective outliers. Isolation forest exhibits great performance in detecting contextual outliers when there are zones affected by bad-hole conditions. Isolation forest also proved efficient in detecting outliers when there is mixture of outliers due to noise (point outlier) and bad-hole conditions (contextual outliers) in the presence of an infrequently occurring but relevant and distinct subgroup (which should not be considered as outlier due to its rare occurrence and distinct characteristics). Isolation forest is by far the most robust and reliable in detecting outliers and inliers in the log data. Performance of unsupervised ODTs depends on selection of features, especially when detecting contextual outliers, which will require hyperparameter tuning for optimum performance. For example, shallow-sensing logs improve the detection of depths where logs are adversely affected by bad holes. Local outlier factor is computationally expensive and needs careful hyperparameter tuning for reliable and robust performance; by far, LOF is the worst-performing unsupervised ODT.

Appendix A Popular methods for outlier detection

See Fig. 1.A1.

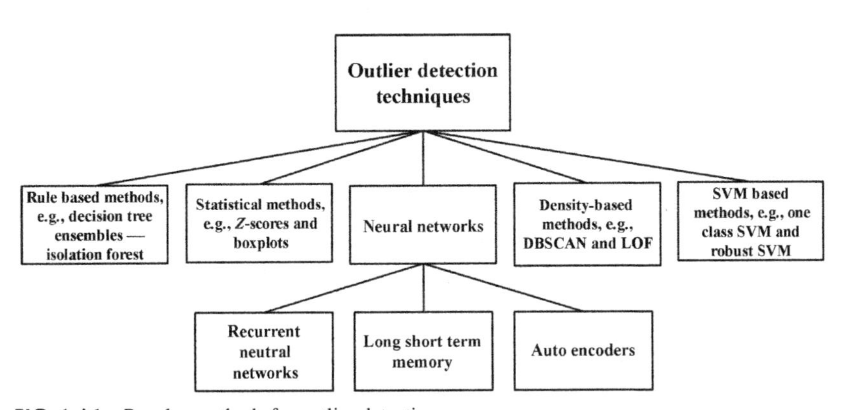

FIG. 1.A1 Popular methods for outlier detection.

Appendix B Confusion matrix to quantify the inlier and outlier detections by the unsupervised ODTs

See Fig. 1.B1.

	Predicted inliers	Predicted outliers
Actual inliers	3962	75
Actual outliers	24	176

(A)

	Predicted inliers	Predicted outliers
Actual inliers	3437	600
Actual outliers	0	91

(B)

	Predicted inliers	Predicted outliers
Actual inliers	553	151
Actual outliers	15	55

(C)

	Predicted inliers	Predicted outliers
Actual inliers	9080	252
Actual outliers	258	396

(D)

FIG. 1.B1 Confusion matrices for (A) DBSCAN applied on the subset FS1 of Dataset #1, (B) IF applied on the subset FS4 of Dataset #2, (C) LOF applied on the subset FS1 of Dataset #3, and (D) OCSVM applied on the Dataset #4. IF applied on the subset FS4 of Dataset #2 has the best performance in detecting outliers. OCSVM applied on the Dataset #4 has the worst performance in detecting outliers.

Appendix C Values of important hyperparameters of the unsupervised ODT models

Model	Hyperparameters
Isolation forest	n_estimators = 100, max_samples = 256, contamination = 'auto'[a], max_features = 1 (default value in scikit learn)
One-class SVM	gamma = 'auto'[b], nu = 0.1
Local outlier factor	n_neighbors = 20, metric = 'euclidean', contamination = 'auto'
DBSCAN	eps = 0.5, min_samples = 5, metric = 'euclidean'

[a]*Contamination refers to the fraction of outlier samples in the dataset; when set at 'auto', the model uses its default threshold. When contamination is set (0 < x < 1), the model selects x of the number of samples in the dataset as outliers based on their anomaly scores.*
[b]*Gamma set at 'auto' simply means the gamma value is 1/(number of features).*

Appendix D Receiver operating characteristics (ROC) and precision-recall (PR) curves for various unsupervised ODTs on the Dataset #1

See Figs. 1.D1–1.D3.

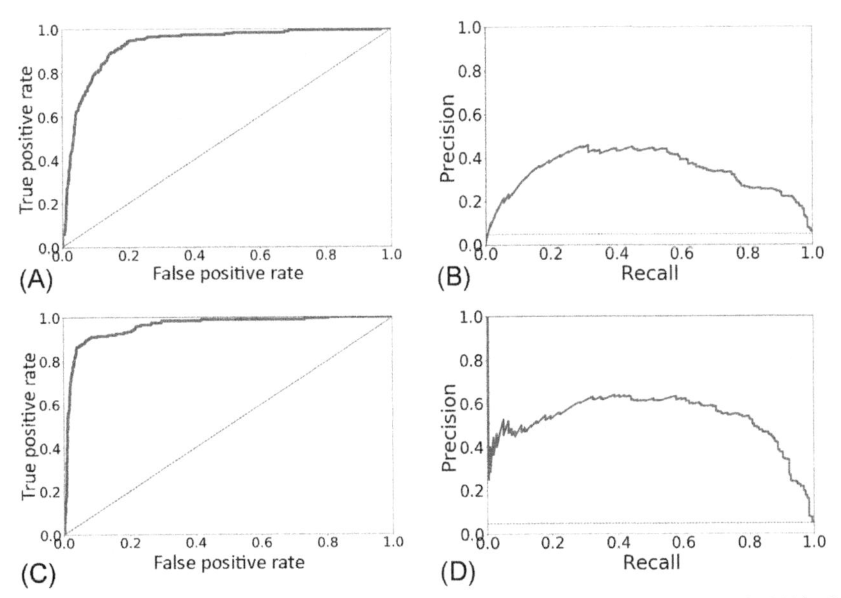

FIG. 1.D1 ROC curves for the isolation forest applied on the subset (A) FS2 and (C) FS2* of Dataset #1. PR curves for the isolation forest applied on the subset (B) FS2 and (D) FS2* of Dataset #1. ROC curve indicates a great performance of the isolation forest model, but PR curve indicates there is room for improvement especially in FS2 dataset. PR curve indicates that the performance on FS2** is better than FS2, which is aligned with other evaluation metric. Compared with ROC curve, PR curve is especially suitable when there is outlier-inlier imbalance in the dataset.

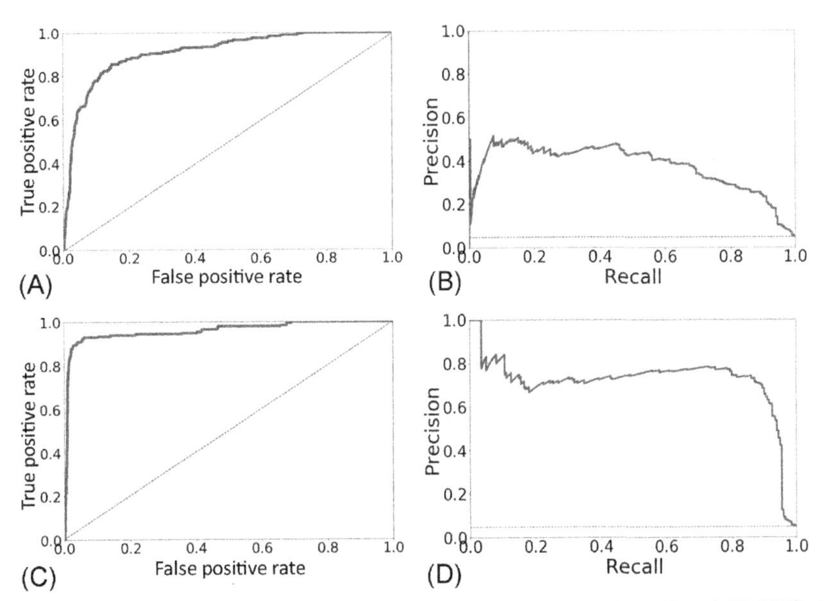

FIG. 1.D2 ROC curves for the one-class SVM applied on the subset (A) FS2 and (C) FS2* of Dataset #1. PR curves for the one-class SVM applied on the subset (B) FS2 and (D) FS2* of Dataset #1. ROC curve indicates similar performances of the isolation forest and one-class SVM, but PR curve indicates the performance of one-class SVM is much better than isolation forest on Dataset #1. Compared with ROC curve, PR curve is especially suitable when there is outlier-inlier imbalance in the dataset.

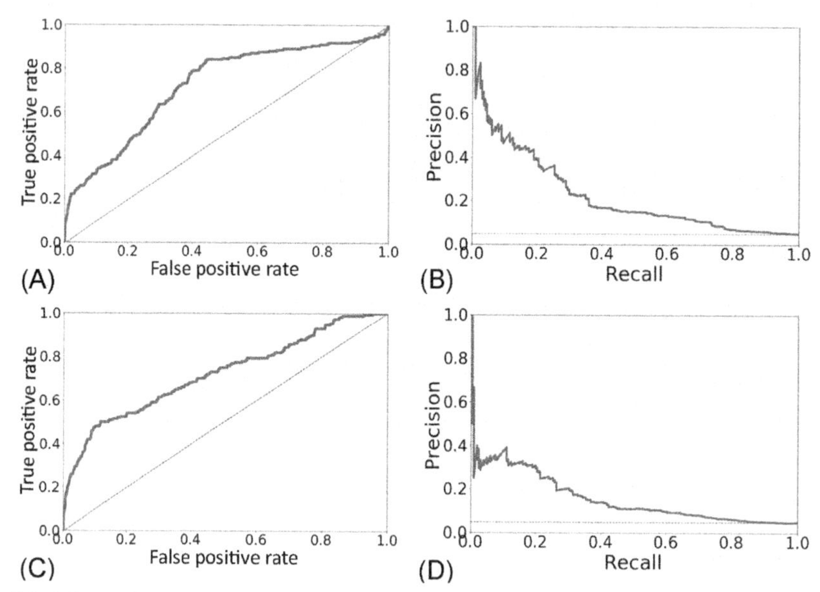

FIG. 1.D3 ROC curves for the local outlier factor applied on the subset (A) FS2 and (C) FS2* of Dataset #1. PR curves for the local outlier factor applied on the subset (B) FS2 and (D) FS2* of Dataset #1. PR curve indicates very poor performance on local outlier factor.

Acknowledgments

Workflows and visualizations used in this chapter are based upon the work supported by the U.S. Department of Energy, Office of Science, Office of Basic Energy Sciences, Chemical Sciences Geosciences, and Biosciences Division, under Award Number DE-SC-00019266.

References

[1] Shin A. Clustering needles in a haystack: an information theoretic analysis of minority and outlier detection. In: Presented at the IEEE International Conference on Data Mining; 2007. p. 13–22.

[2] Zengyou H, Xiaofei X, Shengchun D. Discovering cluster-based local outliers. Pattern Recogn Lett 2003;24(9–10):1641–50.

[3] Chaudhary N, Lee J. Detecting and removing outliers in production data to enhance production forecasting. In: Presented at the SPE/IAEE Hydrocarbon Economics and Evaluation Symposium, Houston, TX, 2016.

[4] Luis M, Nayat S-P, Jose MM, Ana CBG. Anomaly detection based on sensor data in petroleum industry applications. Sensors 2015;15(2):2774–97.

[5] Lewinson E. Outlier detection with isolation forest. Towards Data Science; July 2018.

[6] Liu TF, Ming KM, Zhou Z-H. Isolation forest. In: Presented at the IEEE International Conference, Pisa, Italy; 2008.

[7] Akkurt R, Conroy T, Psaila D, Paxton A, Low J, Spaans P. Accelerating and enhancing petrophysical analysis with machine learning: a case study of an automated system for well log outlier detection and reconstruction. In: Presented at the SPWLA 59th Annual Logging Symposium, London, UK; 2018.
[8] Raschka S, Mirjalili V. Python Machine Learning: Machine Learning and Deep Learning With Python, Scikit-Learn, and TensorFlow. 2nd ed. Packt; 2017.
[9] Ester M, Kriegel H-P, Sander J, Xu X. A density based algorithm for discovering clusters in large spatial database with noise. In: Presented at the International Conference on Knowledge Discovery and Data Mining, Portland, Oregon; 1996.
[10] Wenig P. Local outlier factor for anomaly detection. In: Towards Data Science, 2018.
[11] Pedregosa F, et al. Scikit-learn: machine learning in python. J Mach Learn Res 2011;12:2825–30.
[12] Osborne J, Overbay A. The power of outliers (and why researchers should ALWAYS check for them). Pract Assess Res Eval 2004;9(6):1–8.
[13] Ferdowsi H, Jagannathan S, Zawodniok M. An online outlier detection and removal scheme for improving fault detection performance. IEEE Trans Neural Netw Learn Syst 2014;25(5):908–19.
[14] Orr J, Sackett P, Dubois C. Outlier detection and treatment in I/O psychology: a survey of researcher beliefs and an empirical illustration. Pers Psychol 1991;44:473–86.
[15] He J, Misra S. Generation of synthetic dielectric dispersion logs in organic-rich shale formations using neural-network models. Geophysics 2019;84(3):D117–29.
[16] He J, Misra S, Li H. Comparative study of shallow learning models for generating compressional and shear traveltime logs. Petrophysics 2018;59(06):826–40.
[17] Wu Y, Misra S, Sondergeld C, Curtis M, Jernigen J. Machine learning for locating organic matter and pores in scanning electron microscopy images of organic-rich shales. Fuel 2019;253:662–76.

Chapter 2

Unsupervised clustering methods for noninvasive characterization of fracture-induced geomechanical alterations

Siddharth Misra*, Aditya Chakravarty*, Pritesh Bhoumick[†,a] and Chandra S. Rai[‡]

*Harold Vance Department of Petroleum Engineering, Texas A&M University, College Station, TX, United States, †PricewaterhouseCoopers (PwC), Houston, TX, United States, ‡The University of Oklahoma, Norman, OK, United States

Chapter outline

---

[a] *Present address*: Pricewaterhouse Coopers, Houston, TX, United States.

Machine Learning for Subsurface Characterization. https://doi.org/10.1016/B978-0-12-817736-5.00006-5

1 Introduction

Fractures influence the transport and mechanical properties of materials. Fracture characterization is essential for producing from unconventional hydrocarbon reservoirs, deep mineral resources, and subsurface geothermal energy resources. Fractures are mechanical discontinuities that influence the bulk and shear moduli. Consequently, compressional and shear wave propagations can be used for fracture characterization. When the dimensions of fractures are orders of magnitude smaller than the wavelength, an effective medium approach is valid, wherein the elastic constants can be expressed as a function of the fracture density/intensity [1, 2, 3]. Requirements for the validity of effective medium approach are not met in laboratory-based fracture characterization studies because the fracture dimensions are comparable with wavelength of ultrasonic waves used for various laboratory measurements. Significant gaps still exist in fracture characterization using compressional and shear wave propagations due to heterogeneity of mechanical properties and complexity of fracture geometry in a fractured material.

Experiments have shown that the displacement discontinuity model can capture many of the frequency-dependent and saturation-dependent effects of fracture on a sonic wave propagating through a fractured material. The displacement discontinuity theory predicts that amplitude variation at a given frequency depends on the fracture stiffness [4, 5, 6], which is defined as the ratio between the stress and the magnitude of displacement discontinuity produced by the fracture. A fractured zone in a material will have a lower fracture stiffness compared with an intact zone. A schematic representation of the displacement discontinuity model and the effect of fracture stiffness on the amplitude of a transmitted sonic wave is shown in Fig. 2.1. A fracture is modeled as a zero-width zone between two elastic half spaces, having

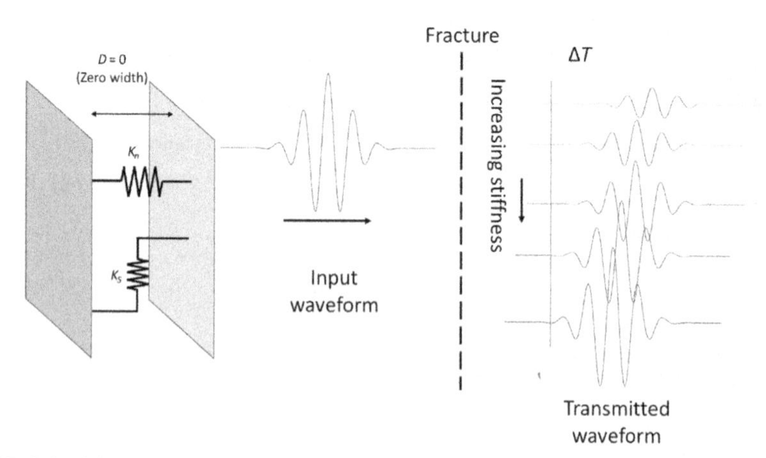

FIG. 2.1 Schematic representation of the displacement discontinuity model, where a fracture is represented as zero-width boundary between two elastic media and the compressional and shear stiffness of the fracture depends on the extent of fracture-induced geomechanical alteration.

compressional and shear stiffness. Upon crossing the fracture the wave experiences attenuation and a time delay, both of which are controlled by the stiffness of fracture and sonic impedance of the intact material. Therefore, a decrease in amplitude along with a delay of the arrival time can be used to identify fractured zones in fractured material.

2 Objective of this study

The study aims to noninvasively map the fracture-induced geomechanical alteration in a hydraulically fractured geomaterial. To that end, unsupervised clustering methods will be applied on the laboratory measurements of ultrasonic shear waveforms transmitted through the fractured geomaterial. The proposed workflow (Fig. 2.2) can be adapted for improved fracture characterization using sonic-logging and seismic waveform data. Several researchers have used machine learning to analyze seismic events, like earthquakes, volcano activity, and rock stability [7, 8, 9]; however, no known reference exists that applies clustering methods to noninvasively visualize the fractured zones and geomechanical alterations in geomaterials.

3 Laboratory setup and measurements

The experiments were performed at the IC^3 laboratory (http://ic3db.ou.edu/home/) at The University of Oklahoma. Present study analyzes the data

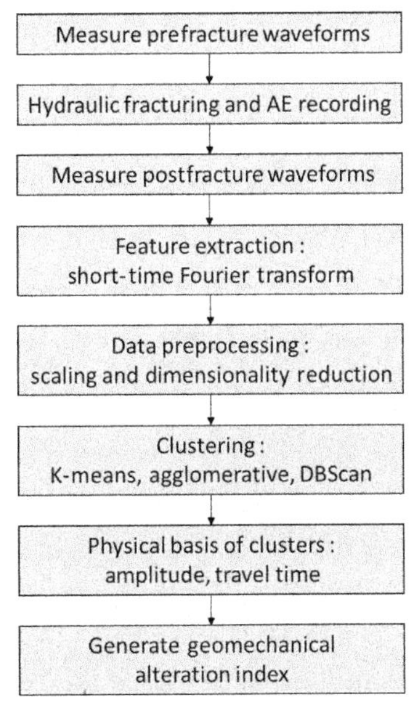

FIG. 2.2 Workflow for the noninvasive mapping/visualization of the fracture-induced geomechanical alterations in a hydraulically fractured geomaterial. Clustering results are converted into geomechanical alteration index using displacement discontinuity theory.

reported by Bhoumick [10]. The hydraulic fracturing, acoustic emissions and shear-waveform measurements were performed on a cylindrical sample of Tennessee sandstone that has a diameter of 152 mm and length of 154 mm. The data acquisition and processing workflow is outlined in Fig. 2.2. The experiment begins with circumferential velocity analysis (CVA), which is the measurement of P-wave velocity across the circumference of the sample at different azimuths. This measurement determines the P-wave velocity anisotropy and hence the direction of fabric in the horizontal plane perpendicular to axis of the cylinder [10]. The maximum azimuthal variation in P-wave velocity was 2.7 %, and the mean P-wave velocity determined from CVA was 3.26 km/s.

The schematic of hydraulic fracturing experiment is presented in Fig. 2.3. To enable the hydraulic fracturing of the Tennessee sandstone sample (Fig. 2.3), a 0.25-in. borehole is drilled into the sample, extending slightly more than half of the sample length (154 mm). A steel tube of 0.24-in. OD and 0.187-in. ID, with two perforation slots placed roughly 180 degrees apart, is placed approximately 0.15 in from bottom of the tube. The tubing is cemented in place using an epoxy. The bottom of tubing is also sealed with epoxy. The epoxy was set for 24 h. Water at room temperature is used as the fracturing fluid. Uniaxial stress of 870 psi oriented at 90 degrees to the direction of fabric was applied on the sample during the hydraulic fracturing. The sample is fractured by pumping fluid at constant rate of 15 cc/min using a syringe pump. Fluid is pumped till breakdown is achieved and continued to pump until the postbreakdown pressure stabilized. The acoustic emissions (AEs) from the hydraulic fracturing are

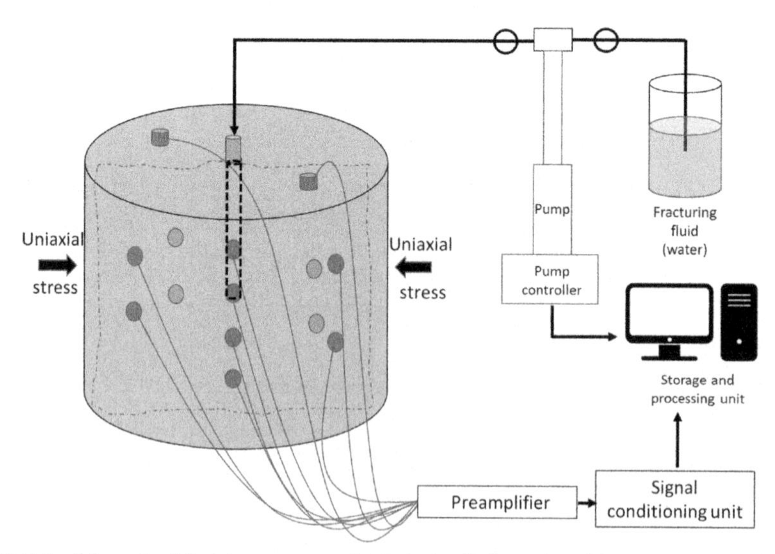

FIG. 2.3 Schematic of the laboratory setup for the hydraulic fracturing of the Tennessee sandstone sample and simultaneous recording of acoustic-emission signals. The boundary of the plane containing the major fracture induced due to the hydraulic fracturing is marked by *dotted lines*.

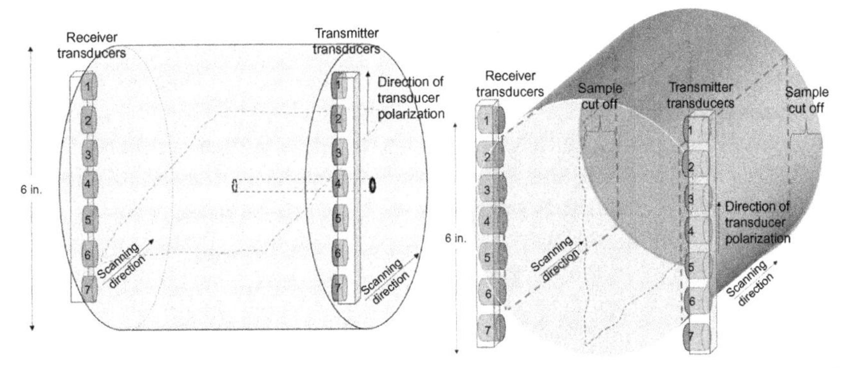

FIG. 2.4 Schematic of shear-waveform data acquisition by moving the sample at an increment of 1 mm with respect to the linear stack of seven transducer assemblies across the axial surface (left) and frontal surface (right). Prefracture waveforms were collected only across the axial surface. Sensors 1 and 7 do not collect data when scanning across the frontal surface. The boundary of the plane containing the major fracture induced due to the hydraulic fracturing is marked by *dotted lines*.

recorded by 16 piezo-electric transducers (Fig. 2.3). A total 1309 AE events could be located within the Tennessee sandstone sample.

The setup for recording shear waveforms consists of seven linearly aligned transducer assemblies, as shown in Fig. 2.4. The spacing between each transducer assembly is 17 mm. Each transducer assembly consists of two linearly aligned piezoelectric transducers (source and sensor pair) placed on front and back frontal/axial surfaces of the sample. The transducer assemblies are scanned across the axial and frontal surfaces of the cylindrical Tennessee sandstone sample. For purposes of attaching the transducers, flat frontal surfaces (Fig. 2.4, right) were created by cutting and removing two portions of the cylindrical sandstone sample from the front and the back (0.5 in. from the sample edge) to obtain smooth flat surfaces. The transducer stacks are pressed on the sandstone surface using air-driven actuators to ensure firm contact between sample and transducers. Honey is used as the coupling agent between the transducers and sample surface. The seven transducer assemblies are moved together over 133 locations separated by 1 mm to separately scan the entire axial and flattened-frontal surfaces of the Tennessee sandstone sample (scanning direction is shown in Fig. 2.4). Hence the lateral resolution of measurements is 1 mm, and the vertical resolution of measurements is 17 mm for both the frontal and axial scans.

For purposes of scanning the axial and frontal surfaces, the bottom plate on which the sample rests is moved at increments of 1 mm. Steel tubing outside sample that connected the sample with the pressurizing system is cut off after fracturing to facilitate the movement of transducers across the surfaces of the sample. When scanning the axial surface (Fig. 2.4, left), all the seven transducer assemblies are in contact, whereas only five inner transducer assemblies are in contact when scanning the frontal surface (Fig. 2.4, right). Sinusoidal pulses of 6 MHz are transmitted into the Tennessee sandstone by each source transducer (one at a time), and the transmitted shear waveforms are measured only by the corresponding sensor, in the same transducer assembly. Each transducer assembly is active one at a time for transmitting the pulses and recording the waveforms. Total number of shear waveforms measured across the axial surface is 133×7 waveforms, whereas that across the frontal surface is 133×5 waveforms. Prefracture shear waveforms (Step 1 in the workflow shown in Fig. 2.2) are recorded only for the axial surface prior to the hydraulic fracturing. Postfracture shear waveforms (Step 3 in the workflow shown in Fig. 2.2) are recorded for both the axial and frontal surfaces after the hydraulic fracturing.

4 Clustering methods for the proposed noninvasive visualization of geomechanical alterations

The goal of the study is to noninvasively visualize the fracture-induced geomechanical alterations in the sandstone sample by first clustering the shear-waveform measurements and then using displacement discontinuity theory to convert the cluster labels/IDs to physically consistent geomechanical alteration index. The interaction of ultrasonic wave propagating through the fractured material with the fractures and geomechanically altered zones influences the amplitude, time delay, and other spatial and frequency-time characteristics of the transmitted waveforms. At 133 locations on the axial surface and 133 locations on the frontal surface, the seven and five transducer sensors, respectively, record the shear waveforms affected by the geomechanical alterations and fractures induced in the Tennessee sandstone due to hydraulic fracturing. Clustering methods are used to group similar shear waveforms, such that waveforms in each group can be assumed to have interacted with similar fracture intensity/density and geomechanical alterations as the wave travels from source to sensor. Therefore, these groups generated using clustering represent the degree of fracture-induced geomechanical alterations. The present study considers three methods for clustering the shear waveforms: K-means clustering, agglomerative clustering, and density-based spectral clustering of applications with noise (DBSCAN). Like any unsupervised learning method, clustering methods do not have a predefined outcome and these methods group samples based on certain similarity/dissimilarity measures.

4.1 K-means clustering

K-means is the simplest and a widely used method for clustering. K-means algorithm begins with a certain predefined number randomly selected samples (data points) as cluster centers. Cluster labels are then assigned to each sample based on the nearest cluster center. Centroid of each newly formed cluster is then treated as a new cluster center, and samples are reassigned cluster labels based on the newly computed nearest cluster centers. This requires calculations of distances of all cluster centers from all the samples in the dataset. The objective of K-means algorithm is to minimize the average sum of the squared Euclidean distances of samples from their corresponding cluster centers. Iterative refinement of cluster centers and cluster labels of samples are then performed to optimize the positions of cluster centers till the algorithm reaches one of the stopping criteria:

1. Convergence of the cluster centers, such that cluster centers negligibly change over several iterations.
2. Number of iterations reaches a maximum value specified by the user.
3. Variance of each cluster changes negligibly over several iterations.
4. Average squared Euclidean distances of samples from their cluster centers reaches local minima.

For each iteration of the algorithm, a new cluster is generated based only on the old clusters. When there are n samples and k clusters to be formed, there are k^n possible clusters. Quality of final clusters, identified using K-means, is quantified using inertia or silhouette score. Few limitations of K-means algorithms include the following:

1. Only suitable for globular, isotropic, well-separated, and equally sized clusters.
2. Number of clusters needs to be predefined by the user. Generally, the elbow method is used to identify optimum number of cluster centers, but it can be a computationally expensive process.
3. A single application of K-means generates nonunique clusters and cluster centers.
4. Being a distance-based method, K-means needs feature scaling and dimensionality reduction.
5. Not suitable for high-dimensional dataset, where each sample has a large number of features/attributes. K-Means is not suitable for high-dimensional dataset because more samples will be equidistant to each other and to cluster centers with increase in dimensions/features. In other words, for higher-dimensional dataset, the concept of distance becomes a weak metric for quantifying similarity between samples.
6. Computationally expensive for large-sized dataset because each adjustment of cluster center requires calculations of distances of all cluster centers from all the samples in the dataset.

4.2 Agglomerative clustering

Hierarchical clustering can be broadly categorized into agglomerative clustering (bottom-up) and divisive clustering (top-down). Agglomerative clustering is a subset of hierarchical clustering. In our study, we use agglomerative clustering, which starts by assuming each sample as a cluster and then groups similar objects (clusters or samples) into higher-level clusters, such that the newly formed higher-level clusters are distinct and the objects within each cluster are similar. Hierarchical clustering builds a hierarchy of clusters (represented as dendrogram) from low-level clusters to high-level clusters based on a user-specified distance metric and linkage criteria. Linkage criteria are used to define similarity between two clusters and decide which two clusters to combine for generating new higher-hierarchy clusters. Few popular linkage criteria are single linkage, complete linkage, average linkage, centroid linkage, and Ward's linkage. Single linkage combines two clusters that have minimum minimally separated samples between the two clusters. Single linkage is suitable for nonglobular clusters, tends to generate elongated clusters, and gets affected by noise. Complete linkage combines two clusters that have minimum maximally separated samples between the two clusters. Complete linkage is not suitable for nonglobular clusters, tends to generate globular clusters, and resistant to noise. Ward's linkage combines two clusters such that the new cluster results in the smallest increase in the variance. Ward's linkage generates dense cluster concentrated toward the middle, whereas marginal samples/points are few and relatively scattered. Both the distance metric and linkage criteria need to be defined in accordance to the phenomena/processes that generated the dataset. For example, when clustering accident sites or regions of high vehicular traffic in a dense urban city, we should use Manhattan distance instead of Euclidian distance as the distance metric. Another example is when dataset is generated due to slowly changing process (e.g., civilization), a single linkage is most suited to group similar clusters (e.g., archeological objects).

Few advantages of agglomerative clustering are as follows:

1. It is suitable when the underlying data have structure, order, and interdependencies (like the correlations in financial markets).
2. It generates a hierarchy that facilitates selection of number of clusters in the dataset.
3. It allows generation of clusters at user-defined granularity by searching through the dendrogram.
4. Unlike K-means, agglomerative (hierarchical) clustering does not require the user to specify the number of clusters prior to applying the algorithm.

Few disadvantages of agglomerative clustering are as follows:

1. It is necessary to specify both the distance metric and the linkage criteria, which are selected without any strong theoretical basis.
2. Compared with K-means that is linear in computational time, hierarchical clustering techniques are quadratic, that is, computationally expensive and slow.

3. Computation time significantly increases with the increase in data size and dimensionality.
4. Compared with K-means, the hierarchical clustering does not provide the best/optimum solution because there is no objective function; consequently, hierarchical clustering is difficult to implement and interpret.
5. Unlike K-means and single/complete linkage, Ward's linkage distorts the feature space and is not space conserving.
6. It does not allow backtracking and object swapping between clusters. Once a certain label is assigned to a sample or to a cluster containing a certain sample, the subsequent labels are assigned to that sample depending on the prior labels and the hierarchy [11].
7. It emphasizes a collection of samples over individual samples when generating new clusters.

4.3 DBSCAN

Density-based spatial clustering of applications with noise (DBSCAN) is based on the assumption that clusters are dense regions in the feature space separated by lower-density regions [12]. DBSCAN uses proximity and density of samples to form clusters. Each sample in the clusters identified by DBSCAN have at least a minimum number of neighboring samples (*nmin*) within a certain distance (depends on the user-specified bandwidth). When implementing DBSCAN, user needs to specify values for *nmin* and bandwidth that are suited for a given dataset. DBSCAN algorithm starts by computing pair-wise distances between all samples. Following that, each sample in the dataset is labeled as either core, border, or noise point based on the user-specified minimum number of neighboring samples (*nmin*) within a certain distance (depends on the user-specified bandwidth) around the sample. In doing so, any sample with at least a certain number of neighbors within a certain distance is marked as core point, and any sample within the neighborhood of a core point but with less than a certain number of neighbors within a certain distance is marked as border point. All points that are neither core or border points are marked as noise. Following that, DBSCAN randomly selects a core point (not assigned to any cluster) and recursively finds all density-connected points, which are assigned to the same cluster as the randomly selected core point. These steps are iterated till all samples are assigned a cluster label or marked as outlier. A user needs to carefully select the optimal values of bandwidth and *nmin*. Small bandwidth and large *nmin* values will result in several sparsely distributed, diffused clusters, where several samples are marked as noise points. Large bandwidth will generate few large clusters. For noisy datasets, it is recommended to have larger values of *nmin*. *K*-distance graph method is used to select the optimal bandwidth.

Few advantages of DBSCAN are as follows:

1. It is suitable for clusters with arbitrary shapes that are not spherical, convex, well separated, and compact.
2. It is robust to noise and outliers.
3. Unlike K-means, DBSCAN infers the optimal number of clusters from the dataset.
4. It is suited for detecting high-density regions and separating them from low-density regions.
5. Compared with K-means and hierarchical clustering, DBSCAN approach is closer to human intuition-based approach to clustering.

Few disadvantages of DBSCAN are as follows:

1. Like K-means and hierarchical clustering, DBSCAN is not suited for high-dimensional large-sized datasets due to the numerous distance calculations required for the assessment of density.
2. It is computationally expensive.
3. It is not suited for datasets that have clusters of large differences in density.
4. Like K-means and hierarchical clustering, DBSCAN requires feature scaling and dimensionality reduction.
5. A user needs to carefully select the optimal values of bandwidth and *nmin*. Choosing these values correctly is important for the performance of the algorithm. Sometimes, domain knowledge is needed to select good values for bandwidth and *nmin*.

5 Features/attributes for the proposed noninvasive visualization of geomechanical alteration

Features are measurable properties or characteristics that describe a system/phenomenon. For example, porosity, permeability, and oil saturation can be considered as features for identifying a good hydrocarbon reservoir. Each feature should be informative, discriminating, and independent to develop robust unsupervised methods. An unsupervised clustering method is as good as the available features and the quality of data. The original data for this study comprise shear waveforms, which are measurements of amplitude at certain time steps (Fig. 2.5, top). Each waveform is made of 1375 signal amplitudes measured at a time step of 40 ns. Sensor is placed opposite to source in a transducer assembly. When scanning the axial surface, waveforms are acquired every 1 mm along the diameter; consequently, 133 waveforms are collected by each of the seven transducer assemblies, which totals to 931 waveforms. The waveform dataset from the axial surface has a size of 931 samples and dimensionality of 1375. Such high dimensionality is not conducive for distance-based or density-based clustering. High dimensionality leads to several adverse effects, also referred as the curse of

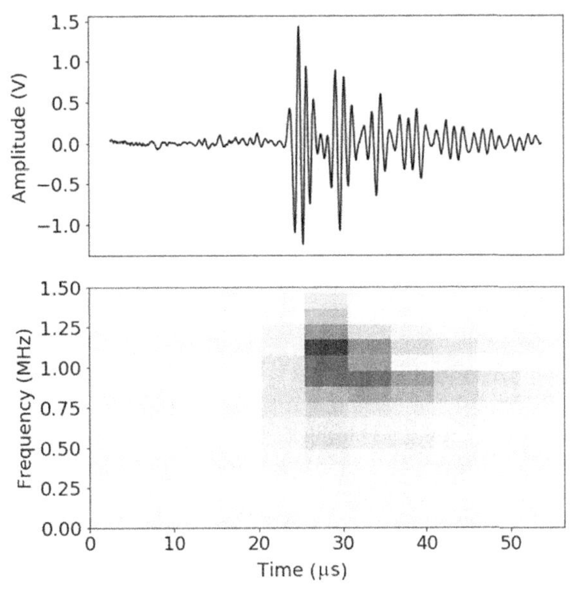

FIG. 2.5 (Upper panel) Original shear waveform and (lower panel) corresponding spectrogram obtained by processing the waveform using short-time Fourier transform (STFT).

dimensionality. Few unwanted challenges due to a large number of features (high dimensionality) are as follows:

- Distance, density, and volume-based machine learning methods fail to find generalizable data-driven model.
- Computationally expensive to develop and test the data-driven models.
- Large memory requirement to store the high-dimensional dataset.
- More data are required to develop generalizable models.
- Nonuniqueness of the model predictions that are more sensitive to noise.
- Data-driven models and their predictions become harder to interpret and explain.
- Exploratory data analysis is challenging because of the difficulty in visualizing the relationships between features and those between the features and targets.

We use feature engineering followed by dimensionality reduction (Steps 4 and 5 in the workflow shown in Fig. 2.2) to convert the high-dimensional shear-waveform dataset to a low-dimensional dataset suitable for clustering methods.

5.1 Feature engineering

Feature engineering is the process of using domain knowledge along with mathematical/statistical transformations to derive new features from the raw data for purposes of improving a specific learning task. The choice of transformations for feature engineering depends on factors like the data type, data structure, data size, learning model, and the desired outcome of the learning. Few examples of feature engineering methods are time series aggregations, image filters, and natural language processing. Popular feature engineering methods tend to be unsupervised and easy to interpret. We implement two distinct feature engineering methods (Step 4 in the workflow shown in Fig. 2.2) to generate two distinct sets of features: The first set comprises statistical measures of a stationary oscillatory signal, and the second set comprises features derived using short-time Fourier transformation of nonstationary signal.

Stationary signals are constant in their statistical parameters over time [13]. To extract the first set of statistical features, each waveform is divided into three equal segments, each having a length of 20 μs, and six features are derived for each segment. This generates a total of 18 features per waveform; therefore, the 1375-dimensional raw waveform data are now transformed into 18-dimensional feature set. Following statistical features were derived for each of the three segments:

- **Energy** is defined as the sum of squares of amplitude of signal. The energy transmitted through a fracture depends on various factors such the fracture stiffness, contact area, and type of fracture filling.
- **Kurtosis** is a descriptor of the distribution of the amplitudes relative to the center of the distribution. It measures whether the data are heavy-tailed or light-tailed compared with a normal distribution. Data sets with high kurtosis tend to have heavy tails indicating lot of outliers.
- **Shape factor** is the ratio of the root mean square (RMS) value to the average (arithmetic mean) of absolute amplitudes. Shape factor is representative of a signal type; for example, sine wave, square wave, and Gaussian white noise have shape factors of 1.11, 1, and 1.15, respectively.
- **Crest factor** is the ratio of peak value to the root mean square (RMS) of a waveform. It indicates how extreme the peaks are in a waveform. Shape factor is representative of a signal type; for example, sine wave, square wave, and Gaussian white noise have crest factors of 1.414, 1, and infinity, respectively.
- **Dominant frequency** is the frequency of maximum amplitude (one that carries the most energy) on the frequency spectrum obtained by applying fast Fourier transform (FFT) on the signal. FFT converts a signal from time domain to frequency domain by decomposing the sequence of values into components of different frequency. Fractures act as low-pass filters [5], that is, transmission of sonic waves through fractures results in attenuation of high frequencies in the signal. On comparing dominant frequencies between fractured and intact rock, a reduction in the dominant frequency indicates wave propagation through a fractured zone.

- **Number of signal zero crossings**: A zero-crossing is a point where sign of mathematical function changes, and it is represented by the intercept of function. This feature represents the number of times the value of signal is zero in a time interval representing the oscillatory nature and wavelength of the signal.

A signal is considered stationary when its statistical measures do not change over time, for example, rotating machinery. Nonstationary signals, on the other hand, have time-varying statistical features as the frequency spectrum of such signals change over time, for example, the speech signal and seismic signal. The features outlined earlier (those used for generating the first set of features) are most useful in detecting changes in the "form" of a time-invariant signals. They are used extensively in machine learning assisted damage detection methods in ball bearings and rotating electrical machinery [14] that produce stationary signals. Seismic signals and ultrasonic shear waveforms, on the other hand, are examples of nonstationary signals and feature engineering methods suited for stationary signals are not designed to capture the time-varying parameters in nonstationary signals [15], other than breaking the signals into time windows (segments) and the deriving the stationary statistical features for each segment.

Few popular feature engineering methods for nonstationary signals are short-time Fourier transform (STFT), the continuous wavelet transform (CWT), and the empirical mode decomposition (EMD). Fourier transform (FT) decomposes a signal into a sum of sinusoids. Unlike the widely implemented Fourier transform, the CWT decomposes the signal into shifted and stretched (or compressed) variations of a wavelet. A wavelet is a wavelike oscillation that is localized in nature [16]. Compressing or stretching a function is termed as scaling. The result of a CWT implementation is not a time-frequency map but a map that is termed *scalogram*. CWT produces a time versus scale variation and not a true time versus frequency variation [17]. In present study, it was observed that the CWT-derived scalogram could not clearly detect the differences between pre- and postfracture waveforms. Nonetheless, it has been applied to analyze seismic attributes [18], vehicle-generated noise analysis [19], and acoustic-emission studies in composite materials [20]. Another feature engineering method for nonstationary signal is empirical mode decomposition (EMD), which decomposes a signal into functions comprising the intrinsic node functions (IMFs). EMD is a completely data-driven algorithm with no predefined decomposition basis functions [21] (such as sinusoids in FT and wavelets in CWT). IMFs are extracted recursively from the signal. The algorithm is based on the identification of local maxima and minima, which are used to define the upper and lower envelopes by fitting a spline curve. The mean envelope is then subtracted from the original signal, and the aforementioned process is repeated on the residual signal. The process stops when the mean envelope is close to zero in the entire time series. The stopping criteria is met when the extracted IMF has a small amplitude or when it becomes monotonic [21]. The drawbacks of the original EMD algorithm include mode mixing (spread of one scale over different IMFs), aliasing (overlapping of IMF functions caused due to insufficient sampling rate), and generation of false modes [22]. Improved EMD techniques have been proposed, namely, ensemble EMD (EEMD) and complete ensemble EMD with adaptive noise (CEEMDAN). The improved versions circumvent the problem of mode mixing; however, the generation of spurious IMFs is not uncommon [24]. EMD-based methods have been extensively applied to nondestructive evaluation. EMD-based methods are more computationally expensive compared with STFT or CWT-based methods. Its application to seismic data is limited. Nonetheless, there have been attempts to apply EMD-based method for seismic attribute analysis [23,24].

In our study, we implement short-time Fourier transform (STFT) on the waveforms as the feature engineering method. Applying FFT over a long time window does not reveal the spectral content change with time. To avoid this problem the FFT is applied over short periods of time. For time windows short enough, nonstationary signals can be considered stationary. Short-time Fourier transform (STFT) is a powerful tool for audio signal processing [25]. STFT is used widely in machine learning assisted speech recognition, music analysis, and automatic transcription of audio [25–27]. To generate the STFT, the first step is to define an analysis window and windowing function to generate segments. FFT is applied on the generated segments to obtain the short-time Fourier transform [26]. Owing to its capabilities in handling nonstationary time series, STFT has been extensively applied to monitor seismicity associated with volcanic activity [7, 8] and seismicity associated with rock stability [9]. Fig. 2.5 shows an example of shear wave used in present study and the corresponding STFT-based spectrogram. The coefficients of a spectrogram express the time-frequency variations and are used as features. Each raw waveform is transformed using STFT to generate a spectrogram having 12 time steps and 15 frequency steps. Each time step is 5 μs, and each frequency step is 100 kHz. Hence, by implementing STFT as the feature engineering method, 180 features are derived from the raw shear waveforms. Robust scaling was then used for scaling the features prior to clustering.

5.2 Dimensionality reduction

After feature engineering, we perform dimensionality reduction (Step 5 in the workflow shown in Fig. 2.2) on the newly derived feature set to obtain fewer informative and nonredundant features that facilitate the subsequent learning and generalization steps. Dimensionality reduction is essential for the processing the engineered dataset represented in terms of 180 STFT-derived features. Dimensionality reduction reduces undesired characteristics in high-dimensional data, namely, noise (variance), redundancy (highly correlated variables), and data inadequacy (features $\gg$ samples). Dimensionality reduction leads to some loss of information. Dimensionality reduction methods can be broadly categorized into feature selection and feature extraction methods. Feature selection methods select the most relevant features from the original set of features based on an objective function. Feature extraction finds a smaller set of newly constructed features, which are some combination of the original features. Features obtained using feature selection retain their original characteristics and meaning as in the original feature set, whereas those obtained using feature extraction are nonphysical transformations of the original features that are different from the original feature set. Popular feature selection methods are variance threshold, recursive feature elimination, and ANOVA F-value and mutual information test. Popular feature extraction methods are principal component analysis, factor analysis, ISOMAP, and independent component analysis. In our study, we use principal component analysis (PCA) for feature extraction as the dimensionality reduction technique.

6 Results and discussions

6.1 Effect of feature engineering

Fig. 2.6 compares the visualizations based on the K-means clustering of shear-waveform dataset transformed using STFT followed by PCA with that transformed using stationary statistical methods followed by PCA. Visualizations in Fig. 2.6 show the geomechanical alterations in the postfracture rock quantified in terms of geomechanical alteration (GA) index. PCA was applied on feature-engineered dataset to reduce the dimensionality for better performance of the clustering methods by avoiding the curse of dimensionality. 180 STFT-derived features were reduced to 67 and 88 PCA-derived components, which account for 98% of variance, for visualizing the geomechanically altered zones in the axial and frontal planes, respectively. Similarly, the 18 stationary statistical features were reduced to 12 PCA-derived components, which account for 98% of variance, for visualizing the altered zones in both the axial and frontal planes.

In the frontal plane, both sets of features (obtained by feature engineering followed by dimensionality reduction) indicate very high geomechanical alteration *[red (dark gray in print version)]* around 100 mm height, which is most likely due to the hydraulic fracturing. Unlike stationary statistical features, STFT-derived GA index indicates symmetrical regions of high alteration *[red (dark gray in print version)]*. Also, unlike the stationary statistical features, the STFT-derived GA index clearly shows a large uniform low-alteration region [*pink (light gray in print version)* and *blue*

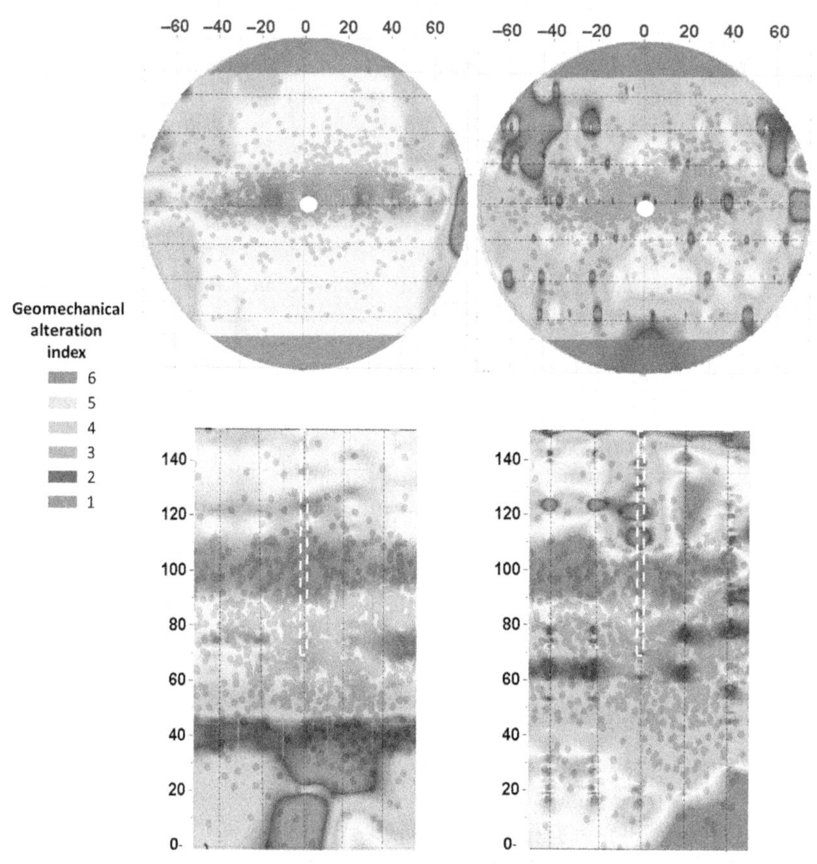

FIG. 2.6 Comparison of the effect of feature engineering. Noninvasive visualization of geomechanical alterations in the postfracture Tennessee sandstone sample in the axial plane (above) and frontal plane (below) obtained by K-means clustering of the shear-waveform dataset transformed using STFT followed by PCA (left) and stationary statistical methods followed by PCA (right). In Figs. 2.6–2.10 the underlying black dots represent the location of acoustic emissions produced during the hydraulic fracturing of the sample under uniaxial stress. In Figs. 2.6–2.10 the thin, *dotted lines* represent the scanning positions of the transducer assemblies, and *white dotted line* represents the drilled borehole in the sample. In Figs. 2.6–2.10 the *gray regions* represent sections not scanned by the transducer assemblies. Hotter colors and larger geomechanical indices indicate larger geomechanical alteration.

(dark gray in print version)] in the entire lower half of the sample. STFT-based visualization exhibits a small region of high alteration at the bottom section of the sample due to issues with sample preparation and sensor placement. In the axial plane the STFT-derived GA index indicates maximum alteration *[yellow (light gray in print version)* and *red (dark gray in print version)]* in the middle of the sample—the same region contains highest density of acoustic-emission hypocenters. Statistical feature-derived clustering shows diffused regions of low alterations [*pink (light gray in print version)* and *blue (dark gray in*

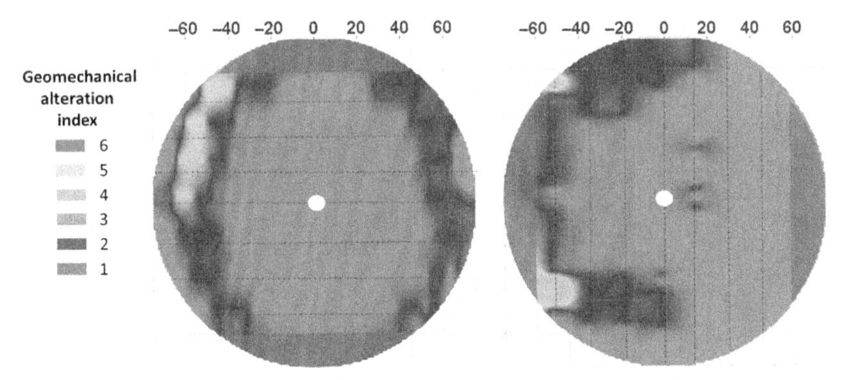

FIG. 2.7 Comparison of the effect of polarization of the shear wave traveling through Tennessee sandstone prior to fracturing. Noninvasive visualization of geomechanical alterations in the prefracture Tennessee sandstone sample in the axial plane obtained by K-means clustering of the shear-waveform dataset measured with direction of transducer polarization parallel (left) and perpendicular (right) to rock fabric. The datasets were transformed using STFT followed by PCA. Hotter colors and larger geomechanical indices indicate larger geomechanical alteration.

print version)]. In the frontal orientation, statistical features show a large region of high alteration in the bottom region of the sample, which is inconsistent. Comparing the figures in Fig. 2.6, clustering of STFT-derived features is more reliable and robust as compared with statistical feature because the STFT-based indices show smooth transition without small zones of abrupt alterations. STFT-derived GA index exhibits smooth transitions in both axial and frontal planes.

Prior to hydraulic fracturing, ultrasonic shear waveforms were measured across the axial surface of the intact sample. Shear waveforms were measured in two perpendicular directions of polarization. For purposes of clustering and visualization, the waveforms are transformed using STFT followed by PCA to obtain 180 features. The dimensionally reduced dataset was clustered using K-means clustering (Fig. 2.7). The two figures in Fig. 2.7 correspond to two directions of polarization of the shear waves. In both cases the geomechanical alteration index is primarily colder colors, which is consistent with the prefracture condition of the rock. Comparing Fig. 2.6 (left, above) with Fig. 2.7 (left), it is evident that maximum geomechanical alteration occurs in the middle portion in the axial plane, and the entire rock has geomechanically altered because of the hydraulic fracturing under uniaxial stress.

6.2 Effect of clustering method

There are numerous clustering algorithms that can be used for the desired noninvasive visualization of geomechanical alteration. In the present study, we implemented three clustering methods: K-means clustering, agglomerative

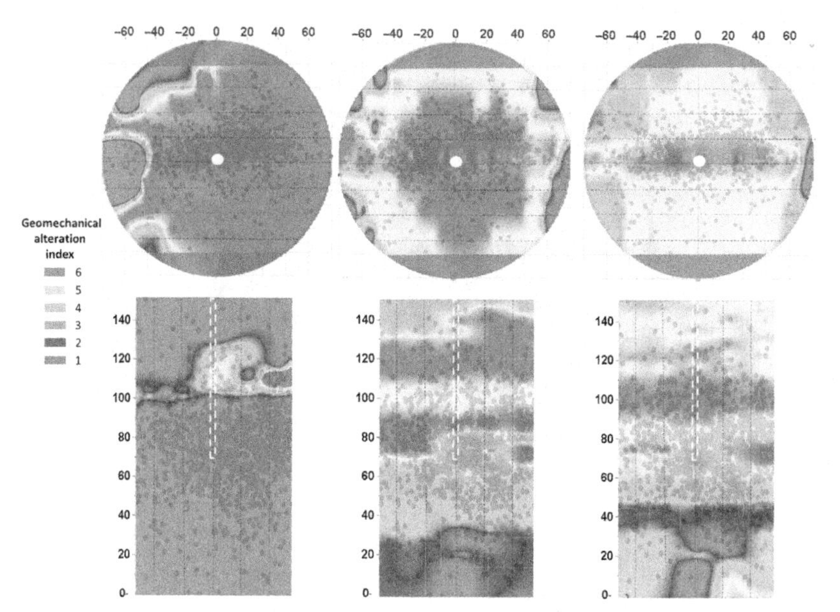

FIG. 2.8 Comparison of the effect of clustering method. Noninvasive visualization of geomechanical alterations in the postfracture Tennessee sandstone sample in the axial plane (above) and frontal plane (below) obtained by DBSCAN (left), agglomerative (middle), and K-means (right) clustering of the shear-waveform dataset transformed using STFT followed by PCA. Hotter colors and larger geomechanical indices indicate larger geomechanical alteration.

clustering, and DBSCAN. Fig. 2.8 shows the geomechanical alteration indices obtained by applying these clustering methods to the postfracture shear-waveform dataset transformed using STFT followed by PCA, similar to that explained in the previous section. In axial orientation, DBSCAN clustering indicates a large region of very high geomechanical alteration *[red (dark gray in print version)]*, whereas the agglomerative clustering shows very high alteration in the middle of the sample and the intensity of alteration tapers away from the sample center toward the sample boundaries. DBSCAN is severely overpredicting the alterations, whereas agglomerative clustering is mildly overpredicting the alterations. In axial plane, K-means clustering shows a large zone of high geomechanical alteration *[yellow (light gray in print version)]* with interspersed small regions of very high alterations *[red (dark gray in print version)]*. K-means-based visualization is the most physically consistent.

In the frontal plane, DBSCAN indicates only a narrow region of alteration in the upper right of the sample; the rest of sample is indicated as intact. Overall the DBSCAN results are the most inconsistent. In frontal plane, both agglomerative and K-means clustering indicate a zone of high alteration extending from the top up to the middle of the frontal section of sample. The lower 40 mm of sample is shown to be relatively less altered. In axial plane, the geomechanical alteration indices generated using agglomerative clustering, as compared with the K-means, have better correlation with the density of acoustic-emission hypocenter. However, K-means generates much better visualization in the front plane that coincides with the acoustic-emission hypocenters. Overall, in comparison with agglomerative clustering, K-means indicates a smaller region of very high alteration in axial and frontal planes and generates more consistent visualization.

6.3 Effect of dimensionality reduction

Dimensionality reduction is used to create a smaller set of informative and relevant features, such that the clustering methods have more generalizable performance. A primary objective of this study is to identify a workflow that generates consistent and repeatable fracture-induced geomechanical alteration index. Therefore, it is important to investigate the effect of dimensionality reduction on the clustering performance. To this end, we compared three different cases of K-means clustering of postfracture shear wave measurements transformed using STFT followed by three different dimensionality reduction approaches (Fig. 2.9). In Case 1, data expressed in

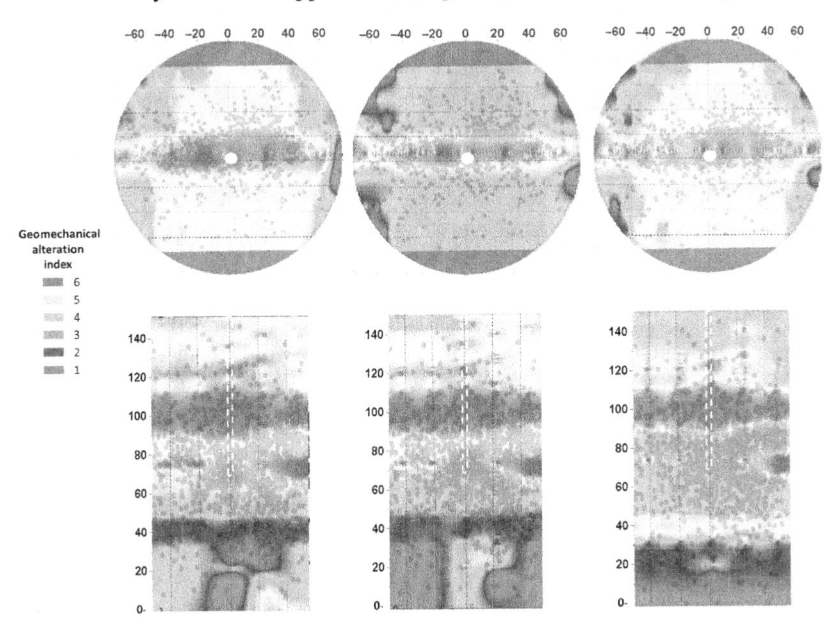

FIG. 2.9 Comparison of the effects of dimensionality reduction. Noninvasive visualization of geomechanical alterations in the postfracture Tennessee sandstone sample in the axial plane (above) and frontal plane (below) obtained by K-means clustering of the shear-waveform dataset transformed using STFT followed by three different dimensionality reduction approaches, namely, (left) Case 1, (middle) Case 2, and (right) Case 3. Hotter colors and larger geomechanical indices indicate larger geomechanical alteration.

terms of 180 STFT-derived features were reduced to 67 and 88 PCA-derived components, which account for 98% of variance, for visualizing the altered zones in the axial and frontal planes, respectively. In Case 2, data expressed in terms of 180 STFT-derived features were reduced to 12 and 15 PCA-derived components, which account for 75% of variance, for visualizing the altered zones in the axial and frontal planes, respectively. In Case 3, feature selection is performed prior to PCA to dimensionally reduce the data expressed in terms of 180 STFT-derived features. In Case 3, two tasks are performed to accomplish the feature selection. First, variance threshold is applied to eliminate STFT-derived features that have variance less than 1.6; then, correlated STFT-derived features having a correlation coefficient higher than 0.9 are removed. The two steps reduce the 180 STFT-derived features to 22 features for axial visualization and 19 features for frontal visualization. After these two steps, PCA is performed to reduce the feature-selected dataset to 18 and 8 PCA-derived components, which account for 98% of variance, for visualizing the altered zones in the axial and frontal planes, respectively. Geomechanical alteration index (Fig. 2.9) for the three cases were generated using K-means clustering. Only postfracture shear wave measurements are considered for this comparison.

In the axial plane, all the three cases show a region of maximum geomechanical alteration around the center of the sample, and the degree of alteration reduces toward the sample boundaries. Case 2 shows unusually large region of low geomechanical alteration; most likely due to the loss of information associated with the loss of 25% of variance. In the frontal orientation, all three cases show a highly altered zone around the height of 100 mm. At the height of 40 mm, Case 1 and Case 2 indicate a slightly altered zone *[blue (dark gray in print version)]*, whereas Case 3 indicates highly altered zone *[yellow (light gray in print version)]*, which is inconsistent. Case 2 seems to be the most inconsistent, while Case 1 is the most consistent with gradual variation in alterations in the vertical and radial directions. In conclusion, shear waveform should be transformed using STFT followed by PCA that retains 98% of the variance for the best visualization.

6.4 Effect of using features derived from both prefracture and postfracture waveforms

Fig. 2.10 shows the effect of combining features derived from prefracture shear waveforms with the features derived from postfracture shear waveforms on the noninvasive visualization. Our hypothesis is that the information from prefracture waveforms may improve the identification of alteration zones. A feature set containing STFT-derived features obtained by transforming prefracture and postfracture measurements has double the number of features than the feature set containing STFT-derived features obtained by transforming only the postfracture waveform. Hence, the dataset containing

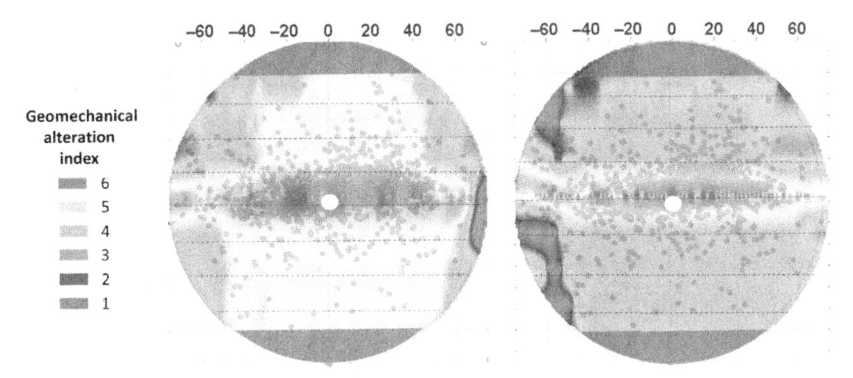

FIG. 2.10 Comparison of the effect of using features derived from both prefracture and postfracture waveforms. Noninvasive visualization of geomechanical alterations in the postfracture Tennessee sandstone sample in the axial plane obtained by K-means clustering of (left) 180 STFT-based features derived from only postfracture shear waveforms and (right) 360 STFT-based features extracted from both pre- and postfracture shear waveforms. For both the cases, STFT-derived features were dimensionally reduced to 120 PCA-derived features, which account for 98% of variance. Hotter colors indicate larger geomechanical alteration.

both pre- and postfracture features has 360 features. This study was done for visualization in the axial plane for which both prefracture and postfracture waveforms were measured. There are no prefracture waveforms for the frontal plane.

In both situations, the clustering produces qualitatively similar alteration indices. The maximum alteration zone is the middle of sample that coincides with the acoustic emission. Surrounding regions show relatively lesser alteration. Visualizations obtained by processing features from both pre- and postfracture waveforms indicate a larger region of low alteration. It appears using prefracture features in this case is equivalent to padding the feature set with constant values. This fact is corroborated by the observation that the same number of PCA-derived components (120) explained 98% variance in both cases. This implies that addition of prefracture data did not introduce any additional variance to the feature set. Overall, using prefracture data may aggravate the curse of dimensionality and reduce the importance of postfracture features for the clustering method.

7 Physical basis of the fracture-induced geomechanical alteration index

According to the displacement discontinuity theory, a lower amplitude of the first arrival and a delayed arrival of the shear waveforms indicate a reduction in the rock stiffness due to fracturing [5]. This principle is used to assign a geomechanical alteration index to the cluster labels/IDs generated by the

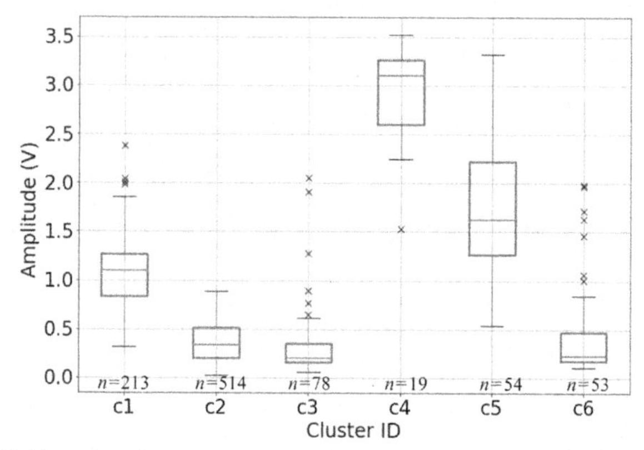

FIG. 2.11 Whisker plot of measured amplitude versus cluster ID assigned by the K-means clustering. Each cluster ID is assigned a geomechanical alteration index based on the amplitude. Higher amplitude indicates higher fracture stiffness or less damage. n indicates the number of waveforms in each cluster out of the total 931 waveforms measured across the axial surface.

clustering methods. In doing so, geomechanical alteration index ranging from 1 to 6 indicates increasing magnitude of alteration in rock stiffness due to the hydraulic fracturing. In this way the data-driven solutions are infused with physics-based factors for improved applications. Fig. 2.11 shows the box plot diagram of the amplitude of first arrival of postfracture shear waveforms measured across the axial surface as a function of the cluster IDs assigned by the K-means clustering. Prior to clustering the dataset was processed using STFT followed by PCA to express the dataset using 67 PCA-derived features, which account for 98% of variance. A box plot is a statistical representation of data wherein the length of the box represents the interquartile range (IQR) of the data, defined as Q3-Q1. The horizontal line within the box indicates the position of the median. The length of the whiskers is proportional to the interquartile range—the upper and lower whiskers extend to Q3 + 1.5*IQR and Q1 − 1.5*IQR, respectively. Data points located outside these bounds are considered outliers and are denoted by *star symbols* above and below the whiskers in the box plot diagram shown in Fig. 2.11.

The geomechanical alteration indices are assigned based on the relative position of the median values. In terms of amplitude of first arrivals of shear waves, highest values of the median are shown by box plot corresponding to cluster C4. Adhering to the tenets of the displacement discontinuity theory, cluster C4 hence is assigned maximum fracture stiffness, that is, low geomechanical alteration index. The lowest value of median is shown by cluster C3, which translates to minimum fracture stiffness, that is, high geomechanical alteration index. Based on the Fig. 2.11, close to 650 of 931 waveforms exhibit interaction with fracture-induced geomechanical alteration.

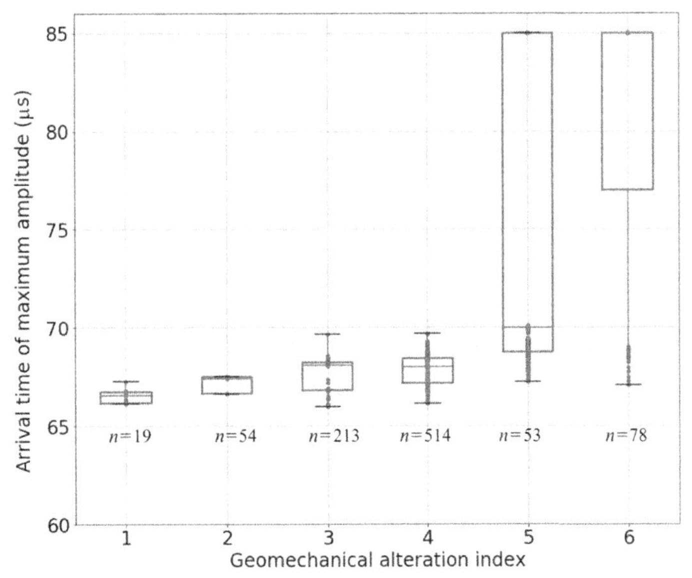

FIG. 2.12 Arrival time of the maximum amplitude of postfracture waveforms as a function of geomechanical alteration index. The alteration index is a surrogate for fracture stiffness. A lower alteration index corresponds to high fracture stiffness and vice versa. The time delay increases with decreasing fracture stiffness, as predicted by the displacement discontinuity theory.

The arrival time of the maximum amplitude of waveforms is plotted as a function of geomechanical alteration index in Fig. 2.12. The figure demonstrates that geomechanical alteration index can be used as a substitute for fracture stiffness. Based on the principles of displacement discontinuity theory, as the alteration index increases, the time delay of first arrival increases. In the absence of a good method to detect first arrival in the presence of fracture, we are using maximum amplitude as the substitute for the first arrival. A smaller index corresponds to high fracture stiffness, and larger alteration index corresponds to lower fracture stiffness, that is, larger alteration. As the amplitude of the first arrival decreases with decrease in fracture stiffness (and increasing geomechanical alteration index), the accuracy of the algorithm to determine the maximum-amplitude arrival decreases. Consequently, we observe wider dispersions in arrival times of the maximum amplitude of waveforms for higher geomechanical alteration index because of the low amplitudes of the waveforms and lower signal to noise ratio. Therefore, the box plots corresponding to high geomechanical alteration indices (5 and 6) show a much greater interquartile range.

8 Conclusions

A data-driven workflow was developed to noninvasively visualize the geomechanical alterations in geomaterials due to hydraulic fracturing. We implemented various unsupervised clustering methods to process ultrasonic shear waveforms dimensionally reduced using short-time Fourier transform followed by principal component analysis. Based on displacement discontinuity theory, each cluster label can be associated with a certain degree of geomechanical alteration (change of stiffness) in geomaterial due to hydraulic fracturing. In this way the data-driven solutions are infused with physics-based factors for improved applications. For an isotropic rock, like Tennessee sandstone, the postfracture shear waveforms (without requiring prefracture waveforms) are shown to be effective in generating reliable noninvasive visualization of geomechanical alterations. Use of short-time Fourier transform followed by robust scaling and principal component analysis ensures that various clustering methods generate relatively similar clustering results. K-means and agglomerative clustering-based geomechanical alteration indices are spatially well correlated to the acoustic-emission events recorded during the hydraulic fracturing; however, K-means clustering generates more consistent visualization. Density-based spectral clustering of applications with noise (DBSCAN) is found be the most ineffective clustering method for purposes of noninvasive visualization of geomechanical alterations.

Acknowledgments

Various workflows and visualizations used in this chapter are based upon work supported by the U.S. Department of Energy, Office of Science, Office of Basic Energy Sciences, Chemical Sciences Geosciences, and Biosciences Division, under Award Number DE-SC-00019266. We thank the Integrated Core Characterization Center (IC^3) and the Unconventional Shale Gas (USG) Consortium at the University of Oklahoma for providing us the shear waveforms and acoustic-emission data.

Declarations

The authors declare that they have no competing interests. AC and SM developed and tested the clustering and feature extraction methods used in the study. AC and SM developed the workflow for the noninvasive visualization/mapping of geomechanical alterations in geomaterials. AC generated all the figures and tables in this study. SM arranged funding to support the development of the workflow for the noninvasive mapping. AC and SM prepared the first complete draft of the chapter. SM wrote various conceptual topics related to data analysis and clustering. PB developed the laboratory setup and performed the shear-waveform measurements at the IC^3 Lab. CR is co-PI of the IC^3 Lab and USG Consortium.

References

[1] O'Connell RJ, Budiansky B. Seismic velocities in dry and saturated cracked solids. J Geophys Res 1974;79:5412–26.

[2] Sayers CM, Kachanov M. A simple technique for finding effective elastic-constants of cracked solids for arbitrary crack orientation statistics. Int J Solids Structures 1991;27(6):671–80.
[3] Hudson JA. Wave speeds and attenuation of elastic waves in material containing cracks. Geophys J Roy Astron Soc 1981;64:133–50.
[4] Schoenberg M. Elastic wave behavior across linear slip interfaces. J Acoust Soc Am 1980;68 (5):1516–21.
[5] Pyrak-Nolte LJ, Myer LR, Cook NGW. Transmission of seismic waves across single natural fractures. J Geophys Res 1990;95(B6).
[6] Myer LR, Hopkins D, Peterson JE, Cook NGW. Seismic wave propagation across multiple fractures. In: Myer LR, Cook NGW, Goodman RE, Tsang P, editors. Fractured and jointed rock masses. Balkema; 1995. p. 105–9.
[7] Bergen KJ, Beroza GC. Earthquake fingerprints: extracting waveform features for similarity-based earthquake detection. Pure Appl Geophys 2019;176(2019):1037–59.
[8] Ibs von-Seht M. Detection and identification of seismic signals recorded at Krakatau volcano (Indonesia) using artificial neural networks. J Volcanol Geotherm Res 2008;176 (2008):448–56.
[9] Weber S, Faillettaz J, Meyer M, Beutel J, Vieli A. Acoustic and microseismic characterization in steep bedrock permafrost on Matterhorn (CH). J Geophys Res: Earth Surf 2018;123.
[10] Bhoumick P, Dang ST, Damani A, Sondergeld C, Rai CS. Stimulated reservoir volume evaluation using shear wave. American Rock Mechanics Association; 2017. p. 1–10 ARMA-17-0409.
[11] Everitt BS, Landau S, Leese M, Stahl D. Cluster analysis. 5th ed. Wiley Publications; 2011.
[12] Ester M, Kriegel HP, Sander J, Xu X. A density-based algorithm for discovering clusters in large spatial databases with noise, In: Proceedings of the 2nd international conference on knowledge discovery and data mining, Portland, OR. AAAI Press; 1996. p. 226–31.
[13] Caesarendra W, Tjahjowidodo T. A review of feature extraction methods in vibration-based condition monitoring and its application for degradation trend estimation of low-speed slew bearing. Machines 2017;5:21. https://doi.org/10.3390/machines5040021.
[14] Hui KH, Ooi CS, Lim MH, Leong MS, Al-Obaidi SM. An improved wrapper-based feature selection method for machinery fault diagnosis. PLoS One 2017;12(12).
[15] Boashash B. Time-frequency signal analysis and processing—a comprehensive reference. Elsevier; 2016.
[16] Meyer Y. Wavelets: algorithms and applications. Philadelphia: Society for Industrial and Applied Mathematics; 1993. p. 13–31. 101–105.
[17] Rioul O, Vetterli M. Wavelets & signal processing. IEEE Signal Process 1991;8(4):14–38.
[18] Sinha S. Time-frequency localization with wavelet transform and its application in seismic data analysis [M.S. thesis]. University of Oklahoma; 2002.
[19] Zhang H, Pan Z, Zhang W. Acoustic–seismic mixed feature extraction based on wavelet transform for vehicle classification in wireless sensor networks. Sensors 2018;18:1862.
[20] Qi G. Wavelet-based AE characterization of composite materials. NDT&E Int 2000;33:133–44.
[21] Huang NE, Shen Z, Long SR, Wu MC, Shih HH, Zheng Q, Yen N-C, Tung CC, Liu HH. The empirical mode decomposition and the Hilbert spectrum for nonlinear and non-stationary time series analysis. Proc R Soc Lond A: Math Phys Eng Sci 1998;454(1971):903–95. https://doi.org/10.1098/rspa.1998.0193.
[22] Colominas MA, Schlotthauer G, Torres ME. Improved complete ensemble EMD: a suitable tool for biomedical signal processing. Biomed Signal Process Control 2014;14:19–29. https://doi.org/10.1016/j.bspc.2014.06.009.

[23] Han J, van der Baan M. Empirical mode decomposition for seismic time-frequency analysis. Geophysics 2013;78(2):O9–O19. https://doi.org/10.1190/geo20120199.1.

[24] Honorio BCZ, Matos MC, Vidal AC. Progress on empirical mode decomposition-based techniques and its impacts on seismic attribute analysis. Interpretation 2017;5(1):SC17–28.

[25] Allen JB. Short term spectral analysis, synthesis, and modification by discrete Fourier transform. IEEE Trans Acoust Speech Signal Process June 1977;ASSP-25:235–8.

[26] Allen JB. Application of the short-time Fourier transform to speech processing and spectral analysis, In: Proc IEEE ICASSP-82; 1982. p. 1012–5.

[27] Klapuri A. Automatic transcription of music, In: Proceedings of the Stockholm musical acoustics conference (SMAC-03)Stockholm: Royal Swedish Academy of Music; August 2003.

Chapter 3

Shallow neural networks and classification methods for approximating the subsurface in situ fluid-filled pore size distribution

Siddharth Misra* and Jiabo He[†,a]

*Harold Vance Department of Petroleum Engineering, Texas A&M University, College Station, TX, United States, †School of Computing and Information Systems, University of Melbourne, Parkville, VIC, Australia

Chapter outline

[a]. Formerly at the University of Oklahoma, Norman, OK, United States

Machine Learning for Subsurface Characterization. https://doi.org/10.1016/B978-0-12-817736-5.00003-X

1 Introduction

NMR log responses acquired in the subsurface are inverted/processed to generate T_2 distribution specific to each depth. In geological formations, T_2 distribution is the transverse relaxation time of hydrogen nuclei of the fluids in the pores. T_2 distribution is represented as a spectra of T_2 amplitudes measured across 64 T_2 time bins. T_2 distribution is a function of the fluid-filled pore volume, fluid phase distribution, and fluid mobility in the pores. NMR T_2 distribution approximates the fluid-filled pore size distribution. T_2 distribution concentrated around small T_2 times are primarily due to small-sized pores. Unlike the conventional logging tools, such as gamma ray, density, neutron porosity, and resistivity, the operational and financial challenges in deploying the NMR logging tool and computing the NMR T_2 distributions impede its use in most of the wells. Deployment of NMR tool is a more severe challenge in shale reservoirs. Well conditions, such as the lateral well section, small-diameter boreholes in deep HPHT reservoirs, and boreholes with large washouts common in carbonates, limit the use of the NMR logging tool. The objective of this chapter is to synthetically generate NMR T_2 distribution from conventional easy-to-acquire logs, so that NMR T_2 distribution can be generated in the wells where NMR log is not available due to well conditions, financial limitations, and operational constraints.

We developed two artificial neural network (ANN)-based predictive models that process the conventional logs to generate the NMR T_2 distribution. The first predictive model implements a generic ANN with fully connected layers that generates T_2 distribution discretized into 64 T_2 bins, identifying relaxation times in the range of 0.3–3000 ms logarithmically split into 64 parts. The second ANN-based predictive model implements two steps: First, the T_2 distribution is fitted with a bimodal Gaussian distribution characterized by six parameters, namely, two amplitudes, two variances, and two T_2 locations of peak amplitude; subsequently, a generic ANN model with fully connected layers is implemented to generate the six parameters for the bimodal T_2 distribution, which are later invoked to generate the T_2 distribution for any depth in the formation. The second approach is based on the observation that the T_2 distribution responses have mostly either unimodal or bimodal distributions.

ANNs have been applied to log-based subsurface characterization. Bhatt and Helle [1] predicted porosity and permeability for wells in the North Sea by processing well logs using committee neural networks. For the porosity prediction, ANN processed sonic, density, and resistivity logs. For permeability prediction, ANN processed density, gamma ray, neutron porosity, and sonic logs. Al-Bulushi et al. [2] predicted water saturation in the Haradh sandstone formation using ANNs to process the density, neutron, resistivity, and photoelectric logs. Recently, Mahmoud et al. [3] predicted TOC for Barnett shale by using ANN to process the resistivity, gamma ray, sonic transit time, and bulk density logs. The model was then applied to estimate TOC for Devonian shale. ANN has been used by several researchers to predict NMR-T_2-derived parameters. Salazar and Romero [4] predicted NMR porosity and permeability in a carbonate reservoir using ANNs that processed gamma ray, resistivity, and neutron logs. Mohaghegh et al. [5] synthesized NMR-derived free fluid, irreducible water, and effective porosity using ANN that processed SP, gamma ray, caliper, and resistivity logs. Later, Elshafei and Hamada [6] predicted permeability using bulk gas model and ANN model. The predicted permeability agreed with permeability measurements on core samples. Labani et al. [7] estimated free fluid-filled porosity and permeability using a committee machine with intelligent systems (CMIS) in the South Pars gas field. CMIS combines the results of fuzzy logic, neuro-fuzzy, and neural network algorithms for overall estimation of NMR log parameters from conventional log data. Recently, in the Asmari formation, Golsanami et al. [8] predicted porosities of eight T_2 bins and T_2 logarithmic mean ($T_{2,LM}$) of NMR T_2 distribution using intelligent models. Notably, coauthors of this book were the first to develop ANN-based predictive models for synthesizing the entire NMR T_2 distribution partitioned into 64 bins.

In relation to the second predictive model implemented in this chapter, Genty et al. [9] fitted NMR T_2 distribution acquired in a carbonate reservoir with multiple Gaussian (or normal) distributions that were parametrized using three parameters (α, μ, σ) for each distribution. In their case, T_2 distributions required three Gaussian components and nine corresponding parameters for purposes of the fitting. Genty et al. [9] utilized these fitted parameters to identify genetic pore types in a carbonate reservoir. Di [10] implemented a method similar to that proposed by Genty et al. [9] to identify different lithofacies in a tight oil reservoir based on the parameters estimated by fitting a Gaussian to the T_2 distribution. We use a similar approach to compute six parameters that describe the NMR T_2 distribution in the shale formation. These six parameters are used for training and testing the second ANN-based predictive model for NMR T_2 prediction.

2 Methodology

2.1 Hydrocarbon-bearing shale system

The shale system under investigation contains conventional and unconventional elements. Conventional intervals consist of middle shale (MS) and conventional reservoir (CR) layers, whereas the source rock intervals include lower shale (LS) and upper shale (US). From the top to bottom, the shale system consists the following intervals: upper shale (US), middle shale (MS), lower shale (LS), and four conventional reservoirs (CR 1–4). These intervals are distinctly different and display highly heterogeneous distributions of reservoir properties. Oil and gas produced in the US and LS got accumulated in the MS and CR intervals. CR 1–4 intervals include siltstone, sandstone, dolostone, and dolo-mudstone intervals. Different minerals as present in these intervals, such as quartz, K-feldspar, plagioclase feldspars, illite, dolomite, calcite, kaolinite, and pyrite.

The US and LS formations are black shale formations deposited during the Late Devonian period to Early Mississippian period. The MS formation displays a range of grain size sorting from poorly sorted, argillaceous siltstone to moderately well sorted fine-grained sandstone and is more complex than US and LS formations. MS can be categorized into upper middle shale (UMS) and lower middle shale (LMS) based on their distinct grain sizes, depositional textures and diagenetic calcite cement contents. UMS contains better reservoir quality with well sorted, fine-grained sandstone, whereas LMS contains more bioturbated, silt-dominated, shallow-marine deposits. CR 1–4 intervals are primarily dolostone with alternating porous dolosiltite facies. The interlaminated CR dolostone is interbedded with clay-rich, conglomeratic dolo-mudstone, which marks stratigraphic intervals that partition CR into four distinct sequences from top to bottom. CR1 is the principal oil-bearing interval of CR. CR2 is also oil-bearing but only locally charged with oil, mainly in the center of the basin. It is rare to find oil in CR3, and the remaining CR4 is nonreservoir dolo-mudstone.

2.2 Petrophysical basis for the proposed data-driven log synthesis

NMR logging tool is generally suited for uniform boreholes of diameter greater than 6.5 inches run at a speed of around 2000 feet/hour. NMR responses of subsurface formation due to the inherent physics have poor signal-to-noise ratio. Highly trained logging engineers and good well conditions are required to ensure high-quality NMR log acquisition in the subsurface borehole environment. After the data acquisition, the NMR logs need to be processed using robust inversion methods to obtain the T_2 distribution. Due to the financial and operational challenges involved in running the NMR logging tool, oil and gas companies do not deploy NMR logging tool in every well. One alternative is to train data-driven models to process conventional logs for predicting the entire NMR T_2 distribution spanning 0.3 ms to 3000 ms. An accurate prediction of NMR T_2 distribution will assist geoscientists and engineers to quantify the pore size distribution, permeability, and bound fluid saturation in hydrocarbon-bearing reservoirs, thereby improving project economics and reservoir characterization capabilities. The use of data-driven methods to predict the entire NMR T_2 distribution without core data is a challenging and novel task.

Generation of NMR T_2 distribution log using conventional "easy-to-acquire" logs is feasible because the various combinations of conventional logs are sensitive to fluid saturations, pore size distribution, and/or mineralogy. For example, resistivity measurements and the mud-filtrate invasion effect on resistivity logs are influenced by the pore sizes, pore throat sizes, tortuosity, and permeability, which also influence the NMR T_2 distribution. Neutron log is sensitive to hydrogen index of fluids in the formation that correlates with porosity of the formation, which also influences the NMR T_2 distribution. GR is sensitive to clay volume that determines the clay-bound water and capillary-bound water, which can be sensed by the NMR T_2 distribution. Density log is sensitive to mineral densities and fluid densities in the formation. Mineral density is controlled by mineralogy that controls the lithology and pore size distribution of the formation. Mineralogy can lead to secondary effects on the NMR T_2 distribution because a certain type of lithology gives rise to a certain type of pore size distribution; for example, sandstone formation with predominantly quartz mineralogy will have a distinct pore size distribution compared with carbonate formation with predominantly calcite mineralogy, which will be distinct from shale formation with predominantly clay mineralogy. NMR T_2 distribution is also affected by surface relaxation, which depends on the mineralogy of the formation. Finally, both the NMR log and the conventional logs are influenced by the vertical distribution of fluid saturations along the formation depth, which is controlled by capillary pressure as governed by the pore size distribution.

In summary, several physical properties that govern NMR T_2 distribution also influence the conventional logs, namely, neutron, density, resistivity, sonic, and GR logs. The petrophysical basis for choosing these conventional logs for predicting the NMR T_2 distribution is that each of these conventional logs are sensitive to pore size distribution, fluid saturation, mineralogy, and capillary pressure. By using artificial neural networks to process the log data, we aim at identifying and extracting the complex and latent statistical relationships that cannot be quantified using any other mechanistic models. The goal of this study is not to develop a precise mechanistic model to describe the physical relationships between the conventional logs and NMR T_2 distribution. Rather, we aim to train neural network models to learn the hidden relationships between the conventional logs and NMR T_2 distribution.

2.3 Data preparation and statistical information

Data preparation and data preprocessing are essential steps prior to the data-driven model development (Fig. 3.1). The first ANN model predicts T_2 distribution discretized into 64 T_2 time bins, whereas the second ANN model predicts the six statistical parameters that parameterize the T_2 distribution as a sum of two Gaussian distributions. Before implementing the second model, we need to compute the six statistical parameters that define the bimodal distribution.

For developing the data-driven model, we used data from XX670 ft to XX990 ft of a shale system comprising seven intervals of a shale reservoir system. Logs acquired in the well include 12 conventional logs and 10 inversion-derived logs (Fig. 3.2). The conventional logs include gamma ray (GR) log sensitive to volumetric shale concentration, induction resistivity logs measured at 10-in. (AT10) and 90-in. (AT90) depths of investigation sensitive to the volumes of connate hydrocarbon and brine, and neutron (NPOR) and density porosity (DPHZ) logs that are influenced by the formation porosity. Other conventional logs include photoelectric factor (PEFZ) log indicating the formation lithology, VCL log measuring the volume of clay, and RHOZ log sensitive

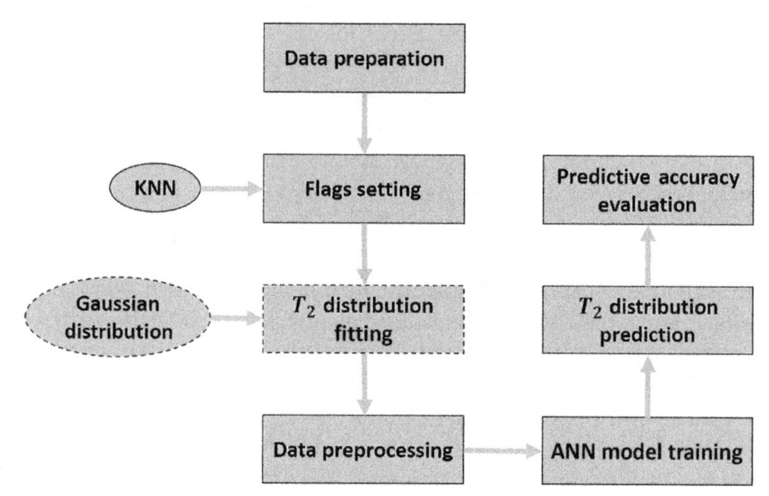

FIG. 3.1 Machine-learning workflow to generate the NMR T_2 distribution for each depth. Actions represented inside the dashed boundaries are involved in the second ANN-based predictive model.

to the formation density. Final set of conventional logs include Delta-T Shear (DTSM), Delta-T Compressional (DTCO) and Shear-to-Compressional Velocity Ratio (VPVS) that are sensitive to geomechanical properties and rock textures. The 12 conventional logs were processed using a commercial log-inversion software to compute 10 additional inversion-derived logs. Six of the ten inversion-derived logs are presented in Tracks 8 and 9 in Fig. 3.2, which are mineral content (of quartz, calcite, and dolomite) and fluid saturation logs. NMR T_2 distributions (Track 10 of Fig. 3.2) are also used to train and test ANN models and to generate the six parameters characterizing the bimodal Gaussian model fit with the T_2 distributions. Environmental correlations and depth corrections were performed on all these logs prior to the data-driven model development.

Figs. 3.A1–3.A3 list the normalized statistical parameters associated with the conventional logs. For example, normalized median (one of the several normalized statistical parameters) is calculated as follows:

1. For each log, its median value for each interval (7 in this case) is calculated.
2. For each log in each interval, the normalized median value is computed by dividing the median value of the log for the interval by the maximum median value of the log for all intervals (7 in this case).

Median GR of US and LS are much higher than those of other intervals (Fig. 3.A1), which indicates the dominance of organic shale in US and LS. AT10 and AT90 are much higher in US and LS than those in other intervals, which indicates high hydrocarbon in these two intervals. DPHZ and NPOR are also the largest in US and LS. Conventional logs in US and LS exhibit

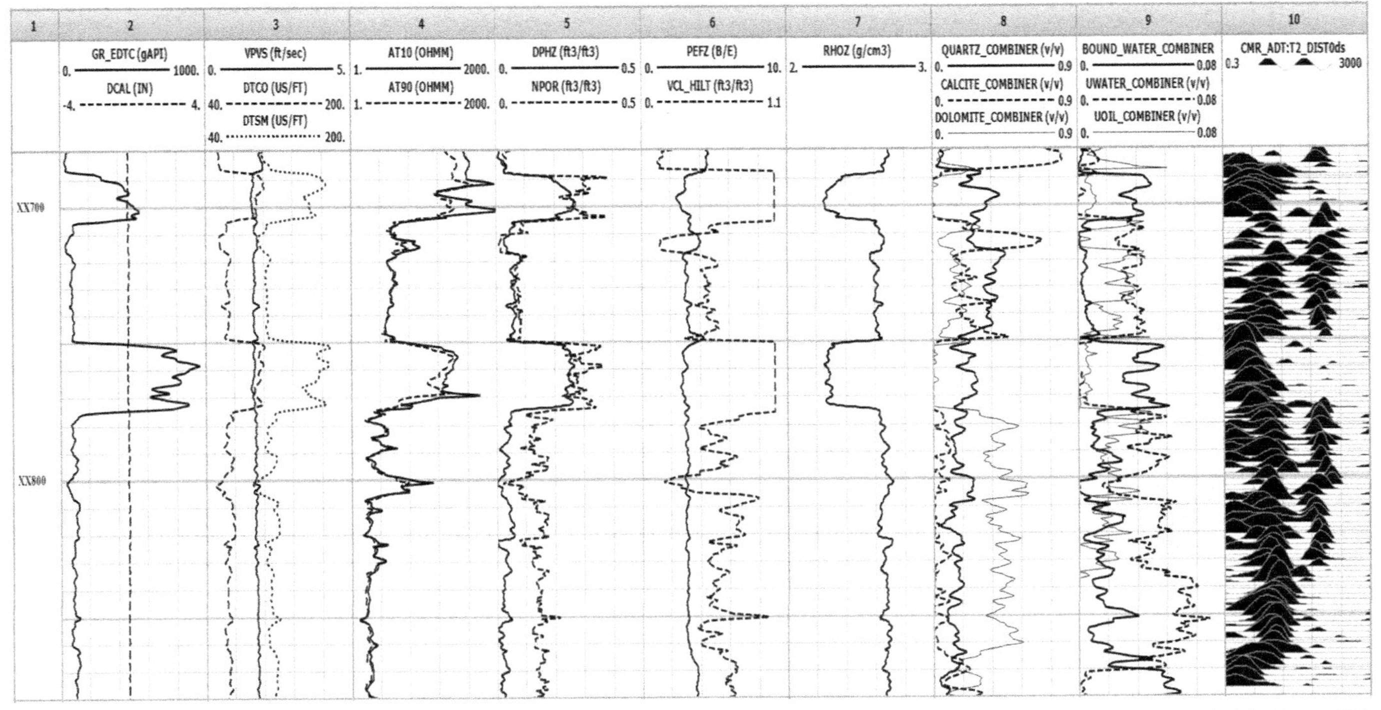

FIG. 3.2 Track 1 is depth; Track 2 contains gamma ray (GR) and caliper; Track 3 contains DTSM, DTCO, and VPVS; Track 4 is induction resistivity logs at 10-in. (AT10) and 90-in. (AT90) depth; Track 5 contains density porosity and neutron porosity; Track 6 is formation photoelectric factor and volume of clay; Track 7 is formation density; Tracks 8 and 9 contain the six inversion-derived logs, namely, mineral content and fluid saturation; and Track 10 is the NMR T_2 distributions.

relatively similar statistical characteristics. Conventional logs acquired in CR1, CR2, CR3, and CR4 also exhibit similar statistical characteristics. Statistical parameters for most of the input logs in MS are closer to those of CR intervals rather than those in US and LS intervals. Median values of VPVS, PEFZ, and RHOZ are similar throughout the seven intervals. Median values of GR, AT10, AT90, mineral compositions, bound water content, and fluid saturations vary significantly across the seven intervals (Fig. 3.A1).

Coefficient of variation of a feature (log) is the ratio of standard deviation over mean (S_d/μ), which is used to measure the dispersion of the feature (Fig. 3.A2). AT10 and AT90 show large dispersion in bottommost CR4 interval. DPHZ shows abnormally high dispersion for the CR3 interval. RHOZ, feldspar, and dolomite show abnormally high dispersion for US interval as compared with other intervals. Skewness is the measurement of asymmetry of data about its mean. A few conventional logs, such as AT10 and AT90, have large values of coefficient of variation and skewness in all intervals. Most of the logs in LS, CR3, and CR4 exhibit higher skewness compared with other intervals. There is a large variability in dispersion of logs for the seven intervals under investigation (Fig. 3.A3).

2.4 Categorization of depths using flags

After selecting the conventional logs and data preprocessing, we generate five synthetic categorical features (referred as Flags) for each depth point. These categorical features (flags) will be used as synthetic discrete-valued logs because their implementation improves the performance of the ANN models in synthesizing the NMR T_2 distribution. Flag-1 is an integer ranging from 1 to 7 identifying the seven distinct intervals intersected by the well, namely US, MS, LS, CR1, CR2, CR3, and CR4. Fig. 3.B1 provides qualitative schematic descriptions of the Flags 2–5. These flags are computed based on certain characteristics of the T_2 distribution, and the underlying pore size distribution. Flag-2 is either 0 or 1 identifying unimodal and bimodal pore size distribution, respectively, at a certain depth. Flag-3 is an indicator of pore sizes in a bimodal system, such that its value is -1, 0, or 1 identifying the abundance of small pores, comparable volumes of small and large pores, and abundance of large pores, respectively. Similar to Flag-3, Flag-4 indicates the relative abundance of pores of certain pore size in a bimodal system, such that Flag-4 is assigned a value of 1 when certain pore size (either small pores or large pores) is negligible or else it is assigned to be 0. Depths for which Flag-4 is assigned a value of 1 can be regarded as unimodal distributions. Flag-5 defines the deviation/spread of pore sizes around the two dominant pore sizes of a bimodal distribution, such that 1 indicates that the spreads around the two peaks are wide and 0 indicates either a unimodal distribution or a narrow spread around the two dominant pore sizes. In brief, Flag-1 classifies intervals based on lithology, Flag-2 identifies number of peaks in the pore size distribution, Flag-3 identifies the dominant pore sizes in bimodal pore systems, Flag-4 checks if certain pore sizes can be neglected, and

Flag-5 captures difference in the deviation of pore size distributions about a dominant pore size. These flags help improve the prediction performance as they provide additional relevant information to the predictive models about each depth.

An expert geologist provided us formation tops and dominant lithology along the well length. We used this information to assign integer values to Flag-1 for each depth. Values for Flags 2–5 were manually assigned to each depth by examining the NMR T_2 distribution for that depth. After manually creating the flags, which are categorical features, we trained a k-nearest neighbor (KNN) classifier (supervised learning) to correctly predict the categorical values of the five flags by processing a specific combination of conventional logs and inversion-derived logs. KNN training should be done only on the training dataset used for training the ANN models for NMR T_2 synthesis. This avoids contaminating the performances of the ANN models when synthesizing the NMR T_2 distributions for the testing dataset. Once the KNN is trained, new conventional and inversion-derived logs from new wells without the NMR T_2 distributions can be processed to generate the Flags 2–5. The KNN predictions of Flags 2–5 along with Flag 1 obtained from the Geologist are fed into the ANN models along with the conventional and inversion-derived logs for purposes of NMR T_2 synthesis in the new wells. Our hypothesis is that the Flags 1–5 act as additional informative features that help improve the performances of the ANN models. Table 3.B1 lists the substantial improvement in predictive performance of the first ANN model when the flags were generated and then used as additional features when training and testing the ANN model. Fig. 3.C1 shows that Flags 1–4 are very important features for the proposed NMR T_2 synthesis task.

When creating the training dataset, Flags 2–5 can be easily generated by examining the available T_2 distributions acquired along the well length and then assigning specific values to the Flags for each training depth depending on the characteristics of the T_2 distribution at that depth. However, for testing data or any new data, T_2 distribution needs to be predicted and is unavailable; consequently, we cannot manually assign specific values to these flags. So, for the testing data and new data, we first train a KNN classifier on the training dataset with manually assigned flags to relate the available conventional logs to the flags. Following that, the trained KNN classifier is used to process the available conventional logs in the testing dataset and new dataset to generate the flags. Flags 2–5 need to be predicted prior to the primary objective of generating the T_2 distribution. The goal of KNN classifier is to first learn to relate the available "easy-to-acquire" conventional and inversion-derived logs to the Flags 2–5 in the training dataset, in the presence of NMR T_2 distribution. The trained KNN is then used to process the easy-to-acquire conventional logs to generate the Flags 2–5, in the absence of NMR T_2 distribution. Twenty-two conventional and inversion-derived logs are used as features to predict the Flags 2–5 one by one with four separate KNN classifiers.

TABLE 3.1 Accuracy of flag generation using KNN classifiers for the testing dataset.

Flag	2	3	4	5
Accuracy	88%	86%	85%	88%

KNN algorithm classifies new samples (without class labels) based on a similarity measurment with respect to the samples having predefined/known class labels. Similarity between a testing/new sample and samples with predefined/known labels is measured in terms of some form of distance (e.g., Euclidean and Manhattan). In KNN algorithm, k defines the number of training samples to be considered as neighbors when assigning a class to a testing/new sample. First, k nearest training samples (neighbors) for each of the testing/new sample are determined based on the similarity measure or distance. Following that, each testing/new sample is assigned a class based on the majority class among the k neighboring training samples. Smaller k values result in overfitting, and larger k values lead to bias/underfitting. We use $k=5$ and Euclidean distance to find nearest neighbors. Distance is expressed as

$$D(\bar{x}, \bar{y}, p) = \sqrt[p]{\sum_{n=1}^{k} (\overline{x_n} - \bar{y})^p} \tag{3.1}$$

where k is number of neighboring training samples to consider when testing/applying the KNN classifier, n indicates the index for a neighboring training sample, $\overline{x_n}$ is the feature vector for the nth training sample, $\bar{y}$ is the feature vector for the testing sample, $p=2$ for Euclidean distance, and $p=1$ for Manhattan distance. Table 3.1 presents the accuracy of flag generation for the testing dataset. After the prediction of four flags (Flags 2–5), 22 logging data (conventional and inversion-derived logs) and 5 flags (Flags 1–5) are used together to predict the NMR T_2 distribution.

2.5 Fitting the T_2 distribution with a bimodal Gaussian distribution

Out of 416 discrete depths, 354 randomly selected depths are used for training, and 62 remaining depths are used for testing. Fitting the original T_2 distribution using a bimodal Gaussian distribution is crucial for developing the second ANN model implemented in the chapter. Genty et al. [9] found that NMR T_2 distribution can be fitted using three Gaussian distributions expressed as

$$f(T_2') = A \sum_{i=1}^{3} (\alpha_i) g_i (\mu_i, \sigma_i, T_{2i}') \tag{3.2}$$

where i is an index that identifies the Gaussian distribution, $T_2' = \log(T_2)$; g_i is the probability distribution function of a Gaussian distribution with mean μ_i and standard deviation σ_i; α_i represents the proportion of pore volumes representing the constituent Gaussian distribution with respect to total pore volume, such that $\alpha_1 + \alpha_2 + \alpha_3 = 1$; and A is the amplitude parameter. In our study, the shale system exhibits NMR T_2 distributions having either one or two peaks. We fit the T_2 distributions using a modified version of Eq. (3.2) expressed as

$$f(T_2') = \sum_{i=1}^{2} (\alpha_i) g_i(\mu_i, \sigma_i, T_2') \tag{3.3}$$

Compared with Eq. (3.2), Eq. (3.3) does not implement the amplitude parameter A and $\alpha_1 + \alpha_2 \neq 1$. When using Eq. (3.3), six parameters are required to fit the T_2 distribution response at each depth. The six parameters are μ_i, σ_i, and α_i, for $i = 1$ and 2. The reliability of the fitting is expressed in terms of the coefficient of determination R^2 formulated as

$$R^2 = 1 - {}^{RSS}/_{TSS} \tag{3.4}$$

where

$$RSS = \sum_{i=1}^{n} \left[f_{i,fit}(T_2') - f_i(T_2')\right]^2 \tag{3.5}$$

and

$$TSS = \sum_{i=1}^{n} \left[f_i(T_2') - \overline{f(T_2')}\right]^2 \tag{3.6}$$

where $n = 64$ is the number of bins into which the original T_2 distribution (corresponding to a depth) is discretized, $f_i(T_2')$ represents the ith discretized T_2 distribution measurement, $f_{i,\ fit}(T_2')$ represents the fit to the ith discretized T_2 distribution computed using the Eq. (3.3), and $\overline{f(T_2')}$ is the mean of the 64 discretizations of the original T_2 distribution for the given depth. RSS is the sum of squares of the residuals, and TSS is the total sum of squares proportional to the variance of the data. T_2 distributions acquired at 416 depth points in the shale system were fitted with Eq. (3.3) to estimate the six characteristic fitting parameters for each depth point. In doing so, the 64 bins of NMR T_2 are transformed to six logs, which were used for training and testing the second ANN-based predictive model. For T_2 distribution with single peaks, $\alpha_2 = \mu_2 = \sigma_2 = 0$. Figs. 3.3 and 3.4 show the results of fitting for randomly sampled depth points. T_2 distributions were fitted at median R^2 of 0.983 (Fig. 3.3). Only 12% of the depths were fitted with R^2 lower than 0.95.

Normalized root mean square error (NRMSE) in synthesizing a specific T_2 bin across all the depths is used together with R^2 of synthesizing all the 64 bins at a specific depth to assess the accuracy of fitting and predicting the NMR T_2 distributions. NRMSE for any specific discretized NMR T_2 bin is expressed as

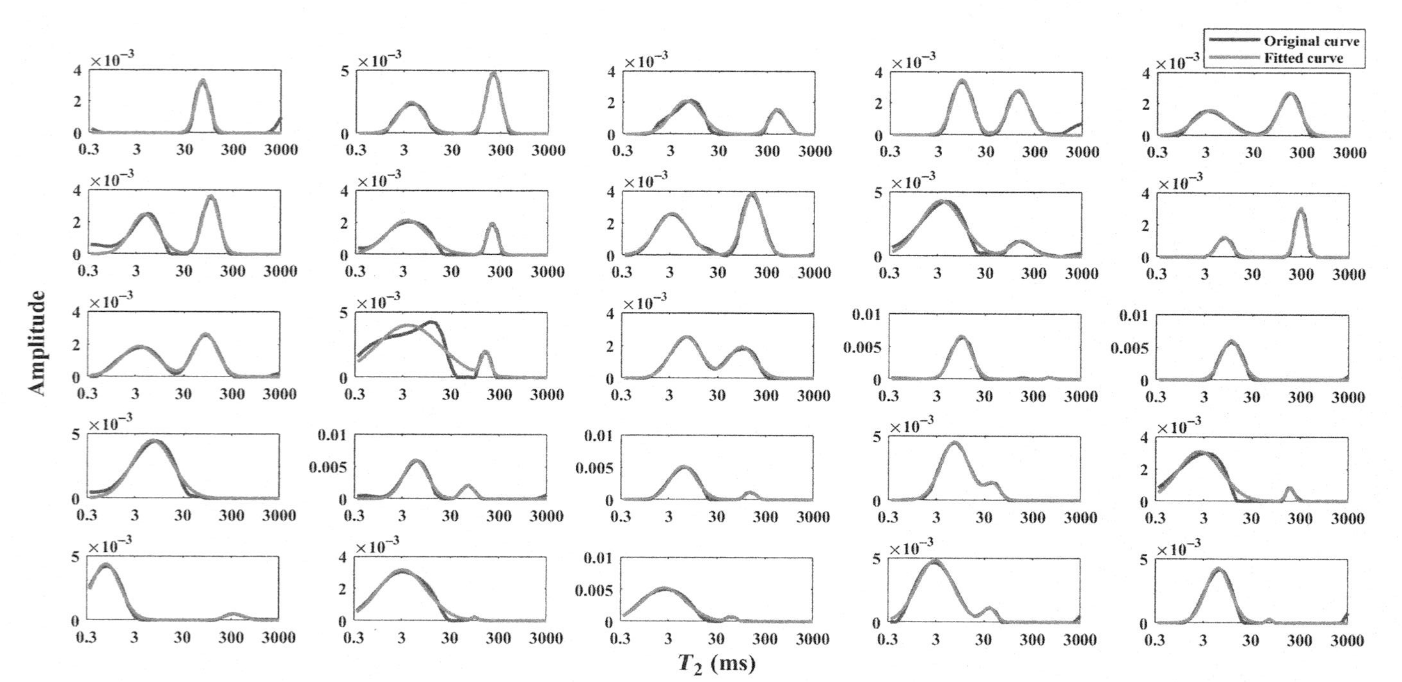

FIG. 3.3 *Blue (dark gray in print version)* curves are original T_2 distributions, and the *red (light gray in print version)* curves are the best-fitting T_2 distributions obtained using Eq. (3.3).

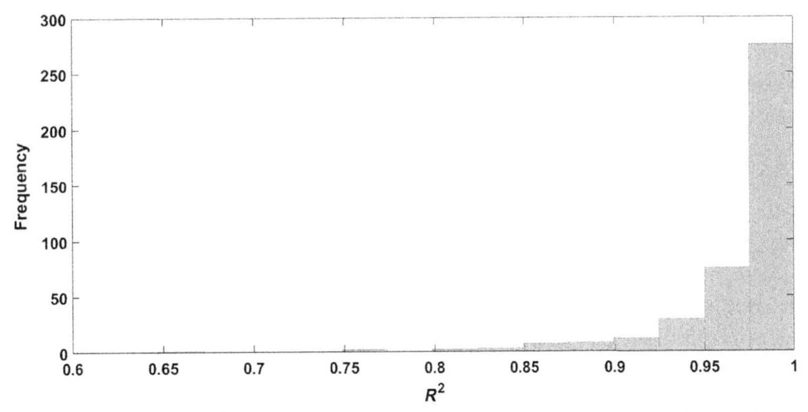

FIG. 3.4 Histogram of R^2 of fitting the NMR T_2 distribution using bimodal Gaussian distribution.

$$RMSE = \sqrt{\frac{\sum_{i=1}^{n} (\hat{y}_i - y_i)^2}{n}} \tag{3.7}$$

$$NRMSE = \frac{RMSE}{y_{\max} - y_{\min}} \tag{3.8}$$

where y_i is the original discretized T_2 measurement for the bin at depth i, $\hat{y}_i$ is the predicted/fitted T_2 for the bin at depth i, n represents the total number of depth points for which the data are available, $y_{\max}$ and $y_{\min}$ indicate the maximum and minimum values for the T_2 bin for the entire depth, and RMSE stands for root mean square errors. NRMSE close to 0 indicates good prediction performance.

2.6 Min-max scaling of the dataset (features and target)

In supervised learning, a data-driven model first learns to relate the features with targets for all the samples in the training dataset. Then, the trained data-driven model is tested on all the samples in the testing dataset. Finally, a well-evaluated, generalizable data-driven model is deployed on new samples to predict/synthesize the targets. In this study, each depth along the length of the well is treated as a sample. For each depth (sample), the 22 easy-to-acquire conventional and inversion-derived logs along with 5 the categorical flags are used as the features, whereas the targets comprise the 64 T_2 amplitudes measured across the 64 T_2 bins. It is recommended that the data (especially features) be scaled prior to training the model. For ANN model, the use of min-max scaler is recommended. Min-max scaler will transform the features and targets to values ranging from -1 to 1 or 0 to 1, which markedly improves the speed of convergence and also improves the reliability and robustness of the ANN-based data-driven model. Min-max scaler is well suited when the feature distribution is non-Gaussian in nature and the feature follows a strict bound (e.g., image pixels). Depths exhibiting outlier log responses need to be removed prior to min-max scaling, which is drastically influenced by outliers. Moreover, the existence of outlier log responses adversely affects the weights and biases of the neurons learnt during the training of an ANN model. In this study, those depths are considered as outliers, where one of the logs has abnormally large or small value as compared with the general trend of the log. At some depths of the shale system under investigation, the log responses are unrealistic and abnormally high; for example, Gamma ray larger than 1000 API unit or DTSM larger than 800 μs/ft. Such outliers are referred as global outliers, which require simple thresholding technique for detection, followed by the removal of such depths exhibiting outlier log responses.

After the removal of depths exhibiting abnormal log responses, each log (feature or target) is transformed to a value between −1 and 1 using min-max scaler. Min-max scaling forces features and targets to lie in the same range, which guarantees stable convergence of weights and biases in the ANN model [11]. Min-max scaling was performed using the following equation:

$$y = 2\frac{x - x_{\min}}{x_{\max} - x_{\min}} - 1 \tag{3.9}$$

where x is the original (unscaled) value of the feature or target and y is the scaled value of x. Scaling is performed for all the features so that all the features have the same influence when training the model. Scaling is essential when using distance, density, and gradient-based learning methods. ANN rely on gradients for updating the weights. For ANN models, unscaled features can result in a slow or unstable learning process, whereas unscaled targets can result in exploding gradients causing the learning process to fail. Unscaled features are spread out over orders of magnitude resulting in a model that may learn large-valued weights. When using traditional backpropagation with sigmoid activation function, unscaled features can saturate the sigmoid derivative during the training. Such a model is unstable exhibiting poor generalization performance. However, when certain variations of backpropagation, such as resilient backpropagation, are used to estimate the weights of the neural network, the neural network is more stable to unscaled features because the algorithm uses the sign of the gradient and not the magnitude of gradient when updating the weights.

2.7 Training and testing methodology for the ANN models

After the feature scaling, the dataset (comprising features and targets) is split into two parts: training data and testing data. Usually, 80% of data are selected as the training data, and the remaining 20% of the original data constitute the testing data. When the size of the dataset available for building a data-driven model increases, we can choose larger percentage of data to constitute the testing dataset, which can be as high as 50-50 split between training and testing datasets. Due to the limited size of data available for building the two ANN-based models, 85% of the dataset is randomly selected to create the training dataset, and the remaining 15% forms the testing dataset. Notably, it is of critical importance that each feature and target in the original dataset, testing dataset, and training dataset should have similar statistical properties and distributions.

We intend to build two ANN models to synthesize the 64 discrete amplitudes corresponding to the 64 T_2 bins. The first model simultaneously predicts all the 64 values of the T_2 bins, whereas the second model predicts 6 parameters that define a bimodal Gaussian distribution, which approximates the 64 discrete values of the T_2 bins. Each ANN model is built with two hidden layers, which is enough for most of the function-approximation problems of similar complexity as our study. Hidden layers lie between the output layer and the input layer. In this study, the input layer takes in 22 conventional and inversion-derived logs and 5 categorical flags. The output layer for first ANN model contains 64 neurons, and that of the second ANN model contains 6 neurons. All neurons of output and hidden layers are connected to all the neurons of the preceding layer; such a connection of neurons is also referred as a dense connection.

Neurons are the computational units of an ANN model. In a densely connected network, each neuron in a hidden layer is connected to all the neurons in the immediately previous layer. A neuron is a mathematical transformation that first sums all the inputs multiplied by weights of the corresponding connections and then applies a nonlinear filter or activation function (generally having a sigmoid shape) on the summation to generate an output that is a value ranging between 0 and 1 or -1 and 1. When a feature vector is fed to the ANN model, the input values are propagated forward through the network, neuron by neuron and layer by layer, until the sequence of mathematical transformations applied on the feature vector reach the output layer.

During the training, the output of the network (i.e., final values obtained at the output layer) is compared with the desired/expected/known output, using a loss function. The error values for each neuron in the output layer are then propagated from the output layer back through the network, until each neuron has an associated error value that reflects its contribution to the original output. Backpropagation algorithm uses these error values to calculate the gradient of the loss function. In the second phase, this gradient is fed to the optimization method, which in turn uses it to update the weights/bias of each neuron, in an attempt to minimize the loss function. ANN training leads to learning the weights and bias for each neuron. ANN training involves feedforward of data signals to generate the output and then the backpropagation of errors for gradient descent optimization. During the process of training an ANN model, first, the weights/biases for the neurons in the output layer are updated; then, the weights/biases of the neurons of each preceding layer is iteratively updated, till the weights/biases of neurons of the first hidden layer are updated.

There are no specific equations or procedures to calculate the number of neurons (computational units) in each hidden layer. Different combinations of neurons in each layer were tried for the NMR T_2 prediction. An arithmetic sequence of the number of neurons in each hidden layer is suggested to generate high prediction accuracy [12]. Consequently, for the first predictive model that takes 27 inputs/features and generates 64 outputs/targets, 39 and 51 neurons were set in the first and second hidden layers of the ANN model, respectively, because the numbers 27, 39, 51, and 64 approximately form an arithmetic sequence. This architecture requires 6460 parameters, including the weights and biases, to be computed/updated during each training step. Following the same logic, for the second predictive model that has 27 inputs/features and 6 outputs/targets, requires 20 and 13 neurons in the first and second hidden layers of the ANN model, respectively, such that the sequence 6, 13, 20, and 27 is nearly an arithmetic sequence. This architecture requires 917 parameters, including the weights and biases, to be computed/updated during each training step.

Training functions are the methods to adjust weights and biases to converge the target functions of ANNs. Target functions quantify errors in learning process. ANN learns by minimizing the target functions based on the training functions. Levenberg-Marquardt (LM) backpropagation [13] and conjugate gradient (CG) backpropagation [14] are two most widely used algorithms for approximating the training functions, which are the relationships between the features and targets with the weights and biases of every neuron in the ANN model. LM backpropagation is suitable for a small number of weights and biases, whereas CG backpropagation can be applied on large neural networks implementing a large number of weights and biases. We use CG backpropagation as the training function for both the ANN-based prediction models. Training time of the ANN model with LM backpropagation was 10 times more than that with CG backpropagation. To be specific, scaled conjugate gradient algorithm [14] was used for our study.

Target function of ANN model adjusts the weights and biases of all neurons to minimize the errors to ensure the best prediction performance during the model training. Overfitting is a challenging problem when minimizing the target function [15]. ANN models cannot learn a generalizable relationship between targets and features due to overfitting. A simple target function is the regularized sum of squared errors (SSE) function expressed as

$$\mathrm{SSE} = \sum_{i=1}^{n} (y_i - \hat{y}_i)^2 + \lambda \sum_{j=1}^{P} \sigma_j^2 \tag{3.10}$$

where n is the number of samples/depths in the training/testing dataset and P is the number of outputs (64 for the first model and 6 for the second model), λ is penalty parameter, y_i is original target at depth i, $\hat{y}_i$ is estimated target at depth i, and σ_j^2 is the variance of predicted target j. Regularization is a popular method to mitigate overfitting [16]. Regularization introduces a penalty term in the target function that penalizes additional model complexity. Penalty parameter is set at 0.7 in the first model and 0.5 in the second model based on extensive numerical experiments to find the optimal value of the penalty term [16, 17]. In place of regularization, cross validation and dropout can be performed to avoid overfitting of ANN models. However, cross validation generally requires larger dataset.

3 ANN model training, testing, and deployment

3.1 ANN models

Twenty-two conventional and inversion-derived logs and five categorical flags are used as features for the first ANN model to predict the 64 discretized amplitudes of T_2 distributions and for the second ANN model to predict the six parameters of the bimodal Gaussian distribution that fits the T_2 distribution. Logs from 416 different depths are randomly split into 354 (85%) depths to be used as the training dataset and 62 (15%) depths as the testing dataset. Normalized root mean square error (NRMSE) for each T_2 bin across the entire dataset is used as a score to evaluate the accuracy of T_2 distributions synthesized by the ANN models. The formulation of normalized root mean square error (NRMSE) is described in Eqs. (3.7) and (3.8).

3.2 Training the first ANN model

Fig. 3.5 presents the prediction performance of the first ANN model for 25 randomly selected depths from the training dataset. Prediction performance on the training dataset is also referred as the memorization performance. For a learning model, the memorization performance cannot be used in isolation as evaluation metric. Memorization performance in conjunction with generalization performance can indicate the level of over-fitting in the learning process. Bimodal T_2 distributions with small dispersions are difficult to predict. Noise in the conventional logs and in the T_2 distributions negatively affects the performance of the trained model. Predictions are in good agreement with true T_2 distributions for unimodal and bimodal distributions with high dispersivity around the dominant T_2. The median NRMSE for the training dataset is 0.1201, which indicates a good prediction performance. Histograms of NRMSE for the 354 depths in the training datasets are plotted in Fig. 3.6. Most NRMSE values are lower than 0.2, implying a good prediction performance of the first ANN model.

Four examples of prediction performance of the first ANN model are illustrated in Fig. 3.7. Each example shows the original and predicted T_2 distributions. Fig. 3.7 aids the qualitative understanding of the training performance. The first subplot at the top left with $R^2 = 0.99$ is the case with best performance. During the training, all depths with single-peak T_2 distribution are trained at a

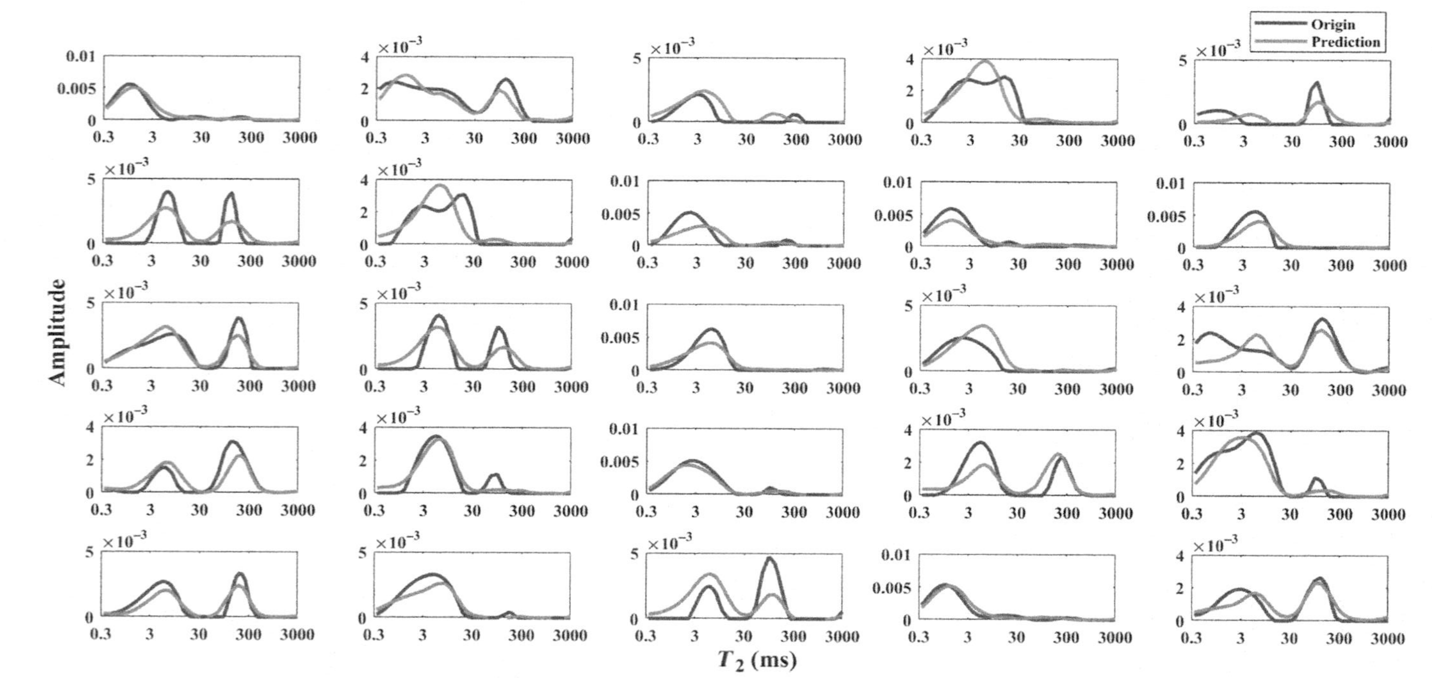

FIG. 3.5 Comparison of original T_2 distributions with those predicted using the first ANN model on the training dataset.

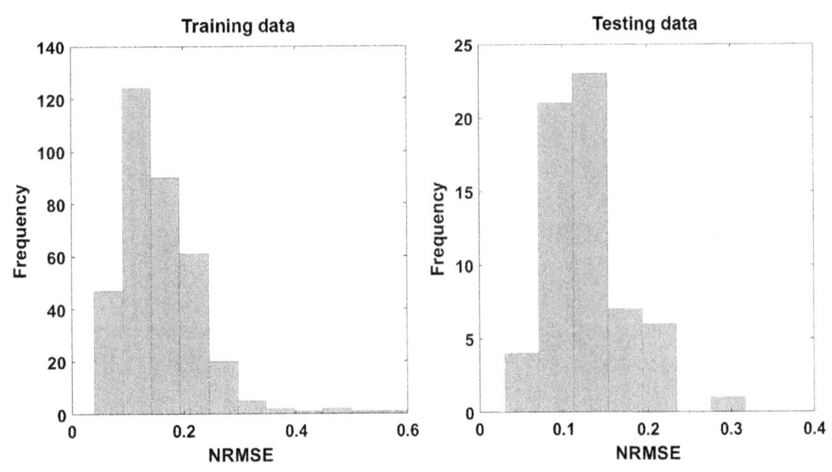

FIG. 3.6 Histograms of NRMSE of predictions of the first ANN model on the training (left) and testing (right) datasets.

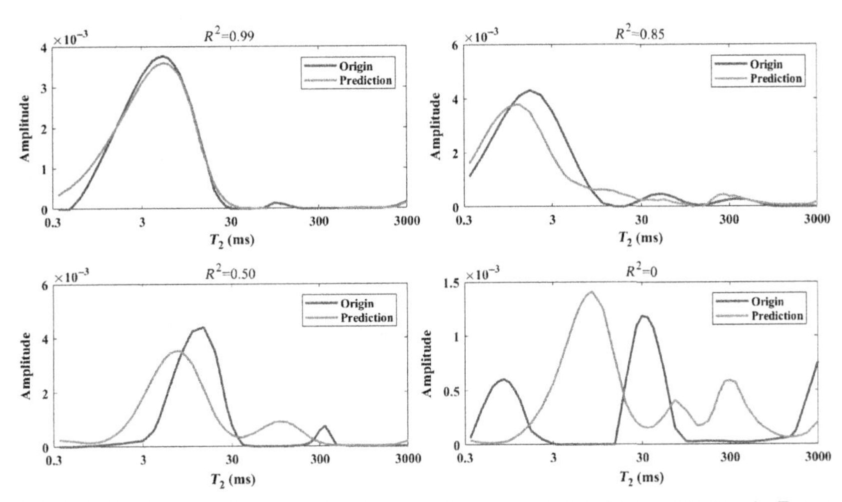

FIG. 3.7 Prediction performances of the first ANN model on the training dataset when the T_2 synthesis has R^2 of 0.99, 0.85, 0.5, and 0.

high prediction performance compared with those with two peaks. The subplot on top right with $R^2 = 0.85$ is the median performance case, and more than half of all depths will perform better than the one shown for $R^2 = 0.85$. $R^2 = 0.50$ is a bad prediction performance, and about 7% of all depths are trained at lower prediction performance than this one. The bottom right subplot with $R^2 = 0$ is an example of poor performance. Less than 3% depths will perform as bad as the one shown for $R^2 = 0$.

3.3 Testing the first ANN model

There are only 62 testing depths, which are comparatively fewer than the 354 depths for training the ANN model. The prediction performance on the testing dataset (also referred as the generalization performance) is similar to that on the training dataset. Fig. 3.6 presents the prediction performance (in terms of NRMSE) of the first ANN model on the testing dataset (Fig. 3.6). The median R^2 and median NRMSE of predictions on the testing dataset are 0.8549 and 0.1218, respectively. This testing performance is remarkable given the hostile subsurface borehole conditions when acquiring the logs, which result in low signal-to-noise ratio, and the limited size of the dataset available to build the model, which gives rise to overfitting and poor generalization. Fig. 3.B2 shows the histograms of NRMSE for training and testing datasets without the implementation of the five categorical Flags (1–5) as additional features. Comparison of Fig. 3.B2 with Fig. 3.6 highlights the necessity of Flags as categorical features to achieve good generalization performance.

Notably, Fig. 3.C1 lists all the features in terms of their importance to the data-driven task of T_2 synthesis. Feature importance was performed to find the most important features out of the 27 features, which include 10 conventional logs, 12 inversion-derived logs, and 5 categorical features. Importance of a feature for a machine-learning task depends on the statistical properties of the feature and on the relationship of the feature with other features, targets, and the machine-learning algorithm used to develop the data-driven model. Feature importance indicates the significance of a feature for developing a robust data-driven model. Feature importance helps us understand the inherent decision making process of a data-driven model and helps in evaluating the consistency of a data-driven model by making the model easy to interpret.

3.4 Training the second ANN model

The second ANN model involves a two-step training process: (1) parameterizing the T_2 distribution by fitting a bimodal Gaussian distribution and (2) training the ANN model to predict the six parameters governing the bimodal Gaussian distribution fit to the T_2 distribution. By following the two-step training process, a trained ANN model can generate the six parameters of the bimodal Gaussian distribution. Prediction performance of the second model is affected by the errors in fitting the T_2 distribution with a bimodal Gaussian distribution (listed in Table 3.1). Fig. 3.8 presents the prediction performance of the second ANN model for 25 randomly selected depths from the training dataset.

The median R^2 and median NRMSE of predictions of the second ANN model on the training dataset are 0.7634 and 0.1571, respectively, as compared with 0.8574 and 0.1201, respectively, for the first ANN model. Consequently, the prediction performance on the training dataset (also referred as the memorization performance) of the first ANN model is superior to that of the second model, but the computational time of the first ANN model is 30% more than that of the second model. Histograms of NRMSE of predictions for the 354 depths

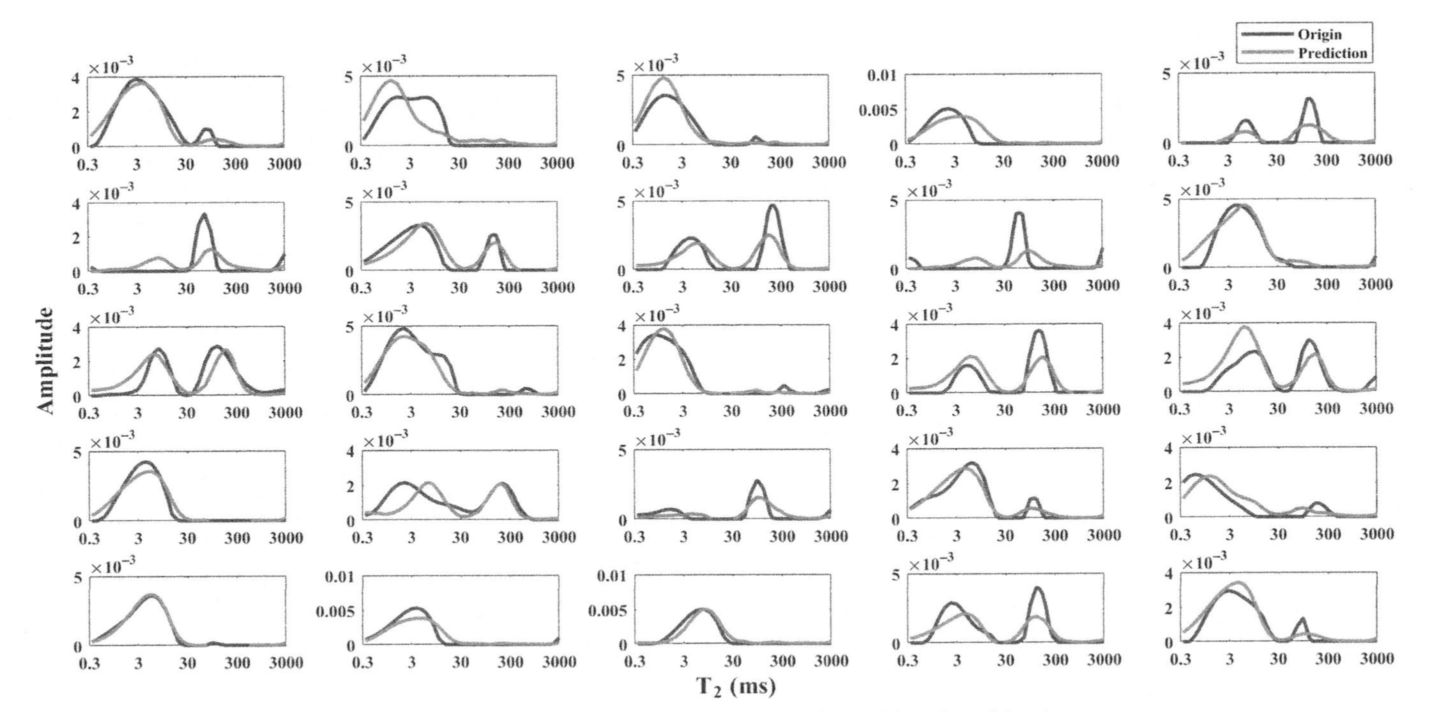

FIG. 3.8 Comparison of original T_2 distributions with those predicted using the second ANN model on the training dataset.

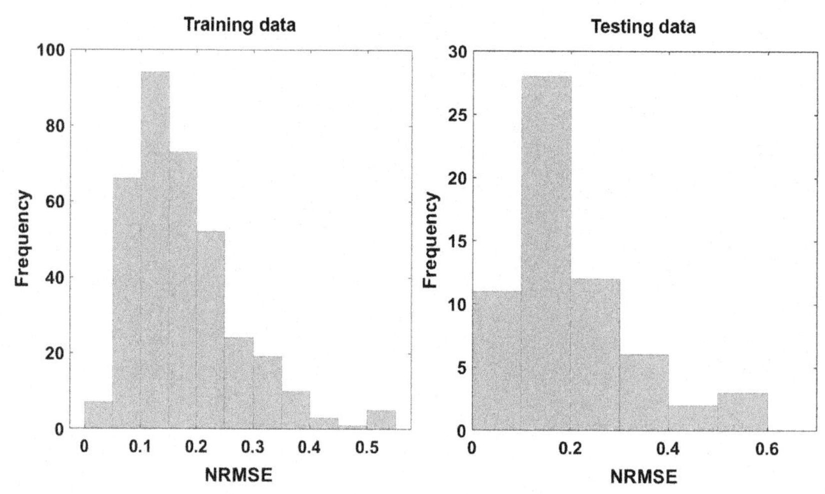

FIG. 3.9 Histograms of NRMSE of predictions of the second ANN model on the training and testing datasets.

are plotted in Fig. 3.9. R^2 and NRMSE of most of the predictions are larger than 0.7 and smaller than 0.25, implying an acceptable prediction performance of the second ANN model. For 22% of all training depths, the memorization performance of the second ANN model is lower than R^2 of 0.5, which indicates a worse prediction performance as compared with the first ANN model, for which only 7% of the training depths were trained at R^2 lower than 0.5. Second ANN model performs at R^2 of 0 for less than 5% of the training depths, whereas the first ANN model performs at R^2 of 0 for less than 2% of the training depths.

3.5 Testing the second ANN model

The prediction performance on the testing dataset (also referred as the generalization performance) is similar to that attained on the training dataset. The median R^2 of testing is 0.7584, and the median NRMSE is 0.1609. This testing performance is remarkable given the hostile subsurface borehole conditions when acquiring the logs, which result in low signal-to-noise ratio, and the limited size of the dataset available to build the model, which gives rise to overfitting and poor generalization. Fig. 3.9 presents the prediction performance (in terms of NRMSE) of the second ANN model on the testing dataset. As shown in Fig. 3.9, 29% of testing depths have prediction performance lower than $R^2 = 0.5$ (NRMSE > 0.25) and 37% of testing depths have prediction performance higher than $R^2 = 0.8$ (NRMSE < 0.15).

3.6 Petrophysical validation of the first ANN model

NMR T_2 distributions are generally used to estimate the formation porosity and permeability, which are the two most important hydrocarbon-reservoir parameters. In this section, we derive few reservoir properties from the predicted and original NMR T_2 distribution to test the robustness of the predicted NMR T_2. The first ANN model being the better performing predictive model, as compared with the second ANN. In this section, we focus on the first ANN model. Petrophysical validation of the first model is demonstrated by comparing ϕ_N and $T_{2,gm}$ derived from the original NMR T_2 distribution with those derived from the ANN-based predictions of T_2 distribution. ϕ_N is the sum of all amplitudes (values) of the 64 bins of a T_2 distribution at a single depth. $T_{2,gm}$ is the 64th root of the product of the 64 discretized T_2 amplitudes at a single depth. Details of the calculation procedures are in Appendix D. Schlumberger-Doll Research (SDR) model is a popular model for the estimation of permeability based on ϕ_N and $T_{2,gm}$, which is expressed as

$$k_{SDR} = C * T_{2,gm}^2 * \phi_N^4 = C * \left(T_{2,gm}\phi_N^2\right)^2 \tag{3.11}$$

where k_{SDR} is the permeability computed using the SDR model and C is a constant. We derived the SDR-model term, $T_{2,gm}\phi^2{}_N$, in Eq. (3.11) using the original and predicted NMR T_2 distribution and then compare them to test the accuracy of the ANN-based predictions of NMR T_2 for purposes of permeability estimation based on the SDR model. Comparison results are presented in Fig. 3.10. Table 3.2 indicates that the ANN-based predictions of NMR T_2 can be reliably used to compute the three reservoir parameters of interest with good accuracy.

Fig. 3.C1 can serve as another petrophysical validation of the ANN-based T_2 synthesis model because it indicates lithology, chlorite content, sonic travel time logs, isolated porosity, rock consolidation, clay content, fluid saturations, and porosity to be important for the NMR T_2 synthesis. The importance of features as shown in Fig. 3.C1 is consistent with physical dependence of NMR logs response on pore size, mineralogy, fluid distribution, and bound fluids.

3.7 ANN-based predictions of NMR T_2 distribution for various depth intervals

Prediction accuracy is the lowest in the MS formation (Table 3.3). The grain size in MS ranges from poorly sorted siltstone to moderately well sorted sandstone. Varying grain size and depositional texture in MS result in more complex pore size distribution that deteriorates the correlations between the feature logs and NMR T_2 distribution in MS. Features in MS tend to have unusually high coefficients of variation (Fig. 3.A2). Furthermore, prediction accuracy in CR1 is lower than those in CR2, CR3, and CR4 formations (Table 3.3). Although all intervals of CR have relatively similar lithology and mineral composition, CR1 is the thinnest, and CR4 is the thickest among the four (Fig. 3.2). The difference of thickness of intervals and the resulting limited training data led to the differences in the prediction accuracies in the various CR intervals. Features in CR1 tend to have unusually high coefficients of variation as compared with CR2, CR3, and CR4 intervals (Fig. 3.A2). Prediction accuracy is the highest in CR4 because CR4 is a thick interval with predominance of dolosiltite. Simple grain size and

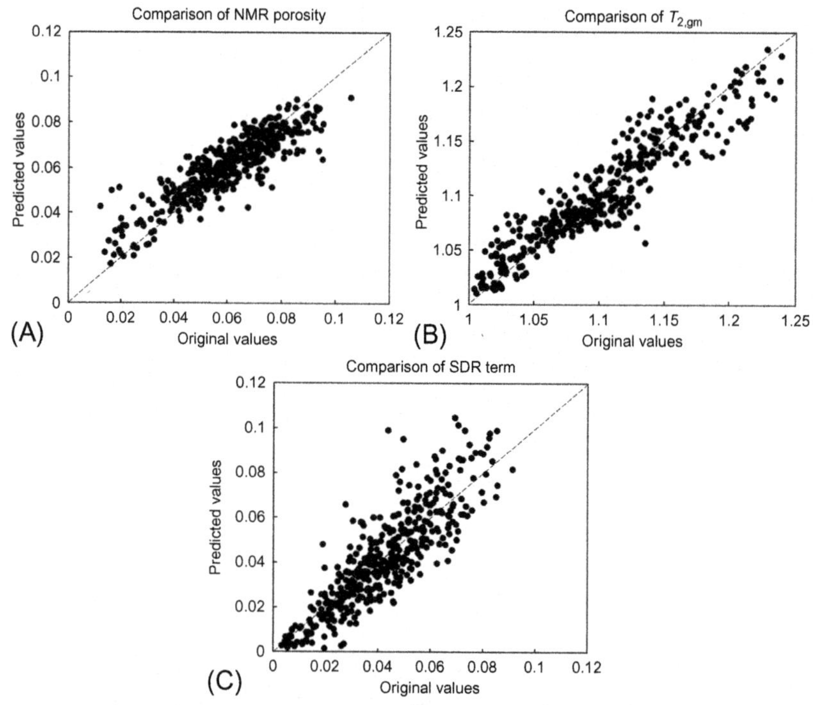

FIG. 3.10 Comparisons of ϕ_N, $T_{2,gm}$, and $T_{2,gm}\phi^2_N$ (referred as SDR-model term) computed from the original NMR T_2 distributions with those computed from the ANN-based predictions of NMR T_2 distributions.

TABLE 3.2 Accuracies of ϕ_N, $T_{2,gm}$, and $T_{2,gm} * \phi^2_N$ derived from the ANN-based predictions of NMR T_2 distribution.

	ϕ_N	$T_{2,gm}$	$T_{2,gm}\phi^2_N$
R^2	0.7685	0.8664	0.7587
NRMSE	0.0909	0.0840	0.0854

TABLE 3.3 Mean values of prediction performances in various intervals using the first ANN-based predictive model.

		US	MS	LS	CR1	CR2	CR3	CR4
Prediction accuracy	R^2	0.850	0.739	0.818	0.792	0.854	0.847	0.877
	NRMSE	0.113	0.153	0.132	0.138	0.112	0.121	0.105

depositional texture with sufficient thickness result in good prediction performance. Median values of the log do not affect the prediction performance; however, extreme values of coefficient of variation and skewness will have adverse effects on the prediction performance of the ANN model.

4 Conclusions

Twelve conventional logs, ten inversion-derived logs, and five log-derived categorical flags in a shale system were processed by two distinct artificial neural network (ANN) models with fully connected layers to synthesize the NMR T_2 distribution responses, which approximate the in situ fluid-filled pore size distribution. The first predictive model generates T_2 distribution discretized into 64 bin amplitudes, whereas the second predictive model generates 6 parameters of a bimodal Gaussian distribution that fit the T_2 distributions. The first predictive model is a better performing method exhibiting median R^2 of 0.8549 on the testing dataset. However, the first model takes 30% more computational time as compared with the second model.

The median R^2 and median NRMSE of predictions of T_2 distributions on the testing dataset are 0.8549 and 0.1218, respectively. This testing performance is remarkable given the hostile subsurface borehole conditions when acquiring the logs, which result in low signal-to-noise ratio, and the limited size of the dataset available to build the model, which gives rise to overfitting and poor generalization in the absence of cross-validation. All testing depths have prediction performance higher than R^2 of 0.6, and 90% of testing depths have prediction errors lower than NRMSE of 0.2.

A few reservoir properties, such as ϕ_N, $T_{2,gm}$, and $T_{2,gm}\phi_N^2$, were derived from the synthetic T_2 distribution at reasonable accuracies. Therefore, ANN-based predictions of NMR T_2 can be reliably used to estimate permeability based on the SDR model. Complex pore size distribution caused by complex grain size distribution and textures can impede the robust ANN-based synthesis of NMR T_2 distribution. Moreover, ANN did not perform well for thin beds due to the lack of data (i.e., statistically significant information) corresponding to the thin beds, which hinders the accuracy of data-driven models. This study provides a workflow to generate in situ fluid-filled pore size distribution, approximated as NMR T_2 distribution, in hydrocarbon-bearing shale reservoirs using neural network models. Notably, the workflow uses features constructed using the k-nearest neighbor classifier to significantly improve the predictive performance of the ANN model. The proposed workflow holds value in the absence of NMR logging tool due to financial and operational challenges.

Appendix A Statistical properties of conventional logs and inversion-derived logs for various depth intervals

See Figs. 3.A1–3.A3.

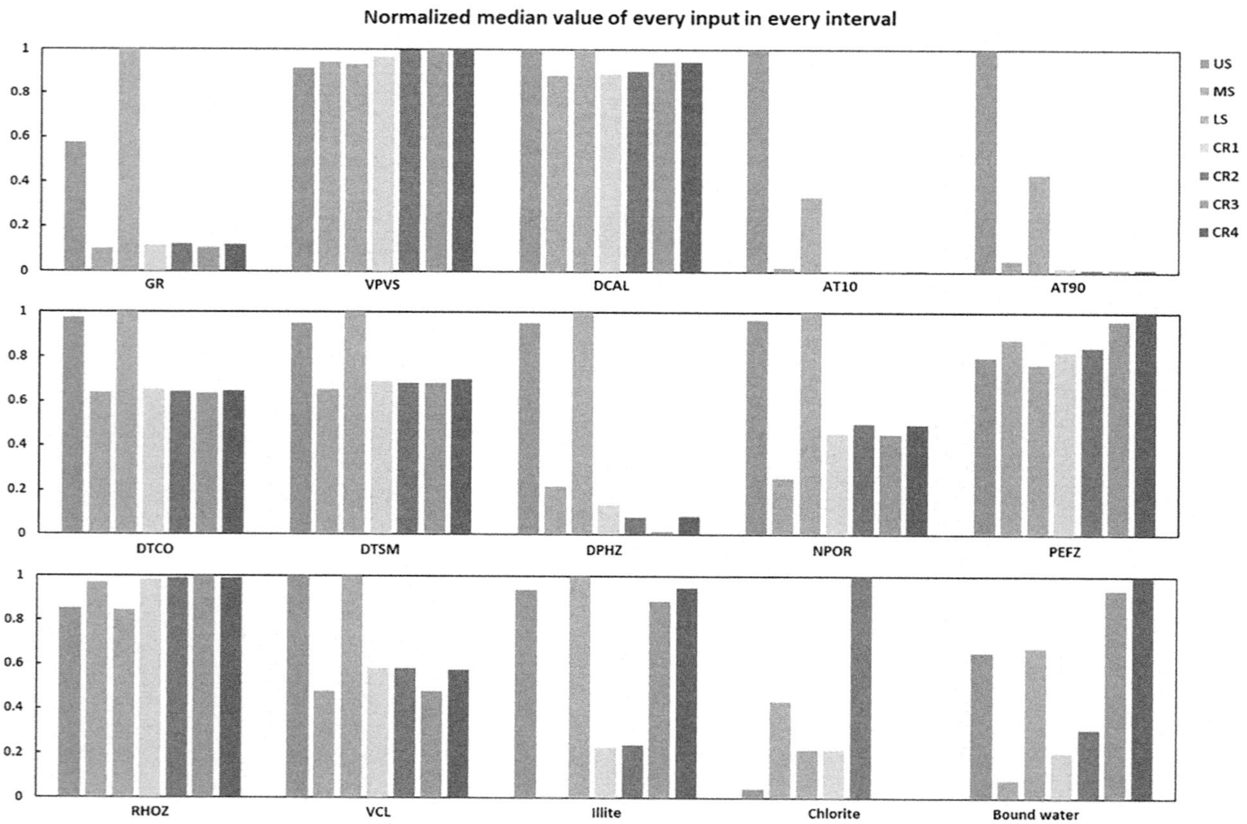

FIG. 3.A1 Normalized medians of conventional and inversion-derived logs for the seven distinct depth intervals, namely, US, MS, LS, CR1, CR2, CR3, and CR4.

Continued

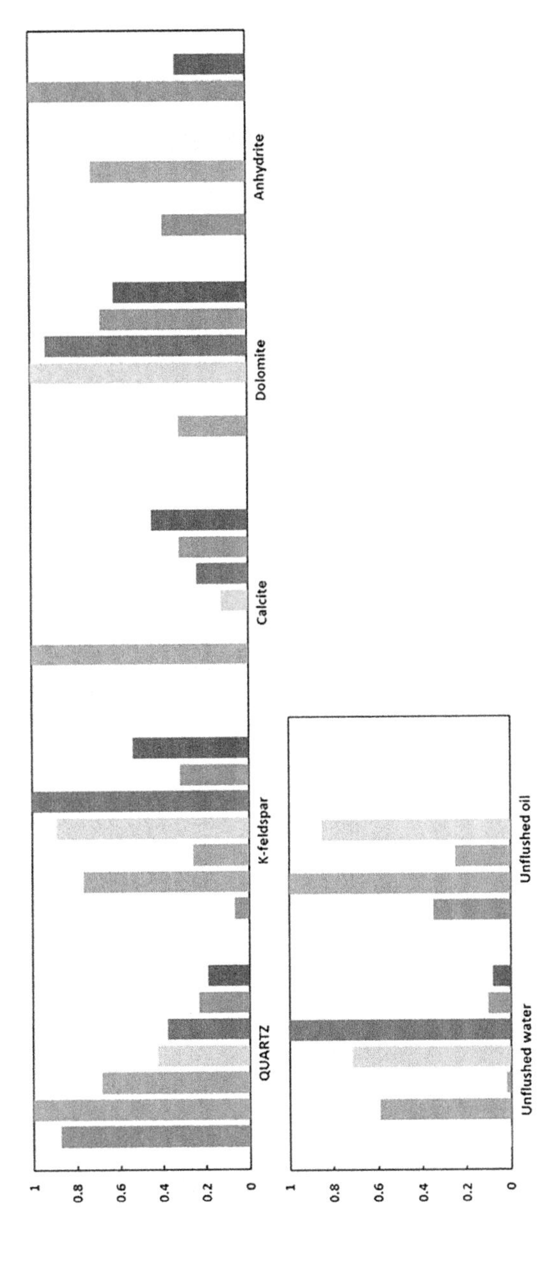

FIG. 3A1, CONT'D

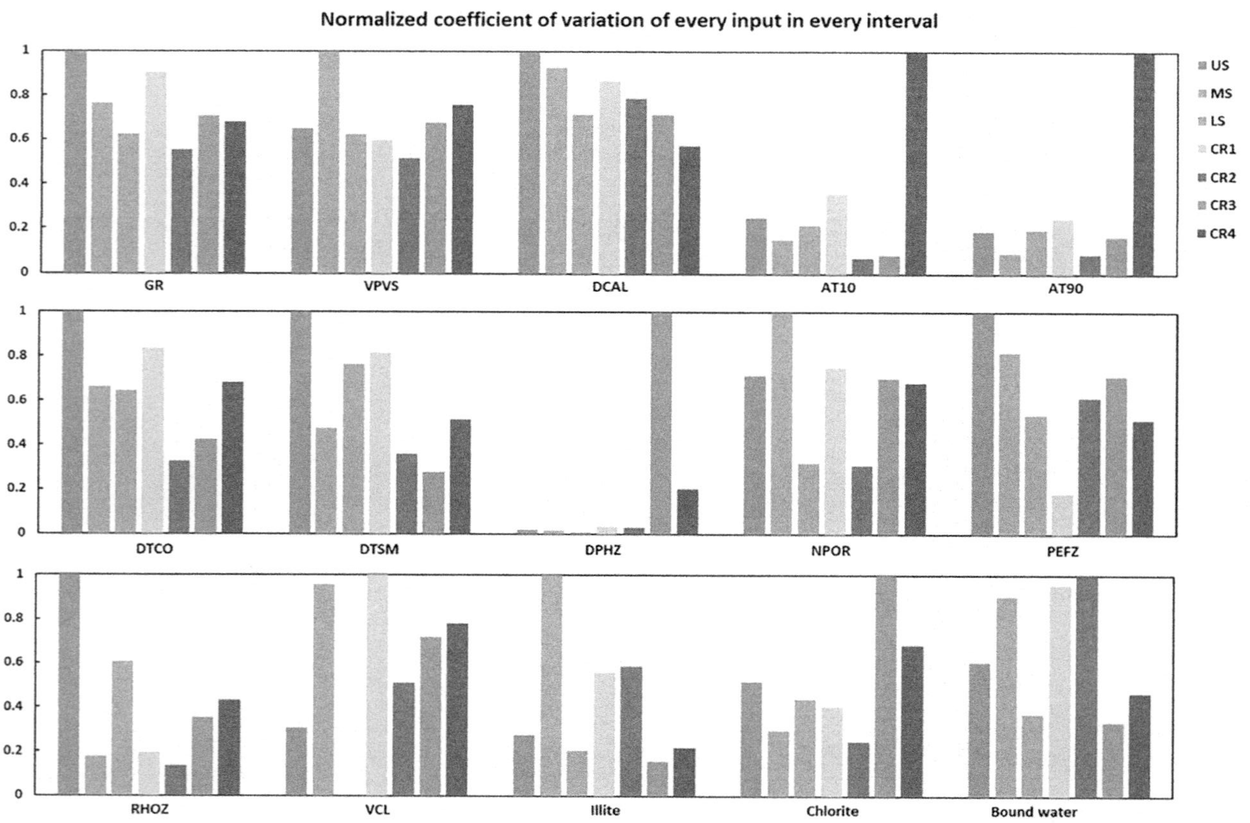

FIG. 3.A2 Normalized coefficients of variation of conventional and inversion-derived logs for the seven distinct depth intervals, namely, US, MS, LS, CR1, CR2, CR3, and CR4.

Continued

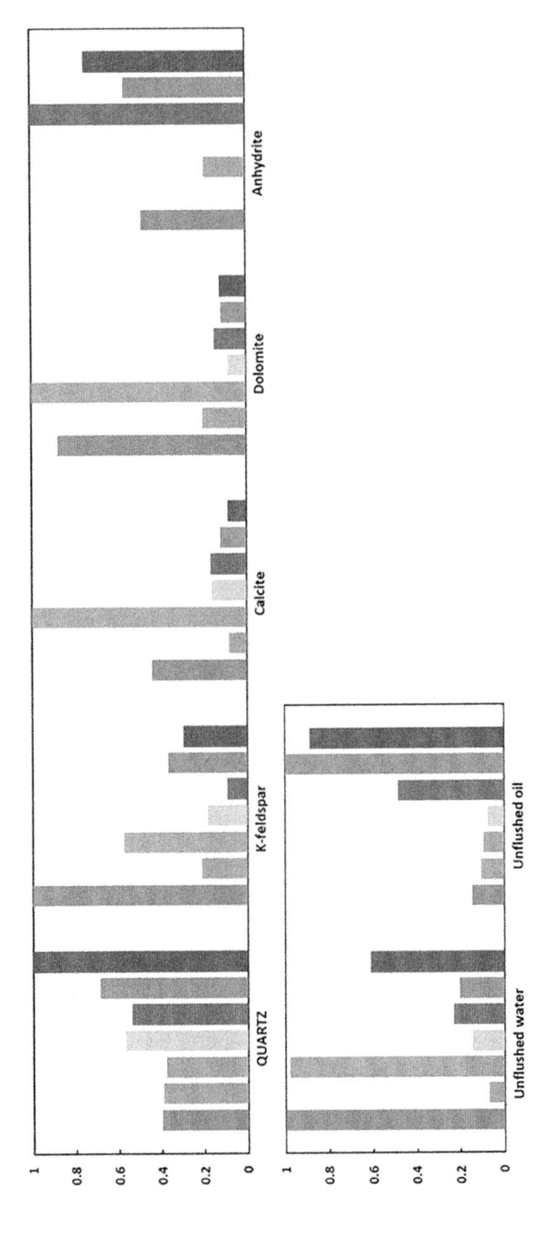

FIG. 3A2, CONT'D

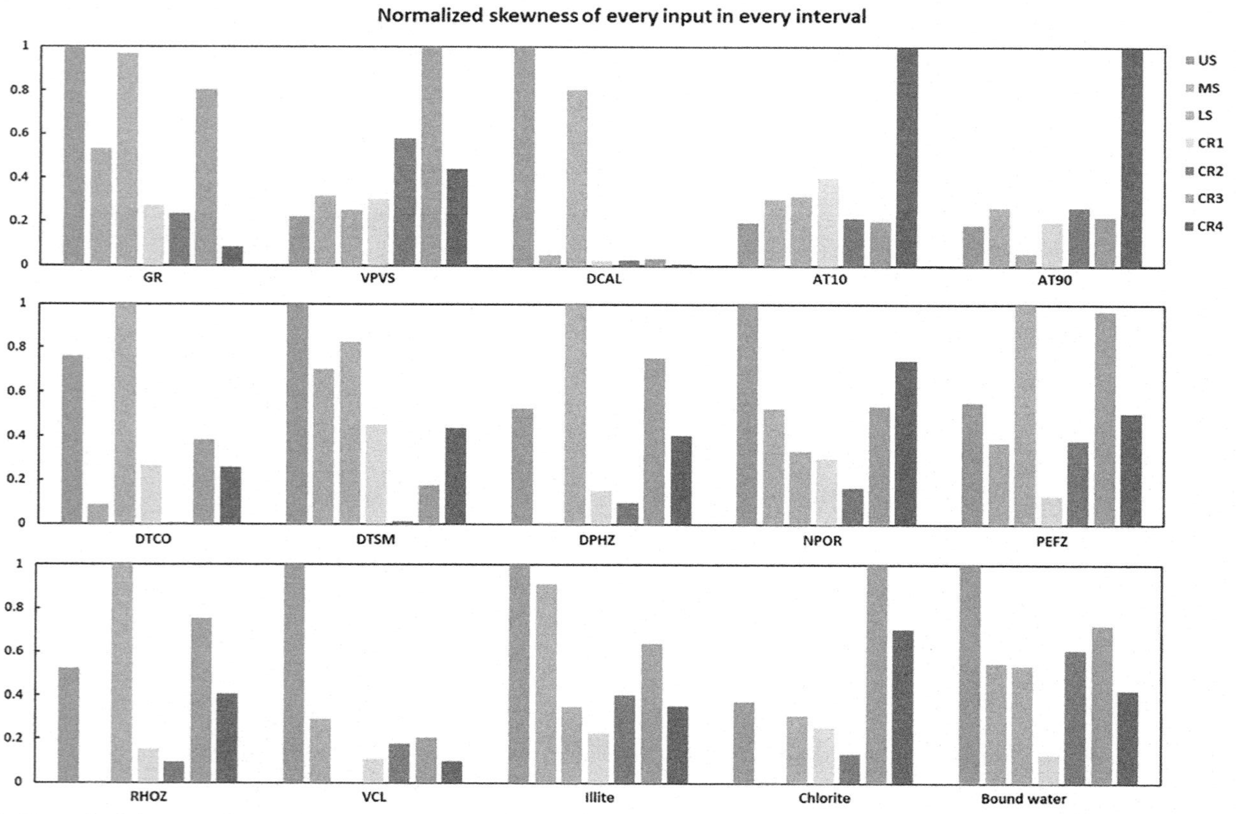

FIG. 3.A3 Normalized skewness of conventional and inversion-derived logs for the seven distinct depth intervals, namely, US, MS, LS, CR1, CR2, CR3, and CR4.

Continued

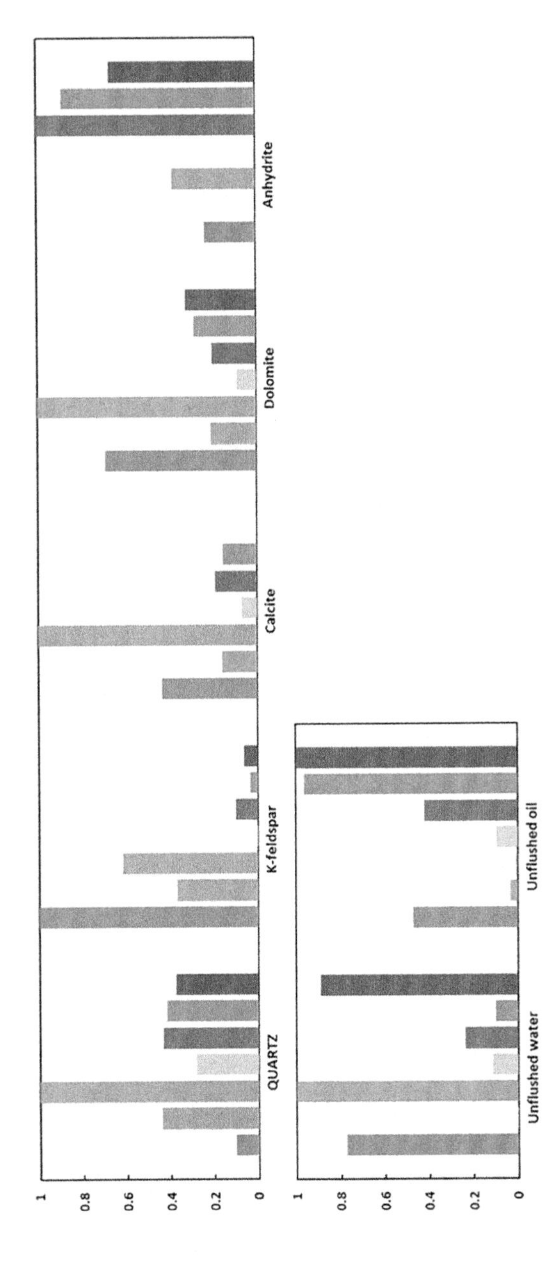

FIG. 3A3, CONT'D

Appendix B Categorization of depths using flags (categorical features)

See Figs. 3.B1 and 3.B2 and Table 3.B1.

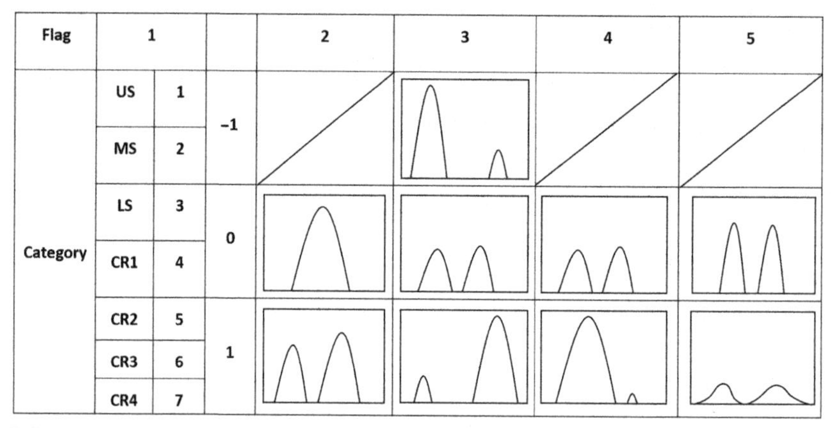

FIG. 3.B1 Qualitative schematic descriptions of the Flags 1–5. Flag 1 has integer values ranging from 1 to 7; Flags 2, 4, and 5 have integer values of either 0 or 1; and Flag 3 has integer values ranging from −1 to 1.

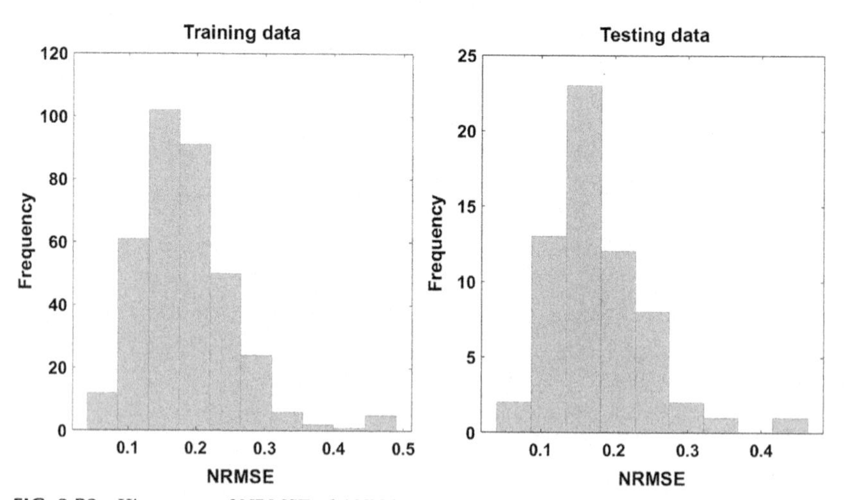

FIG. 3.B2 Histograms of NRMSE of ANN-based predictions of the NMR T_2 distributions when processing the training and testing datasets without flags.

TABLE 3.B1 Comparison of predictive performance, in terms of R^2 and NRMSE, of the first ANN model that processes 12 conventional logs and 10 inversion-derived logs with and without the 5 categorical flags. The use of features/flags constructed using the KNN classifier drastically improves the prediction performance of the ANN model.

	Training		Testing	
	R^2	NRMSE	R^2	NRMSE
With flags	0.8574	0.1201	0.8549	0.1218
Without flags	0.7100	0.1760	0.7152	0.1690

Appendix C Importance of the 12 conventional logs and 10 inversion-derived logs

Machine-learning algorithms process large datasets to develop a data-driven model. The outcomes of a data-driven model for a given set of features/attributes are primarily governed by the importance of the features. Feature importance indicates the significance of a feature for developing robust data-driven model. Importance of a feature for a machine-learning task depends on the statistical properties of the feature and on the relationship of the feature with other features, targets, and the machine-learning algorithm used to develop the data-driven model. Feature ranking helps us understand the inherent decision making process of a data-driven model and helps in evaluating the consistency of a data-driven model by making the model easy to interpret.

There are several techniques for ranking the features. The most popular techniques are the permutation importance method and feature perturbation ranking method. In this study, we are using an alternative feature importance method, wherein the importance of each feature is quantified as the difference between the prediction performance (generalization performance) of the model trained on all the features and that trained on all but the one feature to be ranked. The contribution of one feature is removed by replacing the measurements of that feature for all samples with zeros. A large drop in the prediction performance when one feature is removed (i.e., replaced with zeros) indicates that the feature has a lot of significance when developing (training) and deploying (testing) the data-driven model.

Fig. 3.C1 shows the feature importance for 27 features, which include 10 conventional logs, 12 inversion-derived logs, and 5 categorical flags. Flag 1 (lithology) is the most important feature for the desired synthesis task. Flags 2, 3, and 4 also have high feature importance. This explains the drastic improvement in the ANN-based prediction when using the categorical flags. Among the conventional logs, DTCO, DTSM, GR, NPOR, and DPHZ exhibit high

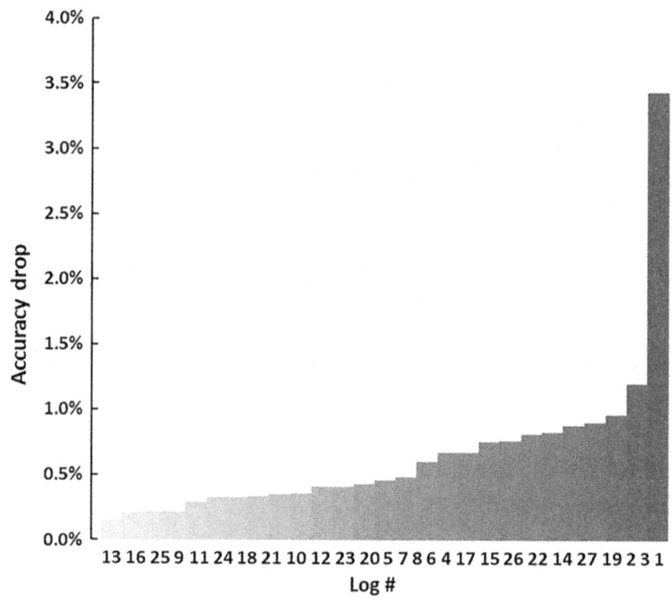

FIG. 3.C1 Feature importance for the first ANN model. The index numbers on the *y*-axis indicate the various logs and flags, namely, #1–#5, Flags 1–5; #6, GR; #7, DPHZ; #8, NOPR; #9, PEFZ; #10, RHOZ; #11, VCL; #12, AT10; #13, AT90; #14, DTCO; #15, DTSM; #16, VPVS; #17, total Sw; #18, illite; #19, chlorite; #20, bound water; #21, quartz; #22, K-Feldspar; #23, calcite; #24, dolomite; #25, anhydrite; #26, unflushed water; and #27, unflushed oil.

importance for the desired synthesis task. Sonic travel logs, which are sensitive to effective porosity and rock consolidation, are the most important conventional logs. Logs sensitive to porosity and clay content are also important for the NMR synthesis. Inversion-derived logs exhibit higher feature importance in comparison with the conventional logs. Among the inversion-derived logs, the chlorite, unflushed oil, K-feldspar, unflushed water, and total water saturation are the most important features, in order from high to low. AT90, VPVS, anhydrite, PEFZ, VCL, dolomite, and illite have low ranking, indicating these logs are not important for the proposed ANN-based predictive task. Deep sensing logs, such as AT90, and logs that have correlation with other important logs tend to have lower feature ranking.

Appendix D Estimations of specific reservoir parameters from NMR T_2 distributions

The first parameter ϕ_N is calculated by integrating the NMR T_2 distribution; in other words, ϕ_N is the summation of the 64 discrete amplitudes for the corresponding 64 T_2 bins at each depth. The T_2 distribution is discretized into 64

T_2 amplitudes as a function of T_2 times ranging from 0.3 ms to 3000 ms. ϕ_N is formulated as

$$\phi_N = \int_{0.3}^{3000} A(T_2)dT_2 = \sum_{i=1}^{64} A(T_{2,i}) \Delta T_2 \tag{3.D1}$$

where $A(T_{2,i})$ is the amplitude A of the ith T_2 bin, and ΔT_2 can be obtained by knowing the range of T_2 such that

$$64 \Delta T_2 = \log_{10}\left(\frac{3000\,\text{ms}}{0.3\,\text{ms}}\right) = 4 \rightarrow \Delta T_2 = \frac{1}{16} \tag{3.D2}$$

Now, the Eq. (3.D1) can be written as

$$\phi_N = \frac{1}{16} \bullet \sum_{i=1}^{64} A(T_{2,i}) \tag{3.D3}$$

The second parameter $T_{2,gm}$ is the 64th root of the product of the 64 T_2 bin amplitudes formulated as

$$T_{2,gm} = \left(\prod_{i=1}^{n} T_{2,i}\right)^{1/n} \tag{3.D4}$$

where $n = 64$ in our case and i indicates the ith T_2 bin. The geometric mean cannot be calculated if any value is equal to 0. Considering the variation in amplitudes for different T_2 bins, weighted $T_{2,gm}$ is calculated at each depth that is expressed as

$$T_{2,gm} = \left(\prod_{i=1}^{64} T_{2,i}^{w_i}\right)^{1/\Sigma w_i} \tag{3.D5}$$

where $w_i = \frac{A_i}{A_T}$, A_i is amplitude of the ith T_2 bin and A_T is the sum of all amplitude values. Here, $\sum_{i=1}^{64} w_i = 1$. The product of weighted T_2 is equivalent to the sum of weighted $\log(T_2)$. Consequently, T_2 geometric mean can also be expressed as T_2 logarithmic mean:

$$\log\left(T_{2,gm}\right) = w_1 \log\left(T_{2,1}\right) + w_2 \log\left(T_{2,2}\right) + \ldots + w_{64} \log\left(T_{2,64}\right) \tag{3.D6}$$

Abbreviations

ANN artificial neural network
AT10 induction resistivity logs at 10-in.
AT90 induction resistivity logs at 90-in.
CMIS committee machine with intelligent systems
CG conjugate gradient

DPHZ	density porosity log
DTCO	Delta-T Compressional
DTSM	Delta-T Shear
GR	gamma ray log
KNN	*k*-nearest neighbor algorithm
LM	Levenberg-Marquardt algorithm
NMR	nuclear magnetic resonance
NPOR	neutron porosity log
NRMSE	normalized root mean square error
SSE	sum of squared errors
TOC	total organic carbon
VPVS	Shear-to-Compressional Velocity Ratio

References

[1] Bhatt A, Helle HB. Committee neural networks for porosity and permeability prediction from well logs. Geophys Prospect 2002;50(6):645–60.

[2] Al-Bulushi N, Araujo M, Kraaijveld M, Jing XD. Predicting water saturation using artificial neural networks (ANNS), In: SPWLA middle east regional symposium. Society of Petrophysicists and Well-Log Analysts; 2007.

[3] Mahmoud AAA, Elkatatny S, Mahmoud M, Omar M, Abdulraheem A, Ali A. Determination of the total organic carbon (TOC) based on conventional well logs using artificial neural network. Int J Coal Geol 2017;179:72–80.

[4] Salazar JP, Romero PA. NMR measurements in carbonates samples and core-logs correlations using artificial neural nets, In: SPE annual technical conference and exhibition. Society of Petroleum Engineers; 2001.

[5] Mohaghegh S, Richardson M, Ameri S. Use of intelligent systems in reservoir characterization via synthetic magnetic resonance logs. J Petrol Sci Eng 2001;29(3):189–204.

[6] Elshafei M, Hamada GM. Petrophysical properties determination of tight gas sands from NMR data using artificial neural network, In: SPE western regional meeting. Society of Petroleum Engineers; 2009.

[7] Labani MM, Kadkhodaie-Ilkhchi A, Salahshoor K. Estimation of NMR log parameters from conventional well log data using a committee machine with intelligent systems: a case study from the Iranian part of the South Pars gas field, Persian Gulf Basin. J Petrol Sci Eng 2010;72(1):175–85.

[8] Golsanami N, Kadkhodaie-Ilkhchi A, Sharghi Y, Zeinali M. Estimating NMR T2 distribution data from well log data with the use of a committee machine approach: a case study from the Asmari formation in the Zagros Basin, Iran. J Petrol Sci Eng 2014;114:38–51.

[9] Genty C, Jensen JL, Ahr WM. Distinguishing carbonate reservoir pore facies with nuclear magnetic resonance measurements. Nat Resour Res 2007;16(1):45–54.

[10] Di J. Permeability characterization and prediction in a tight oil reservoir, Edson Field, Alberta [Doctoral dissertation]. University of Calgary; 2015.

[11] Genty C. Distinguishing carbonate reservoir pore facies with nuclear magnetic resonance as an aid to identify candidates for acid stimulation [Doctoral dissertation]. Texas A&M University; 2006.

[12] Hagan MT, Demuth HB, Beale MH, De Jess O, et al. Neural network design. ACM; 2014.

[13] Chamjangali MA, Beglari M, Bagherian G. Prediction of cytotoxicity data (CC50) of anti-HIV 5-pheny-l-phenylamino-1H-imidazole derivatives by artificial neural network trained with Levenberg-Marquardt algorithm. J Mol Graph Model 2007;26(1):360–7.
[14] Cheng C, Chau K, Sun Y, Lin J. Long-term prediction of discharges in Manwan Reservoir using artificial neural network models, In: Advances in neural networks—ISNN 2005; 2005. p. 975.
[15] Kuhn M, Johnson K. Applied predictive modeling. New York: Springer; 2013.
[16] Li H, Misra S, He J. Neural network modeling of in situ fluid-filled pore size distributions in subsurface shale reservoirs under data constraints. Neural Comput Appl 2019;1–13.
[17] Li H, Misra S. Long short-term memory and variational autoencoder with convolutional neural networks for generating NMR T2 distributions. IEEE Geosci Remote Sens Lett 2018;16 (2):192–5.

Chapter 4

Stacked neural network architecture to model the multifrequency conductivity/permittivity responses of subsurface shale formations

Siddharth Misra* and Jiabo He[†,a]

*Harold Vance Department of Petroleum Engineering, Texas A&M University, College Station, TX, United States, †School of Computing and Information Systems, University of Melbourne, Parkville, VIC, Australia

Chapter outline

1 Introduction

Electromagnetic (EM) properties, such as electrical conductivity, dielectric permittivity, and magnetic permeability, are dispersive in nature, such that the EM properties are functions of the operating frequency of the externally applied EM field. Such frequency dependence is because the polarization phenomenon in a material does not change instantaneously with the applied EM field. EM

[a] Formerly at the University of Oklahoma, Norman, OK, United States

Machine Learning for Subsurface Characterization. https://doi.org/10.1016/B978-0-12-817736-5.00004-1

property is causal in nature; it is mathematically represented as a complex function of the operating frequency because of the phase difference between the material response and the applied EM field. Complex dielectric permittivity (ε^*) and complex conductivity (σ^*) measurements in the frequency range of 10 Hz to 1 GHz are widely used for the oil and gas, hydrological, and several geophysical applications. ε^* and σ^* are fundamental material parameters that influence the transport and storage of electromagnetic (EM) energy in geological materials. ε' is a measure of storage of EM energy due to charge separation and accumulation (also, referred to as polarization), ε'' is a measure of dissipation of EM energy due to friction between polarized structures, and σ is a measure of dissipation of EM energy due to charge transport in a material in response to an external EM field. Consequently, for porous geomaterials and geological formations that are not very conductive, $\sigma^* = -i\omega\varepsilon^* = \sigma - i\omega(\varepsilon' + i\varepsilon'') = (\sigma + \omega\varepsilon'') - i\omega\varepsilon' = \sigma_{\text{eff}} - i\omega\varepsilon_{\text{eff}}$, where subscript 'eff' denotes effective.

Dielectric permittivity exhibits frequency dependence because various polarization phenomena are dominant in various frequency ranges and the effects of polarization lags the applied EM field, that is, polarization does not change instantaneously with the applied EM field. For measurements at low frequencies, the imaginary component of ε^* is much lower than the real component of ε^*. As the frequency increases, the imaginary component of ε^* increases, whereas the real component of ε^* decreases. For fluid-filled porous geomaterials, frequency dispersion is due to various polarization phenomena, each dominant over certain frequency range. Polarization is due to charge separation and subsequent charge accumulation in the presence of externally applied EM field. Few examples of polarization phenomena in fluid-filled geomaterials are orientation/dipolar polarization of dipoles in fluid, Maxwell-Wagner polarization at brine-matrix interface, interfacial polarization of conductive minerals, membrane or double-layer polarization of clays and surface charge-bearing grains, and concentration polarization due to differential mobilities of ions present in fluid. Each polarization mechanism is dominant within a distinct frequency range.

As a consequence of the various polarization phenomena, frequency-dependent ε^* is sensitive to petrophysical properties, such as water saturation, connate water salinity, porosity, tortuosity, wettability, clay content, conductive mineral content, grain sizes, grain texture, and pore size distribution. Multifrequency effective conductivity and permittivity (in the range of megahertz to gigahertz) of a subsurface geological formation are acquired using the wireline dielectric dispersion (DD) logging tool that is run in an open-hole well intersecting the geological formation of interest [1]. DD logging tool has EM transmitters and receivers typically placed on a pad that pushes against the borehole wall to make firm contact with the formation for reliable DD log acquisition. Pad-based transmitters send EM waves of known magnitude and phase generally at four discrete frequencies in the range of 10 MHz to 1 GHz. EM waves travel through the fluid-filled porous formation and reach the pad-based receivers, where the attenuation and phase shift of the waves due to the electromagnetic properties of the formation are recorded. Following that, the wave attenuations and phase shifts at various frequencies are inverted using the tool-physics forward model to compute multifrequency conductivity (σ) and relative permittivity (ε_r) of the formation in the frequency range of 10 MHz to 1 GHz. Relative permittivity is computed as the real component of complex permittivity divided by the vacuum permittivity. Conductivity is computed as the imaginary component of complex permittivity multiplied by the angular frequency. Finally, a physically consistent geo-electromagnetic mixing model or mechanistic polarization-dispersion model is invoked to process the conductivity- and permittivity-dispersion logs to estimate water saturation, bound-water saturation, salinity, clay-exchange capacity, and textural parameters [1]. However, there are several subsurface environments, operational challenges, and project economic scenarios where DD logging tool cannot be run or where DD logs cannot be acquired in the borehole. For such situations that prevent the subsurface deployment of DD logging tool and DD log acquisition, we propose a workflow to generate 8 DD logs, comprising 4 conductivity-dispersion logs and 4 permittivity-dispersion logs, by processing 15 easy-to-acquire conventional logs using a stacked neural network (SNN) model, which combines multiple artificial neural networks (ANNs).

There are several applications of data-driven models and artificial neural networks (ANNs) for subsurface characterization. Neural network can detect conductivity anomalies in sediments [2]. For improved imaging, ANN can reconstruct the 2D complex permittivity profiles in dielectric samples placed in a waveguide system [3]. ANN was used to predict the real and imaginary permittivity components of thermoresponsive materials by processing the magnitude and phase of reflection coefficients measured at various frequencies ranging from 2.5 to 5 GHz [4]. ANN was used to predict dielectric permittivity of epoxy-aluminum nanocomposite at different concentrations [5]. An interesting ANN-assisted application used complex-valued neural network (CVNN) specially designed for generating/synthesizing complex-valued parameters [6]. CVNN was used in landmine detection and classification by processing the ground-penetrating radar data [7]. CVNN model was used in the prediction of soil moisture content [8]. In our study, we designed a novel stacked neural network (SNN) architecture specially designed for the synthesis of eight dielectric dispersion logs acquired at multiple frequencies. The SNN architecture can be generalized for the synthesis/modeling of any dispersive properties that are mathematically represented as real and imaginary (or magnitude and phase) components as function of an independent parameter (e.g., time, frequency, count, and energy). The SNN architecture is suited for dispersive electromagnetic properties where there is direct relationship between the real and imaginary components at any given frequency and there is a strong relationship between the real/imaginary components at various frequencies. Extensive sensitivity analysis and noise analysis are performed on the SNN model to identify the best strategy for generating conductivity and permittivity dispersions. When there are multiple targets (outputs), a general practitioner either uses one ANN to simultaneously predict all the targets or uses one ANN model to individually predict each target by processing all the features (none of the targets) as inputs to the ANN model. In both the cases, the relationships/dependencies between the available targets are not fully used when training the ANN model. In other words, a traditional ANN model is not trained to process few of the available targets along with all the features as inputs to the ANN model for predicting the rest of the targets. The novelty of this study is the SNN architecture that utilizes the inherent dependencies among the targets by using few targets as the inputs/features to synthesize the rest of the targets. Consequently, the SNN model learns to synthesize each target by processing 15 conventional logs along with few other targets as inputs.

2 Method

2.1 Data preparation

The log-synthesis workflow is trained and tested on logs acquired along a 2200-ft depth interval in a well intersecting an organic-rich shale formation comprising six different lithologies. In total, 23 logs were measured during different logging runs; therefore, depth corrections were performed on all the logs prior to any further processing. The 15 easy-to-acquire conventional logs include gamma ray (GR), density porosity (DPHZ), neutron porosity (NPOR), bulk density (RHOZ), volume of clay (VCL), photoelectric factor (PEFZ), delta-T compressional sonic (DTC), delta-T shear sonic (DTS), lithology indicator, and galvanic resistivity at six depths of investigation (RLA0, RLA1, RLA2, RLA3, RLA4, and RLA5). The eight dielectric dispersion (DD) logs consist of four conductivity-dispersion logs (Track 7) and four permittivity-dispersion logs (Track 8) measured at four distinct frequencies in the range of 10 MHz to 1 GHz. The 15 easy-to-acquire conventional logs are processed by the stacked neural network model shown in Fig. 4.1 to generate the 8 DD logs. The stacked neural network uses the 15 conventional logs (Tracks 2–6, Fig. 4.2) as the features (inputs/attributes) and the 8 DD logs (Tracks 7 and 8, Fig. 4.2) as the targets (outputs) to train and test the log-synthesis method. Lithology indicator is a synthetic integer-valued log generated by assigning an integer value in the range of 1 to 6, representing the mineralogy/lithology of the intersected formation.

2.2 Methodology for the dielectric dispersion (DD) log synthesis

The stacked neural network (SNN) model developed in this study was trained and tested on 15 "easy-to-acquire" conventional logs and 8 dielectric dispersion (DD) logs. These logs were acquired in one well intersecting a 2200-ft depth interval of an organic-rich shale formation (Fig. 4.2). The eight DD logs consist of four conductivity-dispersion logs and four permittivity-dispersion logs measured at four distinct frequencies in the range of 10 MHz to 1 GHz. The stacked neural network uses the 15 conventional logs as the features (inputs) and

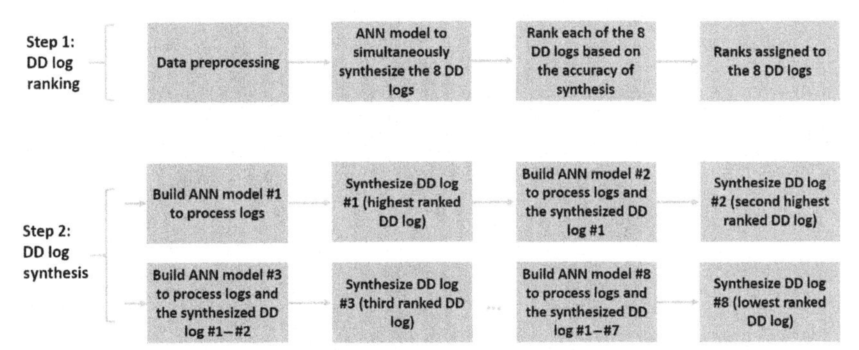

FIG. 4.1 Flowchart of the two-step method, also referred as the SNN model, involving nine neural network models for dielectric dispersion (DD) log synthesis. In the first step, 1 ANN model with 15 inputs and 8 outputs is trained to simultaneously synthesize the 8 DD logs; following that, each of the 8 DD logs is assigned a rank based on the prediction accuracy achieved by the ANN model during the simultaneous synthesis. In the second step, eight distinct ANN models are trained one at a time to sequentially synthesize one of the eight DD logs based on the rank assigned in the first step, starting by predicting the DD log that was synthesized with the highest prediction accuracy and ending by predicting the DD log that was synthesized with the lowest prediction accuracy.

the 8 DD logs are the targets (outputs). The proposed stacked neural network (SNN) architecture combines nine separate artificial neural network (ANN) models with fully connected layers to synthesize the 8 DD logs in two steps. In the first step (Fig 4.1, top), one ANN model with 15 inputs and 8 outputs is trained to simultaneously synthesize the 8 DD logs; following that, each of the eight DD logs is assigned a rank based on the prediction accuracy achieved by the ANN model during the simultaneous synthesis. In the second step (Fig. 4.1, bottom), eight distinct ANN models are trained one at a time to sequentially synthesize one of the eight DD logs based on the rank assigned in the first step, starting with the DD log that was synthesized with the highest prediction accuracy and ending with the DD log that was synthesized with the lowest prediction accuracy. When sequentially training the 8 ANN models implemented in the second step for the sequential synthesis of the 8 DD logs (Fig. 4.1), the ith ANN model that synthesizes the ith ranked DD log is fed with all the previously predicted or measured higher-ranked DD logs (1 to $i-1$) and the 15 conventional logs. In other words, the ith ANN model in the SNN model has $14+i$ inputs and 1 output, where i ranges from 1 to 9.

Various stacked neural network architectures can be designed by changing the number and arrangement of ANN models in the stack, the connection between pairs of ANN models, and the architecture of each ANN model in the stack. In general, the architecture of an ANN model is defined by the number of hidden layers, the number of neurons in a hidden layer, the number and type of inputs and outputs, and the connections between layers and those between the neurons. Similarly, the architecture of a stacked neural network (SNN) model is defined by the number of ANN models in the stack, the number and type of inputs and outputs of each ANN model in the stack, and the connections between the ANN models in the stack.

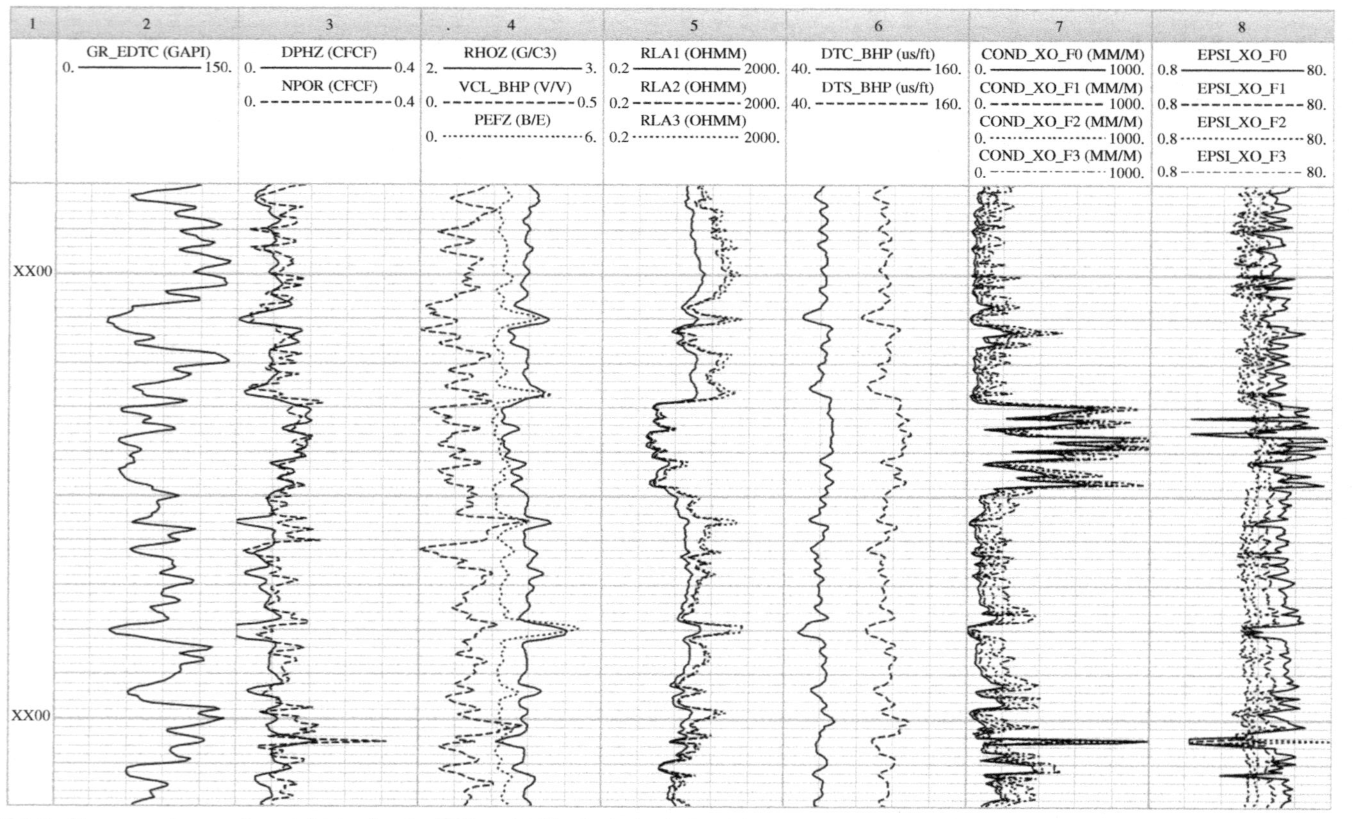

FIG. 4.2 Features and targets for training/testing the SNN model. Track 1 is depth, Track 2 is gamma ray log, Track 3 contains density porosity and neutron porosity logs; Track 4 contains bulk density, volume of clay, and formation photoelectric factor logs; Track 5 contains DTC and DTS logs; Track 6 are laterolog resistivity logs at three out of the six depths of investigation (RLA1, RLA2, and RLA3); Track 7 are conductivity-dispersion logs; and Tracks 8 are permittivity-dispersion logs.

When there are multiple targets (outputs), a general practitioner either uses one ANN to simultaneously predict all the targets or uses one ANN model to individually predict each target by processing all the features (none of the targets) as inputs to the ANN model. In both the cases, the relationships/dependencies between the available targets are not fully used when training the ANN model. In other words, a traditional ANN model is not trained to process few of the available targets along with all the features as inputs to the ANN model for predicting the rest of the targets.

The stacking of ANN models in the SNN model is such that the 8 ANN models implemented in the second step for the synthesis of the 8 DD logs (Fig. 4.1) are sequentially trained, wherein the ith ANN model that synthesizes the ith ranked DD log is fed with all the previously predicted or measured higher-ranked DD logs ($i = 1$ to $i-1$) and the 15 conventional logs. Consequently, the SNN model learns to synthesize each target by processing 15 conventional logs along with few other targets as inputs. Notably, when training the SNN model, the measured values of targets (DD logs) and measured values of features (15 conventional logs) are fed as inputs to each ANN model implemented in the second step, whereas when testing and deploying the SNN model, the predicted values of targets (DD logs) and measured values of features (15 conventional logs) are fed as inputs to each ANN model implemented in the second step. This avoids contamination between training and testing dataset, and ensures that the testing dataset is used in a way that is similar to the deployment/new dataset. Due to the physics of charge polarization, conductivity is related to permittivity at each frequency, and the conductivity/permittivity at one frequency is related to conductivity/permittivity at another frequency. Such relationships are inherent in any dispersive property with a causal behavior. The SNN architecture used in this study is designed to learn these physical relationships as a function of frequency and phase difference.

A simple multivariate linear regression (MLR) model and a traditional ANN model were also trained to synthesize the 8 DD logs one at a time by processing the 15 conventional logs as features/inputs. This evaluation was performed to check the performance of MLR model and ANN model in comparison with the SNN model in the DD log-synthesis task. The prediction performance of the MLR model and ANN model is 20% lower and 10% lower, respectively, compared with that of the SNN model. It can be concluded that the inherent dependencies among the targets can be utilized by using few targets as some of the inputs to synthesize the rest of the targets.

2.3 Evaluation metric/measure for log-synthesis model

A key aspect in the development of data-driven model is the selection and proper interpretation of the evaluation metrics used for measuring and determining the model performances on the training and testing dataset. There are various evaluation metrics for model evaluation. A user should be aware of the limitations and assumptions of a metric and accordingly select a combination of evaluation metrics best suited for a specific predictive/data-driven modeling task. An evaluation metric quantifies the likelihood of a predictive model to correctly predict the future outcomes or that of a data-driven model to correctly quantify the trends/patterns in a dataset. Regression tasks use evaluation metrics that are very different from those used for classification tasks [9]. A good evaluation metric should enable clear discrimination among various models developed on a dataset and should be sensitive to variations in model performances.

A popular evaluation metric for regression tasks is the coefficient of determination, R^2, that measures the fraction of variance in the targets that can be explained by a predictive/data-driven model. In simple terms, R^2 measures how well can the variations in targets/outputs (y) be explained by variations in features/inputs (x) using a certain predictive/data-driven model. R^2 is based on the principle that good models lead to small residuals. R^2 is the square of the correlation coefficient, r, which measures the strength of linear relationship between two variables. Adjusted R^2 is a modification of R^2 to account for the number of features/inputs used in the predictive model. Unlike R^2, adjusted R^2 gives a low score to a model that uses several noninformative, low-importance inputs/features. Few limitations and assumptions of R^2 are as follows:

1. It cannot quantify the bias in the model predictions.
2. It only considers the linear relationships between targets and features.
3. It does not account for the nonlinear relationships between the targets and features unless the targets are appropriately transformed.
4. A large R^2 indicates linear association specific to the model and to the dataset used to develop the model.
5. A large R^2 does not mean causation. It is only an indicator of correlation (association).
6. It over emphasizes large errors versus small errors.
7. It tends to overemphasize errors for samples having large-valued targets versus those having small-valued targets.
8. Though R^2 is scaled between 0 and 1, it is a relative measure and not an absolute measure because it depends on the number of datapoints, selected ranges of the features, and the number and order of the features used to build the model.
9. It does not consider variance in the features.
10. As more features are added, R^2 tends to increase even when the newly added features are not important. This is because of overfitting.
11. It is not suitable when the variance in the target is low, when there are few samples/datapoints, and when the error in data is large.
12. R^2 of models cannot be compared across data sets.

Compared with R^2, mean absolute error (MAE) and root-mean-square error (RMSE) are better evaluation metrics. MAE is the average magnitude of residuals/errors of the predictions for all the samples in the dataset. RMSE is the square root of the average squares of residuals/errors of the predictions for all the samples in the dataset. MAE and RMSE values close to 0 indicate a good model performance. Compared with MAE, RMSE gives more weightage to samples where the model has large errors and is more sensitive to the error distribution and to the variance in errors. Compared with RMSE, MAE is a more stable metric and facilitates easy interpretation, and each sample affects the MAE in direct proportion to the absolute error of prediction for that sample. Large difference between RMSE and MAE indicates large variations in error distribution. MAE is suitable for uniformly distributed errors. RMSE is preferred over MAE when the errors follow a normal distribution. Both MAE and RMSE are not suitable when the errors in model predictions are highly skewed or biased. The sensitivity of the RMSE and MAE to outliers is the most common concern with the use of these metrics. MAE uses absolute value that makes it difficult to calculate the gradient or sensitivity with respect to model parameter.

A modification of RMSE is the normalized root-mean-square error (NRMSE) obtained by normalizing the RMSE with target mean, median, deviation, range, or interquartile range. In our study, we normalize the RMSE of target with the target range, which is the difference between the maximum and minimum values of the target. NRMSE facilitates comparison of model performances when various models are trained and tested on different datasets. Compared with RMSE, NRMSE is suitable for the following scenarios:

1. Evaluating model performances on different datasets, for example, comparing performance of SNN model on training dataset with that on testing dataset or comparing performance of SNN model in Well 1 with that in Well 2
2. Evaluating model performances for different targets/outputs, for example, comparing performance of SNN model for predicting the four conductivity logs with that for predicting the four permittivity logs or comparing performance of SNN model for predicting conductivity log at one frequency with that for another frequency
3. Evaluating performances of different models on completely different datasets, for example, comparing performances of a log-synthesis model with a house-price prediction model
4. Evaluating model performances on various transformations of the same dataset

In our study, we first calculate the normalized root-mean-square error (NRMSE) in SNN model-based synthesis of each of the eight DD logs for various depths. Consequently, for a given dataset, eight NRMSEs are calculated for the eight DD logs. NRMSEs are calculated separately for training and testing datasets. RMSE for log j is expressed as

$$\mathrm{RMSE}_j = \sqrt{\frac{\sum_{i=1}^{n} \left(D_{s,ij} - D_{m,ij}\right)^2}{n}} \tag{4.1}$$

such that n is the number of depth points in the dataset for which DD log synthesis needs to be performed; $j = 1, 2, 3$, or 4 indicates the four conductivity-dispersion logs, and $j = 5, 6, 7$, or 8 indicates the four permittivity-dispersion logs; s indicates synthesized log response, m indicates measured log response; $D_{s,ij}$ is the conductivity σ or relative permittivity ε_r log response synthesized for the depth i; and $D_{m,ij}$ is the σ or ε_r log response measured at depth i. NRMSE for log j is then expressed as

$$\text{NRMSE}_j = \frac{\text{RMSE}_j}{D_{m,j,\max} - D_{m,j,\min}} \tag{4.2}$$

where subscript min and max indicate the minimum and maximum values of the log j, such that $j = 1, 2, 3$, or 4 indicates the four conductivity-dispersion logs and $j = 5, 6, 7$, or 8 indicates the four permittivity-dispersion logs. In our study, high prediction accuracy is indicated by NRMSE less than 0.1. When using NRMSE with range as the denominator, it is crucial to remove outliers from the dataset.

When a model generates the targets, the model performance can be represented as an error/residual distribution by compiling the errors for all the samples in the dataset. A single-valued evaluation metric, like R^2, MAE, RMSE, and NRMSE, condenses the error distribution into a single number and ignores a lot of information about the model performance present in the error distribution. Single-valued metric provides only one projection of the model errors and, therefore, only emphasizes a certain aspect of the error characteristics. When evaluating different models, it is important to consider the error distributions instead of relying on a single metric. Statistical features of the error distribution, such as mean, variance, skewness, and flatness, are needed along with a combination of single-valued metrics to fully assess the model performances. In addition, we should monitor heteroscedasticity of residuals/errors (i.e., difference in the scatter of the residuals for different ranges of values of the feature). The existence of heteroscedasticity can invalidate statistical tests of significance of the model.

2.4 Data preprocessing

In supervised learning, dataset is divided into three parts for purposes of model training, testing, and validation. There should not be any common samples between the three splits. The testing dataset should be treated like a new dataset that should never be used during the model training. Validation data help reduce overfitting, but validation data reduce the size of training and testing datasets. In this chapter, the dataset is split into two parts, namely, training and testing datasets. Instead of using a validation dataset, we use a regularization term in the loss function to reduce overfitting. As a result, more data is available for the training and testing stages, which is beneficial for developing data-driven models under the constraints of data quantity [10].

Before training the ANN models, data preprocessing is necessary to facilitate reliable and robust model predictions. Around 2% of depths in the training and testing datasets exhibited outlier behavior; to be more specific, these outliers are mostly point outliers. Few examples of such outliers are depths where gamma ray responses are close to 1000 API units or shear-wave travel time responses are higher than 800 μs/ft. Depths exhibiting outlier response were removed using simple variance-based method. Outlier removal is essential for obtaining robust NRMSE values. Following outlier detection and removal, the features/inputs and targets/outputs were normalized using MinMax scaler to values within −1 and 1. It is uncommon to scale targets, but features are always scaled. MinMax scaling of the features guarantees stable convergence of the ANN model parameters (i.e., weights and biases) during the neural network optimization [11]. MinMax scaler should not be applied when there are outliers or when a feature has a large variance. In the absence of the limiting conditions, MinMax scaler is well suited for developing robust neural network models. MinMax scaler is performed based on the following equation:

$$D_{sc,i,j} = 2\frac{D_{i,j} - D_{j,\min}}{D_{j,\max} - D_{j,\min}} - 1 \tag{4.3}$$

where $D_{i,j}$ is the original value of log j measured at a specific depth i, $D_{j,\max}$ and $D_{j,\min}$ are the maximum and minimum values of the log j in the entire training dataset, and $D_{sc,i,j}$ is the scaled value of the log j computed for the specific depth i using Eq. (4.3). $D_{i,j}$ can be from training dataset, testing dataset, or new deployment dataset, but $D_{j,\max}$ and $D_{j,\min}$ should always be from the training dataset corresponding to the log j. It should be noted, all data preprocessing methods should be first applied on only the training dataset to learn the statistical parameters or mathematical transformations required for the data preprocessing of the training dataset (e.g., for MinMax scaler operation the parameters are $D_{j,max}$ and $D_{j,min}$). Following that, the statistical parameters or mathematical transformations required for data preprocessing of the training dataset should be used for data preprocessing of the testing dataset or the new deployment dataset. Data preprocessing "should not" be done on the entire dataset prior to split. After split, data preprocessing "should not" be done separately on each split by separately learning the statistical parameters or mathematical transformations required for the data preprocessing of each specific split.

2.5 ANN models for dielectric dispersion log generation

Nine ANN models are implemented in the proposed predictive method that trains and tests the stacked neural network model. The first ANN model with 15 inputs and 8 outputs learns to simultaneously synthesize the 8 DD logs; following that, each DD log is ranked in accordance to the accuracy of simultaneously synthesizing the DD log. The accuracy used for ranking corresponds to the performance of the ANN model on the testing dataset. The second step of training learns to improve accuracy of synthesizing each DD log by sequentially synthesizing the DD logs in accordance to the rank assigned in the first step, such that higher-ranked DD logs along with the conventional logs are used to synthesize the lower-ranked DD logs. Sequential synthesis of one DD logs is done using one ANN; consequently, eight ANN models are implemented to sequentially generate the eight DD logs one at a time. During the training stage, the original higher-ranked DD logs are fed as inputs to learn to generate the lower-ranked DD log, whereas, during the testing and deployment stages, the predicted higher-ranked DD logs are fed as inputs to generate the lower-ranked DD log. For these 8 ANN models used in the second step of prediction, the ith ANN model processes $14+i$ inputs (comprising 15 conventional logs and $i-1$ DD logs) to generate only 1 output DD log, where i varies from 1 to 8. In other words, 1 DD log of a specific rank is synthesized by a corresponding ANN model that processes all the previously predicted or measured higher-ranked DD logs and the conventional input logs. For example, in this second step, the eighth ANN model processes 22 logs to generate the lowest-ranked DD log. All ANN models have two hidden layers and varying number of neurons corresponding to the numbers of inputs and outputs.

Several algorithms can be utilized as the training function to adjust weights and biases of the neurons to minimize target functions of the ANN model. Target function describes error relevant to the prediction performance. We use the sum of squared errors' (SSE) function as the target function expressed as

$$\text{SSE} = \sum_{i=1}^{n} (y_i - \hat{y}_i)^2 + \lambda \sum_{j=1}^{P} \sigma_j^2 \tag{4.4}$$

where n is the number of samples, P is the number of outputs/targets per sample, λ is the penalty parameter, $\boldsymbol{y}_i$ is the original output/target vector, $\hat{\boldsymbol{y}}_i$ is the estimated output/target vector, and σ_j^2 is the variance of predicted output/target j. Regularization method is utilized to introduce the penalty parameter λ into the target function to avoid overfitting and ensure a balanced bias-variance tradeoff for the data-driven model. Overfitting is a critical issue when minimizing the target function. Due to overfitting, ANN model cannot generalize the relationship between inputs/features and outputs/targets and becomes sensitive to noise. Overfitting is evident when the testing performance (referred as the generalization performance) is significantly worse than the training performance (referred as the memorization performance). λ ranges from 0.01 to 0.2 in our models, which is set based on trial and error.

Scaled conjugate gradient (CG) backpropagation [12] is selected as the training function because it requires a small training time. For example, the training time of the ANN model implemented in the first step with Levenberg-Marquardt (LM) backpropagation as the training function is two times more than that with CG backpropagation. Each ANN model is trained for 100 times with different initial weights and biases and different training and testing datasets to avoid local minima during the model optimization. The best model is then selected from among the 100 models.

3 Results

The ranking (Table 4.1) was obtained based on the accuracy of synthesizing each of the eight DD logs when simultaneously synthesizing the eight DD logs using one ANN model. The ranking is in terms of NRMSE, such that lowest NRMSE indicates the most accurate synthetic DD log and is assigned the highest rank. Ranking is accomplished in the first step of SNN model development. In the second step of SNN model development, during the training stage, the ranking facilitates sequential development of eight ANN models that are trained to generate one of the eight DD logs, such that the generation of a lower-ranked DD log by an ANN model uses the higher-ranked DD logs along with the 15 conventional logs.

During the sequential generation of DD logs (Fig. 4.3) in the second step, each of the eight DD logs is predicted one at a time using eight distinct ANN models that process the conventional logs and all the previously predicted dispersion logs as inputs (Fig. 4.1). For example, ANN model #1 generates the highest-ranked σ_{f0} by processing the 15 conventional logs as inputs. Following the generation of σ_{f0}, ANN model #2 generates the second-ranked σ_{f1} by processing the 15 conventional logs and predicted σ_{f0} as inputs. The rest of the lower-ranked DD logs are generated in the similar manner, such that finally the lowest-ranked $\varepsilon_{r,\,f3}$ is generated by ANN model #8 that processes the 15 conventional logs and the 7 previously generated DD logs, namely, σ_{f0}, σ_{f1}, σ_{f2}, σ_{f3}, $\varepsilon_{r,f0}$, $\varepsilon_{r,f1}$, and $\varepsilon_{r,f2}$. This two-step sequential DD log-synthesis method reduces the overall prediction inaccuracy in generating the eight DD logs from 0.705 to 0.637 in terms of NRMSE, which marks a 9.6% relative change with respect to one-step simultaneous DD log synthesis (Table 4.2). For the one-step simultaneous DD log synthesis, $\varepsilon_{r,\,f2}$ and $\varepsilon_{r,\,f3}$ were generated at the highest inaccuracies of 0.098 and 0.112 in terms of NRMSE, respectively. The NRMSE for $\varepsilon_{r,f2}$ and $\varepsilon_{r,f3}$ were lowered to 0.089 and 0.086, respectively, which

TABLE 4.1 Prediction performance of the ANN model implemented in the first step of the two-step DD log synthesis and the ranks assigned to the eight DD Logs based on the accuracy of simultaneous synthesis.

	σ_{f0}	σ_{f1}	σ_{f2}	σ_{f3}	$\varepsilon_{r,\,f0}$	$\varepsilon_{r,\,f1}$	$\varepsilon_{r,\,f2}$	$\varepsilon_{r,\,f3}$
NRMSE	0.068	0.073	0.079	0.087	0.091	0.097	0.098	0.112
Rank	1	2	3	4	5	6	7	8

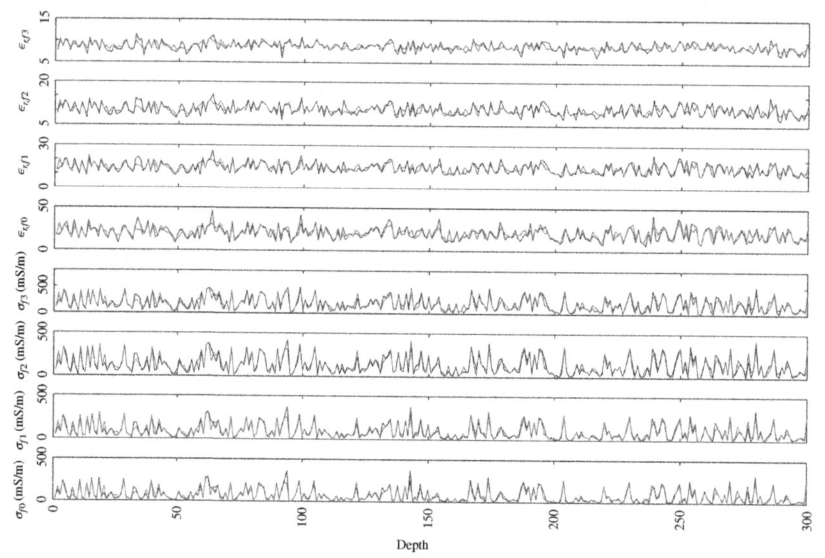

FIG. 4.3 Comparison of the eight original *(dashed)* and eight synthesized *(solid)* dielectric dispersion logs along a 300-ft interval of the shale formations. SNN model successfully synthesized the eight DD logs, comprising four conductivity and four permittivity logs.

TABLE 4.2 Prediction performance the two-step sequential DD log synthesis and relative change in performance as compared with the one-step simultaneous DD log synthesis.

	σ_{f0}	σ_{f1}	σ_{f2}	σ_{f3}	$\varepsilon_{r,\ f0}$	$\varepsilon_{r,\ f1}$	$\varepsilon_{r,\ f2}$	$\varepsilon_{r,\ f3}$
NRMSE	0.067	0.066	0.071	0.077	0.093	0.088	0.089	0.086
Change %	−1.47	−9.59	−10.13	−11.49	2.2	−9.28	−9.18	−23.21

correspond to relative changes in prediction accuracies of 9.8% and 23.21%, respectively (Table 4.2). The proposed predictive method has decent accuracy (Fig. 4.3) despite the low quality and quantity of the dataset. Fifteen conventional and eight DD logs used for training the neural network model were recorded during different logging runs performed in a 5-inch-diameter borehole intersecting the Permian Basin shale formation. Consequently, the logs are prone to noise and adversely affected by the heterogeneity, complexity, and uncertainty of the surrounding borehole and near-wellbore environments. Overall performance of the SNN model in synthesizing the conductivity dispersions is better than that for permittivity dispersions by 0.2.

3.1 Sensitivity analysis

For each case of the sensitivity analysis, 100 cycles with random initial weights and biases and random training and testing datasets are run to ensure that the SNN model optimization avoids local minima. For the purposes of the sensitivity analysis, we investigate the changes in the log-synthesis performance achieved by the SNN model due to changes in inputs/features, outputs/targets, and noise in data. NRMSEs for the synthesis of each conductivity and permittivity-dispersion logs are summed together to compute the accuracy drop in %. An accuracy drop of 1% corresponds to an approximate NRMSE change of 0.007.

The first sensitivity analysis is done to study the importance of the 15 conventional logs, including the one synthetic integer-valued log denoting the lithology type (Fig. 4.4). The SNN model for the proposed log-synthesis method is most sensitive to the removal of sonic (DTC and DTS), neutron porosity (NPOR), integer-valued lithology flag, medium resistivity (RLA3), and clay volume (VCL) logs. The SNN model is the least sensitive to the removal of shallow resistivity (RLA1 and RLA2), density logs (RHOZ), and photoelectric factor (PEFZ). The six resistivity logs sense different formation volumes, such that RLA0 is shallowest sensing measurement and RLA5 is deepest sensing measurement. Among the six resistivity logs, Fig. 4.4 indicates that the DD log generation is the most affected when RLA3 is removed and is least affected when RLA2 is removed. Logs that lead to low accuracy drop when removed can be considered as the least important logs/features for the desired data-driven modeling of DD logs.

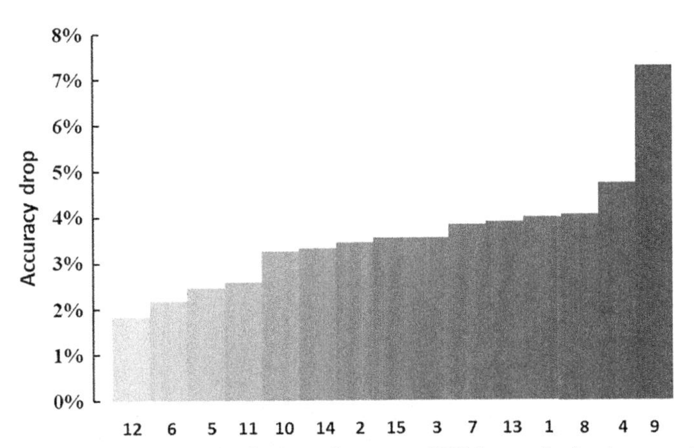

FIG. 4.4 Comparison of reduction in the performance of DD log synthesis when one of the 15 conventional logs is removed one at a time. Each integer on *y*-axis identifies the removal of a specific log: 1, lithology flag; 2, GR; 3, DPHZ; 4, NPOR; 5, PEFZ; 6, RHOZ; 7, VCL; 8, DTC; 9, DTS; and 10–15, resistivity at different depths of investigation (10, RLA0; 11, RLA1; 12, RLA2; 13, RLA3; 14, RLA4; and 15, RLA5).

Next, we determine the smallest set of features/inputs required for maintaining the desirable log-synthesis performance using the SNN model. The 15 conventional logs (features) that are fed into the SNN model may not be available during the real-world deployment of the model. For such scenarios, it is useful to figure out the smallest set of logs that can be fed into the SNN model without significant drop in accuracy of log synthesis. This is determined by deleting features one by one starting with the least important feature, as identified in Fig. 4.4. We conclude that at least 11 most important features should be retained to maintain an accuracy drop less than 10% in the log synthesis as compared to when all the 15 logs/features are used to develop the SNN model. This set of 11 log inputs is obtained by removing RLA2, RHOZ, PEFZ, and RLA1 (Fig. 4.5). As shown by the leftmost bar in Fig. 4.5, a set of features containing only the six most important features (DTC, DTS, NPOR, lithology flag, RLA3, and VCL logs) results in 17% accuracy drop in the log synthesis as compared to when all the 15 logs/features are used to develop the SNN model. Notably, when DTC, DTS, NPOR, and six resistivity logs of different depths of investigation are retained and other low-importance logs (RHOZ, PEFZ, GR, DPHZ, VCL, and lithology) are removed, we observe accuracy drop less than 10% as compared to when all the 15 logs/features are used to develop the SNN model (Fig. 4.6).

To study the sensitivity of the SNN model to noise in training/testing data, 20% Gaussian noise is added one at a time to each feature (inputs and conventional logs), to the six resistivity log inputs together (i.e., 3.33% noise is added to each resistivity log), and to the eight DD log outputs together (i.e., 2.5% noise is added to each DD log). The sensitivity of the SNN model to noise in feature, from the highest to lowest sensitivity, is as follows: resistivity, DTS, GR, RHOZ, NPOR, VCL, DPHZ, DTC, and PEFZ (Fig. 4.7). Twenty percent of overall noise in the six resistivity logs results in maximum reduction in the log-synthesis performance by 4.5%. Log-synthesis performance is also highly

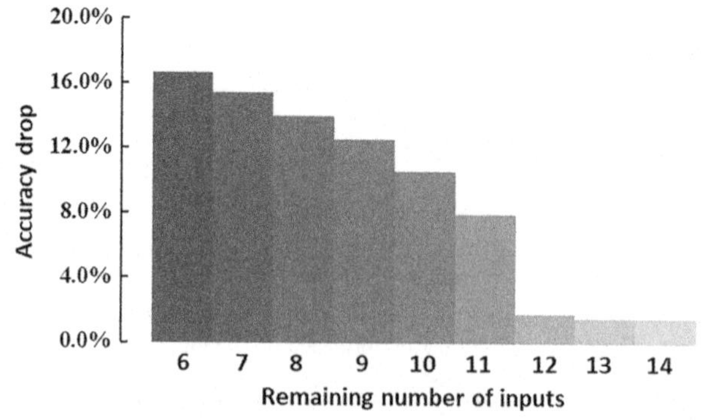

FIG. 4.5 Comparison of reduction in the performance of DD log synthesis by deleting the conventional logs one by one based on the importance of the conventional log as described in Fig. 4.4.

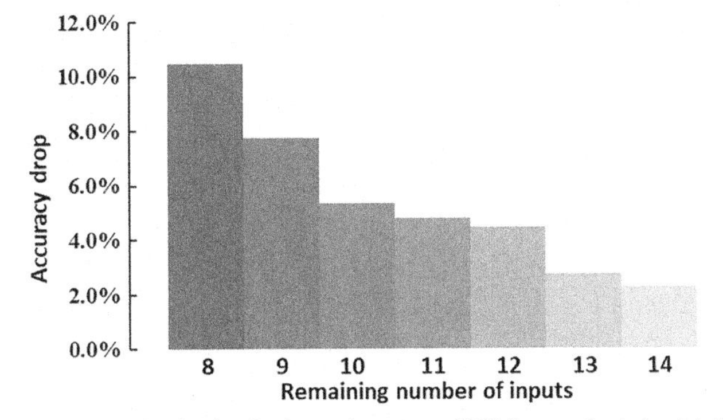

FIG. 4.6 Comparison of reduction in the performance of DD log synthesis by deleting conventional logs (other than the six resistivity logs) one by one based on the importance of the conventional log as described in Fig. 4.4. This preserves the relationship between resistivity and dielectric dispersion measurements.

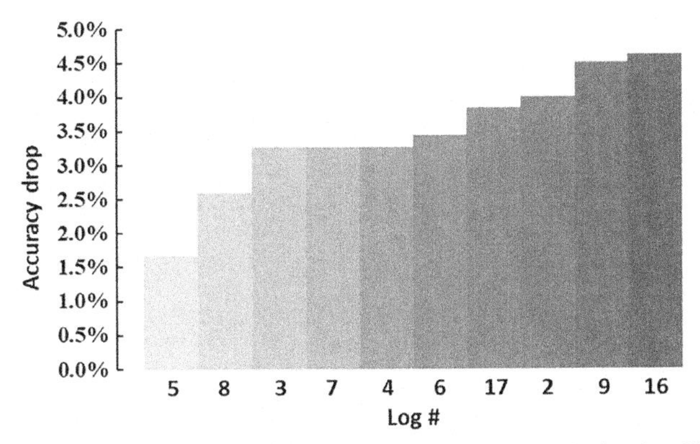

FIG. 4.7 Comparison of reduction in the performance of DD log synthesis when 20% Gaussian noise is added to the conventional logs and DD logs one at a time. Integer-valued log indices are similar to those listed in Figs. 4.4–4.6. Log index #16 represents the six resistivity logs together, and Log index #17 represents the eight dielectric dispersion logs together.

sensitive to noise in GR and DTS. SNN model is least sensitive to noise in PEFZ and DTC. Notably, 20% of overall noise in the eight targets also leads to significant accuracy drop of 3.7%.

Finally, we study the sensitivity of the SNN model to various types of noise [13]. The study is presented in Figs. 4.8–4.10. Gaussian noise is a statistical noise having normally distributed probability density function. It is the most widely studied noise type. Four distinct noise types, namely, exponential, Gaussian, uniform, and Rayleigh noise, were added to the eight DD logs (targets). We simultaneously added 20% noise to each of the eight output logs, such that the noise distribution being added is symmetric around zero. Gaussian and

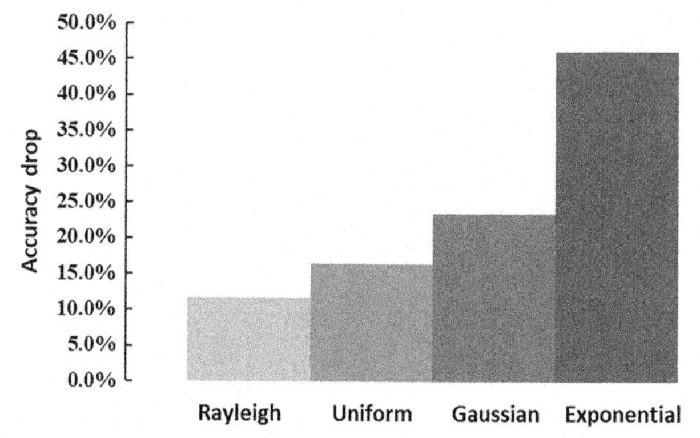

FIG. 4.8 Comparison of reduction in the performance of DD log synthesis when 20% noise of various types is simultaneously added to each of the eight DD logs.

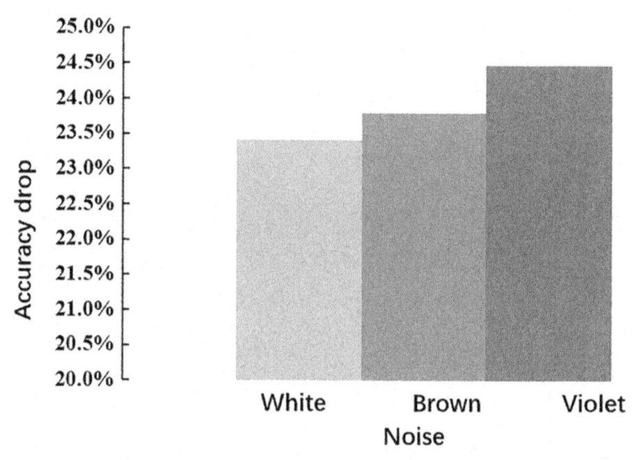

FIG. 4.9 Comparison of reduction in the performance of DD log synthesis when 20% Gaussian noise of a specific color is simultaneously added to each of the eight DD logs.

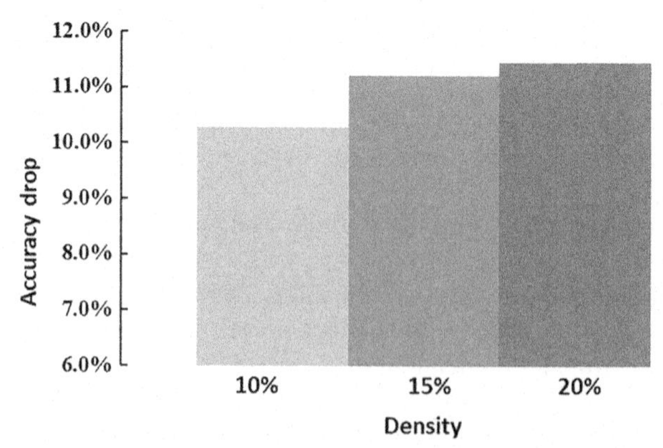

FIG. 4.10 Comparison of reduction in the performance of DD log synthesis when the Gaussian noise of a specific density is simultaneously added to each of the eight DD logs.

uniform noises have a mean of zero, whereas exponential and Rayleigh noises are generated such that the total distribution has a negative portion symmetric to the positive distribution. So, each DD log contains 20% noise. Exponential noise in DD logs has the most adverse influence on the prediction performance, whereas the Rayleigh noise in DD logs has the least influence (Fig. 4.8). Accuracy drop due to the exponential noise is twice that of Gaussian noise.

DD logs are frequency-dependent responses of the geological formation. Such frequency-dependent measurements contain Gaussian frequency-dependent noise of various colors, such as violet, white, and brown noise. White noise has an equal density at all frequencies, violet noise has density that increases 6 dB per octave, and brown noise has density that decreases 20 dB per octave [13]. We simultaneously added 20% Gaussian noise of a specific color to each of the eight DD logs. So, each DD log contains 20% noise. DD log-synthesis performance is not significantly influenced by the color of Gaussian noise (Fig. 4.9). Finally, Gaussian noises of various densities are simultaneously added to each input and output logs. More noise in the training/testing dataset results in higher mean NRMSE, amounting to a reduction in DD log-synthesis performance from 10% to 11.5% for an increase in noise density from 10% to 20% (Fig. 4.10).

Key observations based on the sensitivity study are as follows:

1. Resistivity (RLA3) and shear-wave travel time (DTS) log are important features for the DD log synthesis using SNN model.
2. Noise in resistivity (RLA3) and shear-wave travel time (DTS) log leads to significant deterioration in the DD log synthesis using SNN model.
3. Deep laterolog resistivity logs (RLA3-5) are more important than shallow ones (RLA0-2) for the DD log synthesis using SNN model.
4. Six resistivity logs, one neutron porosity (NPOR), and two sonic logs (DTC and DTS) can be used for the proposed DD log synthesis using SNN model at a prediction performance that is 10% lower than that obtained using all the 15 conventional logs.
5. Exponential noise in target can significantly deteriorate the performance of DD log synthesis using SNN model.
6. Increase in noise density of Gaussian noise in DD logs from 10% to 20% does not cause significant change in the performance of DD log synthesis using SNN model.

3.2 Generalization capability of the DD log synthesis using the SNN model

In this section, we investigate the performance of SNN model when it is trained and tested in one well and deployed in another well; this truly assesses the generalization performance of the data-driven model. For any machine learning or data-driven modeling task, generalization performance has a greater importance as compared to memorization performance. Under data constraints, the training and testing of a data-driven model are generally done in one well and then deployed in another new, unseen well. For purposes of evaluating the model deployment, in the first case, SNN model is trained and tested in Well 1; following that, the SNN model is deployed in Well 2 (row 3 in Table 4.3). In the second case, SNN model is trained and tested in Well 2; following that, the SNN model is deployed in Well 1 (row 4 in Table 4.3). Wells 1 and 2 are in the same field separated by 300 meters. In Wells 1 and 2, the formations have similar sequence of intervals, but the thicknesses of the intervals vary between the two wells. For both the cases, the deployment performance is lower than the testing performance (Table 4.3). In comparison with the testing performance in Well 2 (second row of the Table 4.3), the deployment performances (in terms of NRMSE), when the SNN model is trained/tested in Well 1 and deployed in Well 2, for conductivity-dispersion logs change by 8.3%, 19.5%, 0%, and −3.8%, and that for permittivity-dispersion logs change by 5.1%, −5.0%, 15.5%, and 2.2% for the four frequencies $f0, f1, f2$, and $f3$, respectively. On an average, the deployment performance (rows 3 and 4 in Table 4.3) when synthesizing conductivity-dispersion logs is 6% lower, and that when synthesizing the permittivity-dispersion logs is 4.5% lower as compared with the testing performances (rows 1 and 2 in Table 4.3). These results strongly indicate that the DD log synthesis using the SNN model exhibits good generalization and can be deployed in new wells.

TABLE 4.3 **Comparison of the generalization performances of the DD log synthesis using SNN model when trained/tested in one well and deployed in another well.**

		NRMSE			
		f0	f1	f2	f3
Trained and tested in Well 1 (testing performance)	Conductivity	0.067	0.066	0.071	0.077
	Permittivity	0.093	0.088	0.089	0.086
Trained and tested in Well 2 (testing performance)	Conductivity	0.072	0.077	0.094	0.105
	Permittivity	0.118	0.139	0.129	0.138
Trained and tested in Well 1 and deployed in Well 2 (deployment performance)	Conductivity	0.078	0.092	0.094	0.101
	Permittivity	0.124	0.132	0.131	0.141
Trained and tested in Well 2 and deployed in Well 1 (deployment performance)	Conductivity	0.103	0.092	0.105	0.116
	Permittivity	0.124	0.131	0.141	0.138

3.3 Petrophysical and statistical controls on the DD log synthesis using the SNN model

The conventional and DD logs used to develop the data-driven model are recorded more than 10,000 feet in the subsurface using logging tools that are run at different times in the rugose boreholes filled with drilling fluids. In such conditions, the logging tools sense certain properties of the near-wellbore geological formation volume. The logs are affected by the heterogeneity, complexity, noise, and uncertainty due to the surrounding borehole and near-wellbore environments that adversely affect the accuracy and robustness of the DD log synthesis using the SNN model. In Well 1, relative error (RE) is used to evaluate the performance of DD log synthesis at each depth. RE is formulated as

$$\mathrm{RE} = \frac{|P - M|}{M} \tag{4.5}$$

where P is the predicted value and M is the measured value. Mean RE for the four conductivity-dispersion logs and that for the four permittivity-dispersion logs are calculated at all depths resulting in two REs at each depth. When both the REs are less than 0.2, the depth belongs to the category of good prediction performance (65% for Well 1). When both REs are higher than 0.3 or either of them are higher than 0.4, the depth belongs to the category of poor prediction performance (24% for Well 1). The rest of the depths belong to moderate prediction performance (11% for Well 1). Table 4.4 lists the statistical properties of the formation zones for which the SNN model exhibits good (G), medium (M), and poor (P) performances. Following petrophysical and statistical factors are found to control the performance of the DD log synthesis using the SNN model [14]:

1. Depths with low resistivity exhibit better performance. Low-resistivity depths exhibit larger conductivity dispersion. These depths also exhibit less-noisy dispersion data.
2. Depths having higher conductivity and higher relative permittivity measurements exhibit better performance.
3. Zones exhibiting better performance tend to have lower absolute skewness of DD logs. Lower skewness (symmetric distribution of data) ensures relatively limited bias in the data and uniformly distributed noise in data and that the bounds are equally distant from the mean. Symmetric data are more suited for training the ANN models as compared with asymmetric data.
4. Depths where porosity and water saturation are high exhibit better performance. These depths are related to regions of low resistivity and limited cementation. At a depth of high porosity, dispersion in DD logs is large, the relative permittivity is high, and the dielectric tool response is more sensitive to variations in porosity and pore-filling fluids. The data at such depths are less noisy and more suited for building robust models.

TABLE 4.4 Statistical parameters of conventional and DD logs in Well 1 (G, M, and P indicate zones of good, moderate, and poor log-synthesis performances, respectively).

	Mean (μ)			Coefficient of variation (S_d/μ)			Skewness		
	G	*M*	*P*	*G*	*M*	*P*	*G*	*M*	*P*
GR	82.745	77.859	70.681	0.259	0.326	0.372	0.578	0.527	0.725
DPHZ	0.071	0.061	0.051	0.420	0.548	0.721	−0.538	−0.317	−0.016
NPOR	0.104	0.088	0.074	0.399	0.524	0.642	0.017	0.186	0.440
PEFZ	3.163	3.520	3.831	0.216	0.250	0.235	1.777	1.270	0.897
RHOZ	2.595	2.611	2.627	0.019	0.021	0.023	0.538	0.317	0.017
VCL	0.133	0.114	0.097	0.447	0.580	0.679	0.043	0.124	0.453
RLA0	0.033	0.033	0.033	0.057	0.054	0.053	0.389	0.432	1.284
RLA1	18.592	24.039	26.070	0.619	0.440	0.376	0.382	−0.324	−0.357
RLA2	55.135	87.724	121.421	1.101	0.841	0.830	1.605	1.232	2.003
RLA3	80.439	142.746	219.464	1.314	1.056	0.962	2.053	1.705	1.435
RLA4	103.597	210.112	410.161	1.647	1.488	1.565	3.279	3.375	3.215
RLA5	106.387	221.107	466.832	1.723	1.600	1.804	3.755	3.927	4.038
DTC	66.614	64.011	62.600	0.093	0.103	0.110	−0.516	−0.082	0.036
DTS	111.548	107.945	106.817	0.077	0.076	0.076	−0.103	0.316	0.324
σ_{f0}	106.643	66.775	36.961	0.822	1.077	1.192	0.775	1.994	2.450

σ_{f1}	128.633	84.221	46.973	0.726	0.931	1.036	0.647	1.705	1.993
σ_{f2}	178.156	123.258	69.939	0.615	0.779	0.936	0.460	1.347	1.671
σ_{f3}	243.885	175.763	102.559	0.538	0.686	0.870	0.288	1.111	1.559
$\epsilon_{r,\ f0}$	23.678	20.461	15.792	0.275	0.346	0.404	0.023	0.772	1.429
$\epsilon_{r,\ f1}$	15.981	14.247	11.928	0.206	0.244	0.253	−0.024	0.832	1.749
$\epsilon_{r,\ f2}$	11.226	10.449	9.467	0.149	0.169	0.156	0.069	0.927	1.383
$\epsilon_{r,\ f3}$	9.179	8.840	8.530	0.101	0.109	0.100	−0.047	0.588	0.132

5. Performance tends to improve with increase in clay content, decrease in photoelectric factor, and decrease in bulk density.
6. Zones exhibiting poor performance have higher statistical dispersion for all logs (features and targets). This indicates that the zones of low performance tend to be more heterogeneous compared with the zones of high performance. This is applicable except for the shallow-resistivity measurements and high-frequency permittivity measurements, where there exists an opposite trend.

4 Conclusions

A stacked neural network (SNN) model can process 15 conventional logs to synthesize the 8 dielectric dispersion (DD) logs, comprising 4 conductive-dispersion and 4 permittivity-dispersion logs. The proposed DD log synthesis using the SNN model implements two steps requiring a total of nine neural networks: the first step involves simultaneous DD log synthesis for ranking the eight DD logs, followed by the second step involving rank-based sequential synthesis of the 8 DD logs one at a time by processing the previously predicted, higher-ranked DD logs along with the 15 conventional logs to synthesize a lower-ranked DD log. In the first step, one ANN model with 15 inputs and 8 outputs is trained to simultaneously synthesize the 8 DD logs; following that, each of the eight DD logs is assigned a rank based on the prediction accuracy achieved by the ANN model during the simultaneous synthesis. In the second step, eight distinct ANN models are trained one at a time to sequentially synthesize one of the eight DD logs based on the rank assigned in the first step, starting with the DD log that was synthesized with the highest prediction accuracy and ending with the DD log that was synthesized with the lowest prediction accuracy.

The deployment performance of the SNN model in terms of average normalized root-mean-square errors (NRMSE) is 0.07 and 0.089 for the multifrequency conductivity- and permittivity-dispersion logs, respectively. DD log synthesis using the SNN model exhibits good generalization and can be deployed in new wells sharing similar subsurface characteristics. The two-step DD log-synthesis by the SNN model is 10% better as compared to the simultaneous synthesis of the 8 DD logs using one ANN model. Resistivity logs of various depths of investigation, neutron porosity log, and compressional and shear travel time logs are the most important features (log inputs) for the DD log synthesis. Medium- and deep-sensing resistivity logs are more important than the shallow-sensing resistivity logs. Noise in resistivity, gamma ray, shear travel time, and dielectric dispersion logs adversely influences the performance of the SNN model. Gaussian noises of various colors in the dielectric dispersion logs have similar influence on the performance of SNN model, indicating the robustness of the DD log synthesis to the frequency-dependent noise in the DD logs. Low resistivity, high porosity, high relative dielectric permittivity, large dielectric dispersion, low skewness and large coefficient of variation of conventional logs facilitate better DD log synthesis.

References

[1] Han Y, Misra S, Simpson G. Dielectric dispersion log interpretation in Bakken Petroleum System. In: SPWLA 58th annual logging symposium. Society of Petrophysicists and Well-Log Analysts, June; 2017.

[2] Ko WL, Mittra R. Conductivity estimation by neural network. In: IEEE Antennas and Propagation Society international symposium, AP-S digest, June, vol. 4; 1995.

[3] Brovko AV, Ethan KM, Vadim VY. Waveguide microwave imaging: neural network reconstruction of functional 2-D permittivity profiles. IEEE Trans Microw Theory Tech 2009;57 (2):406–14.

[4] Hasan A, Andrew FP. Measurement of complex permittivity using artificial neural networks. IEEE Antennas Propag Mag 2011;53(1):200–3.

[5] Paul S, Sindhu TK. A neural network model for predicting the dielectric permittivity of epoxy-aluminum nanocomposite and its experimental validation. IEEE Trans Compon Packag Manuf Technol 2015;5(8):1122–8.

[6] Nitta T. Complex-valued neural networks: utilizing high-dimensional parameters. Hershey, NY: IGI Global; 2009.

[7] Yang C, Bose NK. Landmine detection and classification with complex-valued hybrid neural network using scattering parameters dataset. IEEE Trans Neural Netw 2005;16(3):743–53.

[8] Ji R, et al. Prediction of soil moisture with complex-valued neural network. In: IEEE control and decision conference (CCDC), 29th Chinese, May; 2017.

[9] Wu Y, Misra S, Sondergeld C, Curtis M, Jernigen J. Machine learning for locating organic matter and pores in scanning electron microscopy images of organic-rich shales. Fuel 2019;253:662–76.

[10] Li H, Misra S, He J. Neural network modeling of in situ fluid-filled pore size distributions in subsurface shale reservoirs under data constraints. Neural Comput Applic 2019;1–13.

[11] Genty C. Distinguishing carbonate reservoir pore facies with nuclear magnetic resonance as an aid to identify candidates for acid stimulation [M.S. thesis]. Texas, USA: Dept. Petroleum Eng., Texas A&M Univ.; 2006.

[12] Cheng C, et al. Long-term prediction of discharges in Manwan Reservoir using artificial neural network models. In: Advances in neural networks–ISNN 2005. vol. 3498. 2005. p. 1040–5.

[13] Han Y, Misra S. Joint petrophysical inversion of multifrequency conductivity and permittivity logs derived from subsurface galvanic, induction, propagation, and dielectric dispersion measurements. Geophysics 2018;83(3):1–63.

[14] He J, Misra S. Generation of synthetic dielectric dispersion logs in organic-rich shale formations using neural-network models. Geophysics 2019;84(3):D117–29.

Chapter 5

Robust geomechanical characterization by analyzing the performance of shallow-learning regression methods using unsupervised clustering methods

Siddharth Misra*, Hao Li[†] and Jiabo He[‡,a]

*Harold Vance Department of Petroleum Engineering, Texas A&M University, College Station, TX, United States, [†]The University of Oklahoma, Norman, OK, United States, [‡]School of Computing and Information Systems, University of Melbourne, Parkville, VIC, Australia

Chapter outline

Nomenclature

RSS Residual sum of squares
TSS Total sum of squares
SSE Sum of squared errors
OLS Ordinary least squares

[a] Formerly at the University of Oklahoma, Norman, OK, United States

Machine Learning for Subsurface Characterization. https://doi.org/10.1016/B978-0-12-817736-5.00005-3

PLS	Partial least squares
LASSO	Least absolute shrinkage and selection operator
MARS	Multivariate adaptive regression splines
ANN	Artificial neural network
DBSCAN	Density-based spatial clustering of application with noise
SOM	Self-organizing map
GMM	Gaussian mixture model
RE	Relative error
t-SNE	t-Distributed stochastic neighbor embedding
y_i	Original log response at depth i
y'_i	Normalized value of the log response (y) at a depth i
$\hat{y}_i$	Synthesized value of the log response (y) at a depth i
β	Coefficient of OLS model
w	Coefficient/parameter vector
X	Feature vector
Y	Target vector
$\mathcal{N}$	Gaussian distribution
R^2	Correlation coefficient

Subscripts

i Formation (i)

j Formation parameter (j)

1 Introduction

Well logging is essential for oil and gas industry to understand the in situ subsurface petrophysical and geomechanical properties. Certain well logs, like gamma ray (GR), resistivity, density, compressional sonic travel time, and neutron logs, are considered as "easy-to-acquire" conventional well logs and deployed in most of the wells. Other well logs, like nuclear magnetic resonance, dielectric dispersion, elemental spectroscopy, and shear wave travel-time logs, are deployed in limited number of wells.

Sonic logging tools transmit compressional and shear waves through the formation. These waves interact with the formation matrix and fluid. Compressional waves travel through both the rock matrix and fluid, whereas shear waves travel only through the matrix. The time taken by the wave to travel from the transmitter to the receiver, referred as travel time, depends on the geomechanical properties, which are influenced by the matrix composition, fluid composition, and microstructure. Compressional and shear travel-time logs (DTC and DTS, respectively) can be computed from the waveforms recorded at the receiver. Sonic travel-time logs contain critical geomechanical information for subsurface characterization around the wellbore. The difference in the DTC and DTS logs is a function of the formation porosity, rock brittleness, and Young's modulus, to name a few.

Both shear and compressional travel-time logs are not acquired in all the wells drilled in a field due to financial or operational constraints. Under such circumstances, machine learning-generated synthetic DTC and DTS logs can be used to improve subsurface characterization. This chapter has two objectives: (1) develop a workflow to synthesize both DTC and DTS logs from "easy-to-acquire" conventional well logs and (2) apply clustering techniques to the "easy-to-acquire" conventional well logs to simultaneously determine the reliability of the synthetic DTS and DTC logs. Novelty of this study is to use unsupervised clustering methods to generate an indicator of the reliability/accuracy of the DTC and DTS logs synthesized by the shallow-learning regression methods when deployed in new wells.

Easy-to-acquire well logs can be processed using statistical and machine learning methods to synthesize the advanced well logs that are not acquired in most of the wells in a field. Researchers have explored the possibility of synthesizing certain "hard-to-acquire" well logs under data constraint [1–3]. Several studies have tried to implement machine learning techniques to determine sonic log from other well logs. One study used ANNs, adaptive neurofuzzy inference system, and support vector machines to synthesize both compressional and shear sonic travel time by processing GR, bulk density, and neutron porosity [1]. In another study, shear wave velocity (reciprocal of DTS) was predicted using fuzzy logic and ANNs [4, 5]. Apart from machine learning algorithms, DTS and DTC have been predicted using empirical equations [6, 7], empirical correlations [8], or self-consistent models [9].

Our study aims to evaluate the performances of six regression models, developed using supervised learning, for simultaneous synthesis of DTC and DTS logs by processing conventional "easy-to-acquire" logs. The shallow-learning models implemented in this study are ordinary least squares (OLS), partial least squares (PLS), least absolute shrinkage and selection operator (LASSO), ElasticNet (combination of LASSO and ridge regression), multivariate adaptive regression splines (MARS), and artificial neural networks (ANN). The first four models are linear regression models, and the last two models are nonlinear regression models. The 13 "easy-to-acquire" conventional logs together with DTC and DTS logs acquired in two wells are used for training and testing the machine learning methods.

A novel aspect of our study is the implementation of clustering techniques to determine the reliability of the sonic logs synthesized using the regression models. Our study shows that the K-means clustering method can process "easy-to-acquire" logs to group together certain depths such that the group/cluster assigned to any given depth is correlated to "a certain range of accuracy" (referred as reliability) of the shallow-learning model in synthesizing the DTC and DTS logs at that depth. We compared five clustering techniques to find the one that can be best used to determine the reliability of the sonic-log synthesis at any given depth. The five clustering algorithms tested in this study are K-means, Gaussian mixture model, hierarchical clustering, density-based spatial clustering of application with noise (DBSCAN), and self-organizing map (SOM). Dimensionality reduction technique is used to facilitate the visualization of the characteristics of each clustering technique.

In this chapter, 6 shallow-learning models and 5 clustering algorithms process 13 "easy-to-acquire" conventional logs to synthesize the compressional and shear travel time logs (DTC and DTS) along with an indicator of the reliability of the log synthesis. ANN model has the best prediction accuracy among the six regression models for log synthesis. K-means clustering can generate cluster numbers that positively correlate with ANN prediction accuracy. By combining the shallow ANN model and the K-means clustering, we developed a prediction workflow that can synthesize the compressional and shear travel-time logs and simultaneously determine the reliability of the log synthesis. This study will enable engineers, petrophysicists, geophysicists, and geoscientists to obtain reliable and robust geomechanical characterization when sonic logging tool is not available due to operational or financial constraints.

2 Methodology

2.1 Data preparation

Well logs used in this study were acquired from two wells. In Well 1, well logs were measured at 8481 depths across 4240-ft depth interval. In Well 2, well logs were measured at 2920 depths from 1460-ft depth interval. The 13 easy-to-acquire conventional logs used for the proposed log synthesis include gamma ray log (GR), caliper log (DCAL), density porosity log (DPHZ), neutron porosity log

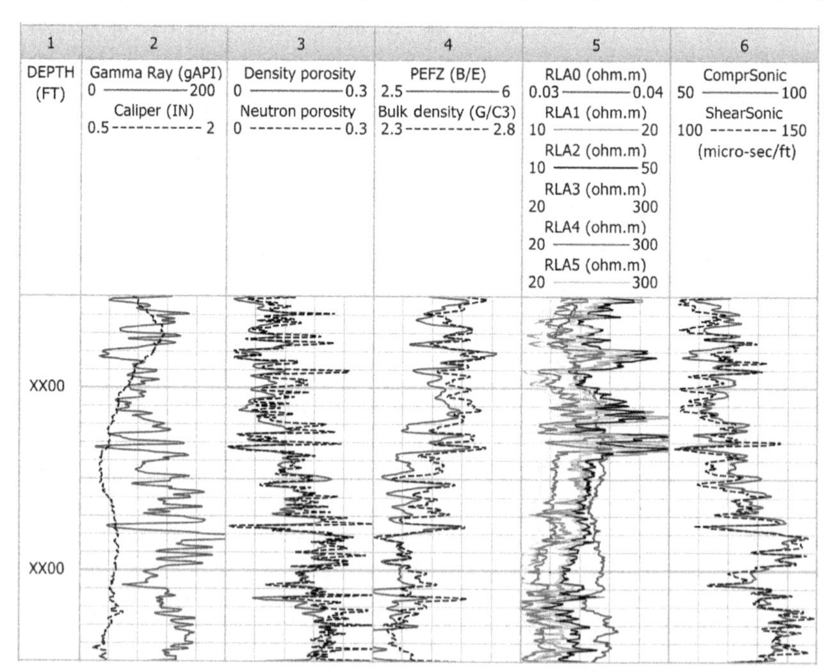

FIG. 5.1 Track 1 is depth, Track 2 contains gamma ray and caliper logs, Track 3 contains density porosity and neutron porosity logs, Track 4 contains formation photoelectric factor and bulk density logs, Track 5 is laterolog resistivity logs at various depths of investigation (RLA0, RLA1, RLA2, RLA3, RLA4, and RLA5), and Track 6 contains shear and compressional travel-time logs for a 200-ft section of the 4240-ft depth interval in Well 1.

(NPOR), photoelectric factor log (PEFZ), bulk density log (RHOZ), discrete lithology flags, and laterolog resistivity logs at 6 depths of investigation (RLA0, RLA1, RLA2, RLA3, RLA4, and RLA5). These logs from Well 1 are shown in Tracks 2–5 of Fig. 5.1. These easy-to-acquire logs are fed into the six shallow-learning models and the five clustering methods. The 4240-ft depth interval in Well 1 has 13 distinct lithologies, and the corresponding discrete lithology flag ranges from 1 to 13 indicating the 13 lithology. DTC and DTS logs (Fig. 5.1, Track 6) are the outputs of the shallow-learning log synthesis models. From a machine learning perspective, the 13 "easy-to-acquire" logs will be referred as features and the 2 DTC and DTS logs being synthesized will be referred as targets.

2.2 Data preprocessing

Data preprocessing aims to facilitate the training/testing process by appropriately transforming and scaling the entire dataset. Preprocessing is necessary before training the machine learning models. Preprocessing removes outliers and scales the features to an equivalent range. We use min-max scaling that ensures fast convergence of the gradient-based learning process, especially for neural network models. Min-max scaling is performed on one feature at a time using the following equation:

$$y'_i = 2\frac{y_i - y_{min}}{y_{max} - y_{min}} - 1 \tag{5.1}$$

where y_i is the original value of a log response (y) and y'_i is the scaled value of the log response (y) at a depth i. y_{min} and y_{max} are the minimum and maximum values of the log response (y), respectively. Min-max scaling is performed only on the 13 "easy-to-acquire" logs, which are considered as features for the shallow-learning task of synthesizing DTS and DTC logs. We do not scale the DTS and DTC logs, which are the targets for the machine learning task. As mentioned in previous chapters, a machine learning workflow first learns from the training dataset, then is evaluated on the testing dataset, and finally deployed on the new dataset. Any data preprocessing step should adopt the following sequence of steps: (1) perform data preprocessing on the training dataset; (2) learn the statistical parameters required for the data preprocessing of the training dataset; and (3) perform data preprocessing on the testing dataset and new dataset by applying the statistical parameters learnt from the preprocessing of the training dataset. In our case, minimum and maximum of each feature (log) is first learnt during the scaling of training dataset, and then those minimum and maximum values are used for scaling the corresponding features in the testing dataset and the new dataset.

2.3 Metric to evaluate the log-synthesis performance of the shallow-learning regression models

The correlation coefficient (R^2) is used to compare the prediction performance of all models, which is formulated as

$$R_j^2 = 1 - RSS_j / TSS_j \tag{5.2}$$

where

$$RSS_j = \sum_{i=1}^{n} \left(y_{pi,j} - y_{mi,j}\right)^2 \tag{5.3}$$

and

$$TSS_j = \sum_{i=1}^{n} \left(y_{pi,j} - \overline{y}_j\right)^2 \tag{5.4}$$

where n is the total number of depths for which DTC and DTS logs need to be synthesized; $j=1$ indicates the DTC log and $j=2$ indicates the DTS log; i represents a specific depth; and $y_{pi,j}$ is the sonic log j predicted at depth i, $y_{mi,j}$ is the sonic log j measured at depth i, and $\overline{y}_j$ is the mean of sonic log j measured at all depths in the training or testing dataset. RSS_j is the sum of squares of the residuals, and TSS_j is the total sum of squares proportional to the variance of the corresponding sonic log j.

2.4 Shallow-learning regression models

Six shallow-learning models synthesize DTS and DTC logs by processing 13 "easy-to-acquire" logs. The trained shallow-learning model captures the hidden relationships between the 13 "easy-to-acquire" logs and the 2 DTC and DTS sonic logs. Machine learning workflow involves the following steps in chronological order:

1. Identify targets and features in the dataset.
2. Split the dataset into training and testing datasets. Perform data preprocessing of training dataset; following that, perform data preprocessing of testing dataset; finally, perform data preprocessing of the new dataset on which the models need to be deployed.
3. Select a set of hyperparameters for the machine learning model.
4. Train the model on the training dataset. Continuously monitor the performance metric to evaluate the memorization error of the model.
5. Stop training the model after the model performance crosses a certain threshold.
6. After the training, test the model on the testing dataset and compute the generalization error of the model.
7. Compare the model performance on training dataset (memorization error) against that on testing dataset (generalization error).
8. Repeat the steps 4, 5, 6, and 7 with another set of hyperparameter till the memorization error (training error), generalization error (testing error), and the difference between these two errors are below certain thresholds.

In our study the DTS and DTC logs are the targets, and the 13 other "easy-to-acquire" logs are the features. The shallow-learning models will learn to synthesize the targets by processing the features. Regularization/penalty parameter in ElasticNet and LASSO models and the number of neurons, activation function, and the number of hidden layers in ANN model are examples of hyperparameters. On the other hand, the weights of neurons in ANN model and coefficients to be multiplied with the features in the OLS and LASSO models are examples of parameters. Hyperparameters control the learning process, and the model parameters are the outcome of the learning process. Parameters are computed during the training, and hyperparameters are set before the training and modified after comparing the memorization error with the training error. During the machine learning, one goal is to reach to the sweet spot where memorization error (training error), generalization error (testing error), and the difference between these two errors are below certain thresholds.

2.4.1 Ordinary least squares (OLS) model

OLS model assumes the target y_i is a linear combination of features x_{ip} and a residual error ε_i at any given depth i. Target can then be formulated as

$$y_i = \beta_1 x_{i1} + \beta_2 x_{i2} + \cdots + \beta_p x_{ip} + \varepsilon_i \tag{5.5}$$

where i represents a specific depth and p represents the number of "easy-to-acquire" logs available as features for training the model to synthesize the target log y. In our case, $p = 13$. In the training phase, the OLS model learns/computes the parameters β that minimize the sum of squared errors (SSE) between the modeled and measured targets. SSE is expressed as

$$\text{SSE} = \sum_{i=1}^{n} (\hat{y}_i - y_i)^2 \tag{5.6}$$

where $\hat{y}_i$ is the synthesized target of the model for a specific depth i and n is the number of samples in training the dataset. In this study the features x_{ip} are the 13 raw logs at depth i, and the target y_i are DTC and DTS sonic logs. OLS models tend to be adversely affected by outliers, noise in data, and correlations among the features. Like other linear models, OLS is suited for small-sized, high-dimensional datasets, where the dimensionality of the dataset is the number of available features for each sample, and the size of the dataset is the number of samples available for training the model.

2.4.2 Partial least squares (PLS) model

Partial least squares regression is an extension of the multiple linear regression for situations when the features are highly colinear and when there are fewer number of samples than the number of features, that is, when the size of the dataset is much smaller than the dimensionality of the dataset. In such cases, OLS model will tend to overfit, whereas PLS model performs much better in terms of building a generalizable model. Scaling of features and targets is crucial for developing robust PLS model. PLS model learns to find the correlations between the targets and features by constructing and relating their latent structures. PLS model learns the multidimensional direction in the feature space that explains the maximum multidimensional variance direction in the target space. Latent structure corresponding to the most variation in target is first extracted and then explained using a latent structure in the feature space. New latent structures are combined with original variables to form components. The number of components m (no more than the number of features) is chosen to maximally summarize the covariance with targets. The requirement to build PLS model with the number of components m fewer than the number of features makes PLS model more suitable than OLS model when there are high correlations among the features. The removal of redundant features that are colinear or not strongly correlated to the variance in the targets generates more generalizable and accurate model.

In applying the PLS model, the most important hyper parameter is the number of components to be generated. In our study, smallest m with the best prediction performance is used to build the model. We select the number of components by testing a range of values and monitoring the change in the model performance. The best performance of the model occurs when the number of components equals 13, and model performance does not change significantly when the number of components is reduced from 13 to 8, which indicates that there are 5 correlated features in the training dataset. A shallow-learning model that can work with fewer number of features avoids the curse of dimensionality, which makes the model development more robust to noise, computationally efficient and improves accuracy of predictions. Unlike OLS, PLS can learn to generate multiple targets in a single model.

2.4.3 Least absolute shrinkage and selection operator (LASSO) model

LASSO learns the linear relationship between the features and targets, such that the correlated features are not included during the model development to prevent overfitting and ensure generalization of the data-driven model. LASSO model implements an L1 regularization term that severely penalizes nonessential or correlated features by forcing their corresponding coefficients to zero. Unlike the OLS, LASSO model learns the linear relationship in the data by minimizing the SSE and the regularization term together to ensure the sparsity of the coefficients. The objective function that is minimized by the LASSO algorithm is expressed as

$$\min_{w} \frac{1}{2n} \|\boldsymbol{X}\boldsymbol{w} - \boldsymbol{Y}\|_2^2 + \alpha \|\boldsymbol{w}\|_1 \tag{5.7}$$

where $\boldsymbol{w}$ is the coefficient vector comprising coefficients β associated with features, which are the parameters of the LASSO model; $\boldsymbol{X}$ is the feature vector; $\boldsymbol{Y}$ is the target vector; n is the number of depth samples in the training dataset; and the hyperparameter α is the penalty parameter that balances the importance between the SSE term and the regularization term, which is the L_1 norm of the coefficient vector.

When α increases, the regularization term punishes the coefficient matrix to be sparser. The α term is the hyperparameter of the LASSO model that is optimized by testing a range of values for α. We select α equals to 4.83 for which the LASSO model achieves the best generalization performance. R^2 values for the DTC and DTS predictions are 0.79 and 0.75, respectively. For α equals to 4.83, the LASSO model learns the value of parameter (coefficient) for each feature (input log), as listed in Table 5.1. LASSO model-derived coefficients for 6 out of the 13 features are 0. Those logs with a coefficient values (Table 5.1) close to zero are either well correlated with other logs or less important for the desired log synthesis as compared with other logs with nonzero coefficients. A similar redundancy in features was noticed with the PLS model. Table 5.1 indicates that shallow-resistivity logs, deep-resistivity logs, and density porosity log are not essential for the desired DTC and DTS logs (Fig. 5.2). Shallow resistivity (RLA0 and RLA1) and deep resistivity (RLA4 and RLA5) logs are correlated with the medium sensing RLA2 and RLA3 logs. Further, RLA2 and RLA3 being medium sensing have similar depth of investigation as the sonic DTC and DTS logs; consequently, RLA2 and RLA3 are used by the LASSO model, while the other resistivity logs are not used for the synthesis of DTC and DTS logs.

2.4.4 ElasticNet model

Similar to LASSO, the ElasticNet algorithm uses a regularization term to penalize the coefficients of correlated and nonessential features. ElasticNet model learns linear relationship between features and targets using a regularization term that is a weighted sum of L1 norm and L2 norm of the coefficients. Unlike the LASSO model, ElasticNet model preserves certain groups of correlated features that improve the precision and repeatability of the predictions. ElasticNet model does not penalize correlated features as severely as LASSO model. ElasticNet algorithm generates more unique model as compared to the LASSO model for high-dimensional data with highly correlated variables. The objective function of the ElasticNet model is formulated as

$$\min_{w} \frac{1}{2n} \|Xw - y\|_2^2 + \alpha_1 \|w\|_1 + \alpha_2 \|w\|_2^2 \tag{5.8}$$

where the penalty parameters α_1 and α_2 are the hyperparameters of the ElasticNet model and determined through optimization to be equal to 4.8 and 0.1, respectively. This is aligned with the findings of the LASSO model because α_2 is a small value and α_1 is almost equal to α of the LASSO model. Dataset used in our study is not a high-dimensional dataset, and the benefits of ElasticNet model in comparison to the LASSO model are only observed for high-dimensional datasets.

TABLE 5.1 Estimates of coefficients β learned by the LASSO model for $\alpha = 4.83$

Lithology	GR	DCAL	DPHZ	NPOR	PEFZ	RHOZ	RLA0	RLA1	RLA2	RLA3	RLA4	RLA5
0.31	0.06	−1.29	0.00	41.63	0.60	−55.51	0.00	0.00	−1.01	−1.05	0.00	0.00

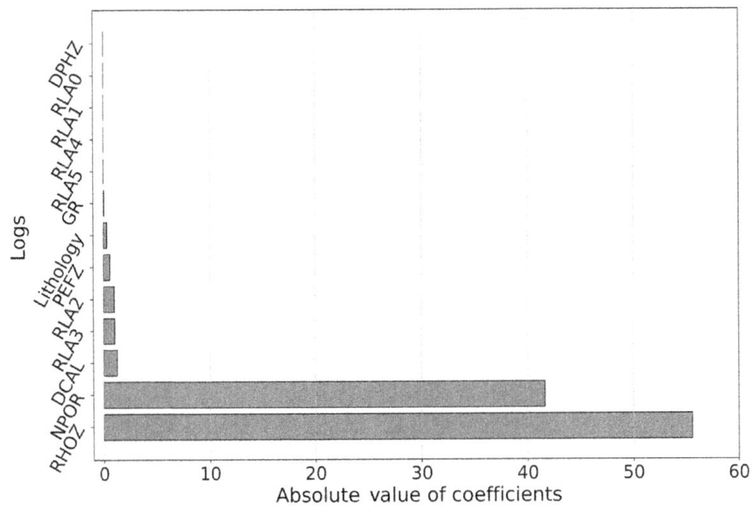

FIG. 5.2 Bar plot of the estimated coefficients β for LASSO model.

2.4.5 *Multivariate adaptive regression splines (MARS) model*

Few advantages of linear models are their ease and speed of computation and also the intuitive nature of interpreting their coefficients/parameters. However, the strong assumption about linearity affects the predictive accuracy of linear models. MARS models the nonlinear relationship between features and targets by splitting the feature space into subspaces and then learns the linear relationship between features and targets for each of the subspaces. MARS uses a divide and conquer strategy in which the training datasets are partitioned into separate piecewise linear segments (splines) of differing gradients (slope). These piecewise linear segments (or curves), also known as basis functions $B_q(x)$, result in a flexible model that can handle both linear and nonlinear behavior. The points of connection C_q between the piecewise segments are called knots. By relating the features and targets using multiple independent linear regressions, the model can capture the nonlinear trends in the dataset. MARS assesses each data point for each feature as a knot to partition the original feature space into two new subspaces. Then, two different linear models with the candidate feature(s) are identified for each subspace that results in the smallest error. This partitioning is continued until many knots are found, producing a highly nonlinear pattern, which is a collection of linear models for individual subspaces. Increase in number of knots allows better fit with the training dataset; however, the learnt relationship may not generalize well to new, unseen dataset. Knots that do not contribute significantly to predictive accuracy can be removed using the process known as "pruning." MARS model formulates the predicted target $\hat{y}$ as a weighted sum of basis function $B(x)$:

$$\hat{y}_i = \sum_{q=1}^{p} \alpha_q B_q(x_{qi}) \quad (5.9)$$

where x_{qi} is the value of qth feature (log) x_q at the ith depth point, $B_q(x_{qi})$ is a basis function, and α_q is the coefficient of B_q. The basis function can have many different forms, but most likely, it is a hinge function. Each knot generates two basis functions (hinge functions): $\max(0, x_{qi} - C_q)$ and $\max(0, C_q - x_{qi})$, where C_q is the knot. Hinge function partitions the feature space into subspaces by using different values of C_q. In our study, we use 10 knots to partition the feature space; consequently, there are 21 parameters that need to be computed in the training phase. The MARS model synthesizes DTC and DTS logs with R^2 of 0.85 and 0.83, respectively. LASSO and ElasticNet models showed that RLA0, RLA4, and RLA5 logs have high correlations with RLA1, RLA2, and RLA3 logs. Therefore, when training the MARS model, we do not use the RLA0, RLA4, and RLA5 logs. In conclusion, the MARS algorithm involves a forward phase and a backward phase. The forward phase places candidate knots at random positions within the range of each feature to define a pair of basis functions (BFs). At each step, the model adapts the knot and its corresponding pair of BFs to give the maximum reduction in sum-of-squares residual error. This process of adding BFs continues until a user-defined threshold. The backward phase involves deleting the redundant knots and corresponding BFs that made the least contribution to the model fitting.

2.4.6 *Artificial neural network (ANN) model*

ANN is a widely used machine learning model suitable for both linear and nonlinear regression tasks. A neural network comprises an input layer, an output layer, and several hidden layers. The capacity of the neural network model to fit data can be adjusted by changing the number of hidden layers and the number of neurons in each hidden layer. Each hidden layer and output layer are made of neurons. In a densely connected neural network architecture, each neuron in a layer is connected to neurons in the immediately previous layer. The neuron in a layer receives the output from the neurons in the immediately previous layer multiplied by weight of the connection between the two neurons. Each neuron sums the weighted outputs from the neurons in the immediately previous layer and feeds the summation through a activation function that generates values between -1 and 1 or 0 and 1. Activation function, also referred as transfer function, controls the nature of output of a neuron. For the neurons in hidden layers, it is recommended to use Rectified Linear Unit as the activation function because it overcomes the vanishing gradient problem of once popular Sigmoid and Tanh activation functions. For the neurons in output layer, softmax is used as the activation function for classification tasks, whereas a 1:1 linear function is used as the activation function for regression tasks. Activation function adds nonlinearity to the computation. The input layer contains neurons without any activation function. Each feature value of a sample is fed to a corresponding neuron in the input layer. Each neuron in the input layer is connected to each neuron in the subsequent hidden layer, where the weights of the connections are few of the several parameters that need to be computed during the training of the neural network and essential for nonlinear complex functional mappings between the features and targets. A neural network without activation function will act as a high-order linear regression model. An important feature of an activation function is that it should be differentiable so as to perform back-propagation optimization strategy while propagating the errors backward in the network for updating the weights/parameters of the connections. In our case, there are 13 features and 2 targets to be synthesized. The number of neurons in the input and output layers of the neural network are 13 and 2, respectively. We use two fully connected hidden layers in the ANN model having nine and five neurons in the first and second hidden layers, respectively. Such a connection results in total of 188 parameters/weights that need to be computed. Out of the 188 parameters, 126 parameters define the connection between the input layer and first hidden layer, 50 parameters define the connection between the first and second hidden layers, and 12 parameters define the connection between second hidden layer and the output layer. The neural network implemented in our study utilizes conjugate gradient back propagation to update the parameters of the neurons. The number of neurons in a layer, number of hidden layers, and activation function serve as the hyperparameters of the ANN model.

2.5 Clustering techniques

The goal of clustering is to group data into a certain number of clusters such that the samples belonging to one cluster share the most statistical similarity and are dissimilar to samples in other clusters. In this study, we implement five clustering techniques: centroid-based K-means, distribution-based Gaussian mixture, hierarchical clustering, density-based spatial clustering of application with noise (DBSCAN), and self-organizing map (SOM) clustering. The clustering methods process the "easy-to-acquire" features to differentiate the various depths into distinct groups/clusters based on certain similarities/dissimilarities of the "easy-to-acquire" features. The goal is to generate clusters/groups based on unsupervised learning technique (i.e., without the target sonic logs) and assess the correlations of the clusters with the accuracies of the regression models for log synthesis. In doing so, the cluster numbers can be used to evaluate the reliability of the log synthesis during the deployment phase, when it is impossible to quantify the accuracy/performance of the log synthesis, unlike what is done for the training and testing datasets.

K-means clustering technique takes the number of clusters as an input from user and then randomly sets the cluster centers in the feature space. The cluster centers are shifted till the distortion/inertia metric converges, that is, with further iteration to find the best clusters, the cluster centers do not shift a lot. Gaussian mixture model (GMM) assumes the clusters in the dataset are generated based on Gaussian processes. The data points in the multidimensional feature space are fitted to multivariate normal distributions that maximize the posterior probability of the distribution given the data. Hierarchical clustering model clusters dataset by repeatedly merging (agglomerative) or splitting (divisive) data based on certain similarities to generate a hierarchy of clusters. For example, agglomerative hierarchical clustering using Euclidian distance as a measure of similarity repeatedly executes the following two steps: (1) Identify the two clusters that are closest to each other, and (2) merge the two closest clusters with an assumption that the proximity of clusters indicates similarity of the clusters. This continues until all the clusters are merged together. DBSCAN clusters the data points based on the density of the data points. The algorithm puts data points with a lot of neighbors into similar groups and recognizes points with fewer neighbors as outliers. DBSCAN needs the user to define the minimum number of points that are required to form the cluster and the maximum distance between two points required for the two points to be part of the same cluster. The fifth clustering technique, SOM, utilizes neural network for unsupervised dimensionality reduction by projecting the high-dimensional data onto two-dimensional space while maintaining the original similarity between the data points. Here, we first apply SOM projection, then use K-means to cluster the dimensionality-reduced data in the lower-dimensional feature space into groups.

We first applied the five clustering techniques on all the "easy-to-acquire" logs (features). The clusters so obtained did not exhibit any correlation with the performances of the shallow-learning regression models for the synthesis of DTS and DTC logs. Clustering methods that use Euclidian distance, for example, K-means and DBSCAN, perform poorly in high-dimensional feature space due to the curse of dimensionality. High dimensionality and high nonlinearity when using all the 13 "easy-to-acquire" logs resulted in complex relationships among the features that were challenging for the clustering algorithms to resolve into reliable clusters. In order to avoid the curse of dimensionality, only three "easy-to-acquire" logs, namely, DPHZ, NPOR, and RHOZ, were used for the desired clustering because these logs exhibit good correlations with the log synthesis performance of the shallow-learning models (Fig. 5.3). We chose these three logs to build the clusters for determining the reliability of log synthesis using the shallow-learning models in new wells. For the five clustering techniques, we processed the three selected features, namely, DPHZ, NPOR, and RHOZ, to generate only three clusters that could show potential correlation with the good, intermediate, and bad log-synthesis performances, respectively, of the shallow-learning models. The 4240-ft formation in Well 1 is clustered into three clusters by processing the DPHZ, NPOR, and RHOZ logs. Following that, the averaged cluster numbers of each 50-ft depth interval and the averaged relative errors in log synthesis for each

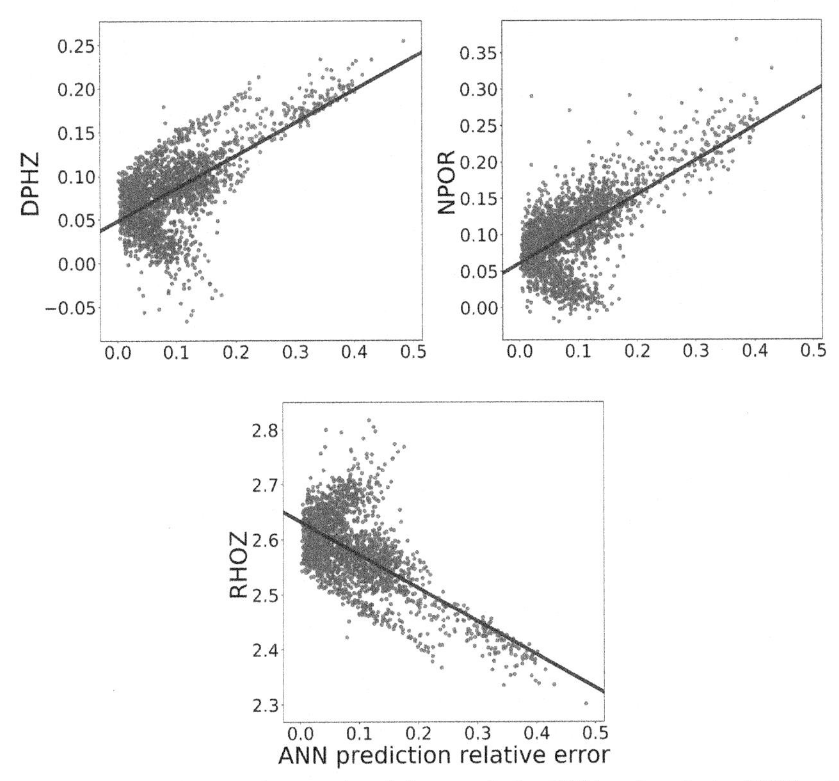

FIG. 5.3 Correlation plots between the relative error in the ANN-based synthesis of DTC and DTS logs and the log measurements of (A) density porosity (DPHZ), (B) neutron porosity (NPOR), and (C) the bulk density (RHOZ) at various depths in the 4240-ft depth interval of Well 1.

50-ft depth interval are calculated for purposes of investigating the correlation between the cluster and log-synthesis performance.

2.5.1 K-means clustering

K-means clustering requires us to manually set the number of clusters to be generated for a given dataset. This may lead to unwanted consequences when the number of clusters actually present in the high-dimensional data is not equal to that set by a user. Fig. 5.4 shows the inertia, or the sum of squared criterion, with respect to the number of clusters. The turning point of the plot (also referred as the elbow) is around the cluster number 3, beyond which the rate of drop in inertia decreases. Till the cluster number 3 marking the elbow, the rate of decrease in the inertia is significantly large. After the elbow, addition of more clusters gives rise to a low rate of decrease in the inertia. Consequently a cluster number 3 is chosen to be the most suitable for the desired K-means clustering.

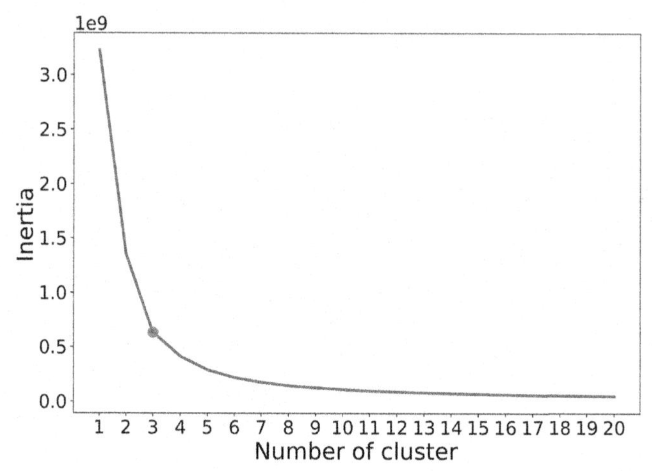

FIG. 5.4 Inertia of the clusterswith respect to the number of clusters implemented in the K-means clustering. The elbow specified by the *red dot* identifies the optimal number of clusters.

2.5.2 Gaussian mixture model clustering

A GMM assumes all the data points are generated from a mixture of a finite number of Gaussian distributions with certain parameters. Expectation-maximization (EM) algorithm is used to fit the GMM to the dataset. GMM learns the representation of a multimodal data distribution as a combination of unimodal distributions. GMM assumes the data in a specific cluster are generated by a specific Gaussian distribution/component. GMM fits K Gaussian components to the dataset by parameterizing the weight, mean, and covariance of each cluster, where i is the cluster number. If there are K clusters in the dataset, Gaussian mixture model fits the dataset by optimizing the following sum of Gaussian distributions/components:

$$p\left(\vec{x}\right)=\sum_{i=1}^{K}\phi_i\mathcal{N}\left(\vec{x}\mid\vec{\mu_i},\Sigma_i\right) \tag{5.10}$$

where $\vec{x}$ is the data point vector, $\mathcal{N}$ is Gaussian distribution, K is the number of clusters, $\vec{\mu_i}$ is the mean of a cluster, Σ_i is the covariance matrix, and ϕ_i is the weight/parameter to be learnt by the GMM algorithm. The sum of the weights of all distributions equals to 1. After fitting the data with multiple Gaussian distributions, the results can be used to cluster any new data point into one of the identified clusters.

Using multiple Gaussian distributions to fit the dataset is reasonable for clean data with limited noise, especially when the data is generated by relatively homogeneous processes/systems. Well logs are sensing geological formations that are heterogeneous and layered. Further, well logs generally contain noise and uncertainties, which may result in a high variance of each cluster. As a result, GMM is not most suited for well logs. In Fig. 5.5A, the GMM-derived cluster number weakly correlates with the relative error with a Pearson correlation coefficient of −0.22. As explained above, each cluster number plotted in Fig. 5.5 is the averaged cluster number for a 50-ft depth interval, whereas each relative error in the log synthesis is the average of relative errors for the corresponding 50-ft depth interval. The points scatter all over the plot and do not show obvious correlation. Gaussian mixture model identifies several clusters by differentiating the formation depths but the cluster patterns learned by the GMM model are completely different from the patterns of relative errors of shallow-learning models used for log synthesis (Fig. 5.8).

2.5.3 Hierarchical clustering

Our application of hierarchical clustering algorithm starts with every data point as a cluster; then clusters are repeatedly merged together based on their similarity (proximity), until the target number of clusters is reached. This clustering ends when each cluster is most distinct from other clusters and the samples within each cluster are most similar. The similarity of two clusters is evaluated based on the sum of squared Euclidean distances for all pairs of

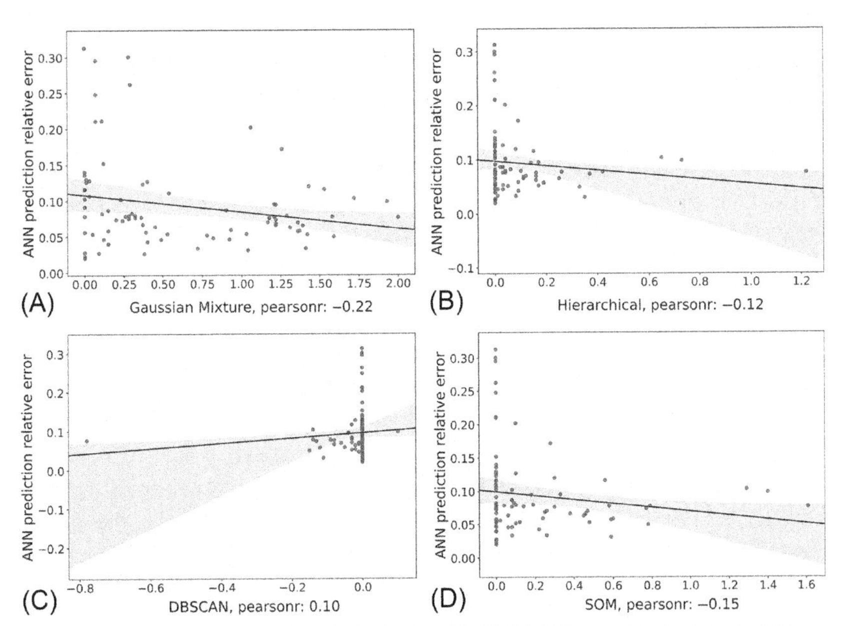

FIG. 5.5 Averaged cluster numbers derived using (A) GMM, (B) hierarchical, (C) DBSCAN, and (D) SOM clustering versus the averaged relative errors in ANN-based synthesis of DTS and DTC logs such that the the averages are computed for each of the 50-ft depth intervals in the 4240-ft depth interval of Well 1, where training and testing were done. The *gray region* indicates 95% confidence interval.

samples, one from each cluster. The merging process forms a hierarchical tree of clusters. In Fig. 5.5B, most of the cluster numbers are 0, which indicates the hierarchical clustering, finds most of the samples to be similar, and groups most of the formation depths into one cluster. The results shown in Figs. 5.5B and 5.8 demonstrate that the hierarchical cluster algorithm does not do a good job in differentiating the formation depths.

2.5.4 DBSCAN clustering

DBSCAN is a density-based clustering method. Unlike the K-means clustering the DBSCAN method do not need the user to manually define the number of clusters. Instead, it requires a user to define the minimum number of neighbors to be considered in a cluster and the maximum allowed distance between any two points for them to be a part of the same cluster. Within a certain user-defined distance around a sample, the DBSCAN will count the number of neighbors. When the number of neighbors within the specified distance (i.e., data density) exceeds a threshold, DBSCAN will identify that group of data points as belonging to one cluster. Based on our extensive study, we set minimum number of neighbors as 100 and the range of distance as 10. Fig. 5.5C shows that DBSCAN clustering method identifies many data points as outliers, which are clustered into cluster number 1 and most of the formation depths are clustered into cluster number 0 (Fig. 5.8).

2.5.5 SOM followed by K-means clustering

Self-organizing map (SOM) is a neural network-based dimensionality reduction algorithm generally used to represent a high-dimensional dataset as two-dimensional discretized pattern. Reduction in dimensionality is performed while retaining the topology of data present in the original feature space. In this study, we perform SOM dimensionality reduction followed by K-means clustering. The clustering method is basically a K-means clustering performed on the mapping generated by SOM. As the first step, artificial neural network is trained to generate low-dimensional discretized representation of the data in the original feature space while preserving the topological properties; this is achieved through competitive learning. In SOM, the vectors that are close in the high-dimensional space also end up being mapped to SOM nodes that are close in low-dimensional space. K-means can be considered a simplified case of SOM, wherein the nodes (centroids) are independent from each other. K-means is highly sensitive to the initial positions of the centroids, and it is not suitable for high-dimensional dataset. The two-stage procedure for clustering adopted in this study first uses SOM to produce the low-dimensional prototypes (abstractions) that are then clustered in the second stage using K-means. This two-step clustering method reduces the computational time and improves the efficiency of K-means clustering. Even with relatively small number of samples, many clustering algorithms—especially hierarchical ones—become intractably heavy. Another benefit of the two-step clustering method is noise reduction. The prototypes constructed by SOM are local averages of the data and, therefore, less sensitive to random variations than the original data. In our study the SOM has a dimension of 50 neurons by 50 neurons, upon which the 8481 samples from Well 1 will be mapped. All "easy-to-acquire" logs were fed as inputs to the SOM model. The weights of SOM were randomly initialized. During training the weight vectors are updated based on the similarity between the weight vectors and input vectors, which results in moving the SOM neurons/nodes closer to certain dense regions of the original data. The similarity between data points and SOM nodes during the weight update is evaluated based on Euclidean distance. The result of the two-step SOM followed by K-Means clustering does not have a strong correlation with the relative error in log synthesis, as shown in Figs. 5.5D and Fig. 5.8.

3 Results

3.1 Prediction performances of shallow-learning regression models

The six shallow-learning regression models discussed in the earlier section were trained and tested on 8481 data points (distinct depths) acquired from the 4240-ft depth interval in Well 1 and deployed for blind testing on 2920 data points acquired from 1460-ft depth interval in Well 2. Model training was on 80% of randomly selected data from Well 1, and model testing was on remaining dataset. The data-driven model is trained to capture the hidden relationships of the 13 "easy-to-acquire" logs (features) with the DTC and DTS sonic logs (targets). The performance of log synthesis is evaluated in terms of the coefficient of determination, R^2. The log synthesis results for Wells 1 and 2 are shown in Table 5.2. The log-synthesis performances for DTC and DTS in Wells 1 and 2 in terms of R^2 are illustrated in Fig. 5.6. OLS and PLS exhibit similar performances during training and testing but not during the deployment (blind testing). LASSO and ElasticNet have relatively similar performances during training, testing, and blind testing. Among the six models, ANN performs the best with R^2 of 0.85 during training and testing and 0.84 during the blind testing, whereas LASSO and ElasticNet exhibit the worst performance with R^2 of 0.76 during the blind testing. Cross validation was performed to ensure the model is trained and tested on all the statistical features present in the dataset, which is crucial for the robustness of the shallow-learning models. As shown in Table 5.2, when the trained models are deployed in Well 2, all models exhibit slight decrease in prediction accuracy. ANN has the best performance during the deployment stage in Well 2. The accuracy of the DTC and DTS logs synthesized using ANN model is shown in Fig. 5.7, where the measured and synthesized sonic logs are compared across randomly selected 300 depth samples from Well 2 for the purpose of blind testing, that is, no data from Well 2 was used for training the model.

TABLE 5.2 Prediction performances in terms of R^2 for the six shallow-learning models trained and tested on data from Well 1 and deployed on data from Well 2. The row corresponding to Well 1 are the testing results, and the row corresponding to Well 2 are the blind-testing results.

Accuracy (R^2)		OLS	PLS	LASSO	ElasticNet	MARS	ANN
Well 1 (testing)	DTC	0.830	0.830	0.791	0.791	0.847	0.870
	DTS	0.803	0.803	0.756	0.753	0.831	0.848
Well 2 (blind testing)	DTC	0.804	0.790	0.778	0.774	0.816	0.850
	DTS	0.794	0.769	0.763	0.755	0.806	0.840

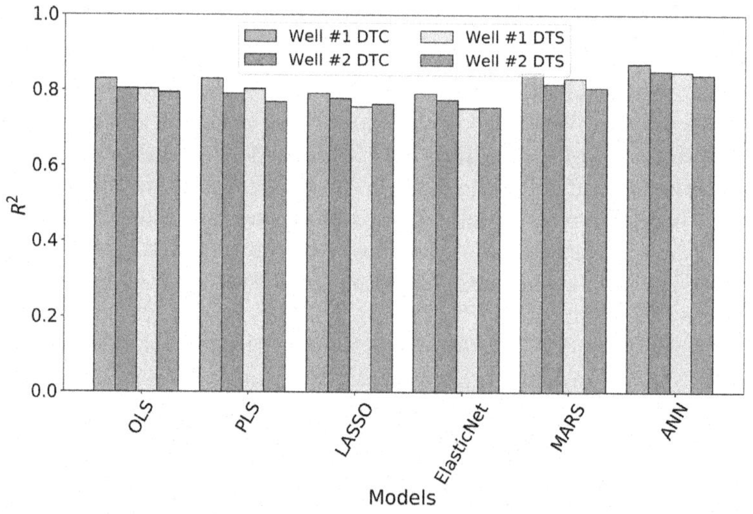

FIG. 5.6 Bar plot of the prediction performances for the six shallow-learning regression models in synthesizing DTC and DTS logs in Well 1 (train-test) and Well 2 (blind).

3.2 Comparison of prediction performances of shallow-learning regression models in Well 1

In this section, relative error (RE) is used to evaluate the prediction performances of the shallow-learning models for log synthesis in Well 1, where DTS and DTC logs are available for quantifying the relative error in the log synthesis. RE for a log synthesis is formulated as

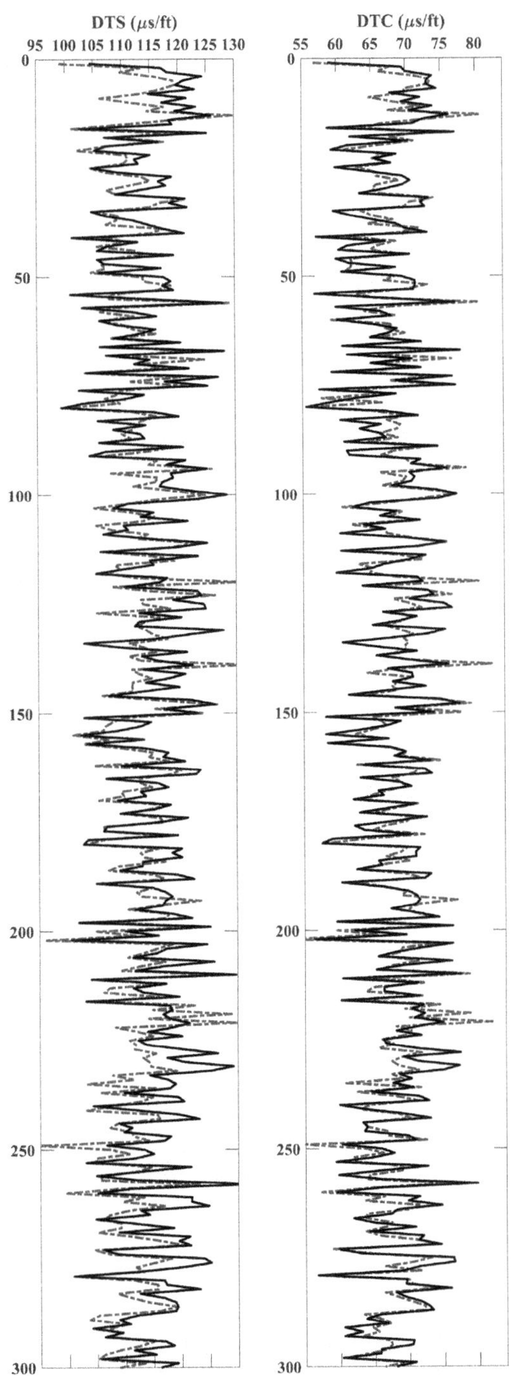

FIG. 5.7 Comparison of measured (*dashed blue*) and synthesized (*solid black*) compressional (DTC) and shear (DTS) travel-time logs in Well 2, when the ANN model is trained and tested in Well 1 and deployed in Well 2 to synthesize the DTC and DTS logs.

$$\mathrm{RE} = \frac{|P - M|}{M} \tag{5.11}$$

where P is the predicted value and M is the measured value of either DTS or DTC log at a depth i. RE values are first separately calculated for the DTC and DTS logs; then the two RE values corresponding to the two targets are averaged at each depth to represent the overall prediction performance of a shallow-learning model for any given depth. Averaged RE at each depth is further averaged over 50-ft depth intervals to reduce the effects of noise, borehole rugosity, and thin layers. The averaged RE will better describe the overall performance of a model in formations with different lithologies. Later, this final form of the averaged RE is compared with the averaged cluster numbers generated by the clustering methods to identify a clustering technique that can be used to indicate the reliability of regression-based synthesis of DTS and DTC logs in new wells without these logs.

Fig. 5.8 contains averaged relative errors in log synthesis for the shallow-learning models (first six columns) and the averaged cluster numbers generated by the clustering methods (last five columns) for the 4240-ft depth interval of Well 1. The first six columns in Fig. 5.8 show the averaged RE of the six shallow-learning log-synthesis models in Well 1. Whiter colors represent higher RE, and darker colors represent lower RE such that RE is

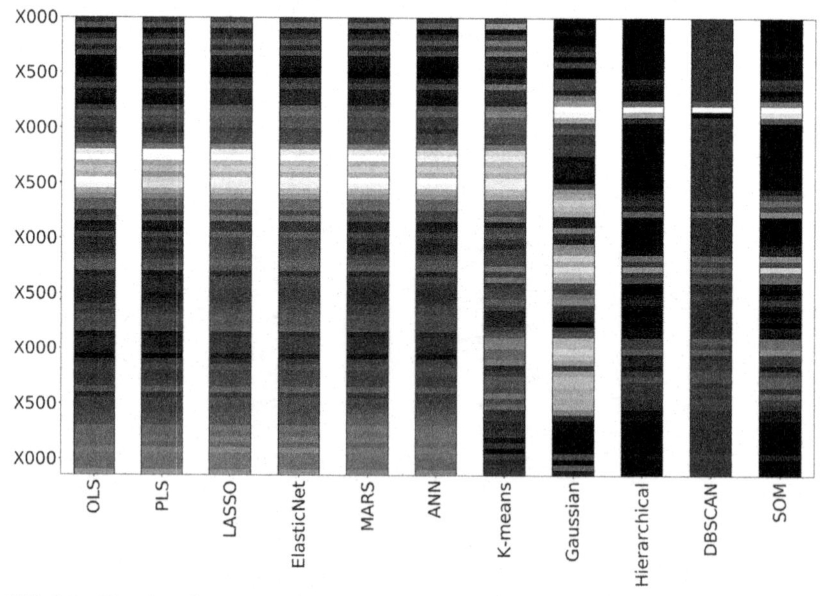

FIG. 5.8 First six columns are the 50-ft-averaged relative errors in synthesizing DTS and DTC logs using the six shallow-learning regression models for the 4240-ft depth interval of Well 1, where darker colored intervals represent depths that represent zones in which the learning models exhibit better prediction performances. The last five columns are the averaged cluster numbers generated using different clustering methods for each 50-ft depth intervals in the 4240-ft depth interval of Well 1.

inversely related to the performance of the log-synthesis model. The six log-synthesis models exhibit very similar patterns of RE over the entire depth interval, as shown in the six columns of Fig. 5.8. The six models perform badly in the upper middle part of the selected formation (around 1250–1800 ft. below the top of the formation depth under investigation). Possibly the zone of poor log-synthesis performance has certain physical properties that are very different from the rest or those where the logs have very distinct statistical features. In the following sections, clustering methods process the "easy-to-acquire" logs to identify clusters that exhibit high correlation with the relative errors of the synthesized DTC and DTS logs.

3.3 Performance of clustering-based reliability of sonic-log synthesis

Similar to the calculations of REs for the synthesized logs, we find the averaged cluster numbers generated by the clustering methods for each of the 50-ft depth intervals in the 4240-ft depth interval of Well 1 to eliminate the effects of noise and outliers. In order to determine whether a clustering method can be used as reliability indicator for the synthesis of DTS and DTC logs, the averaged cluster numbers are compared against the averaged REs across the 4240-ft depth interval of Well 1, as shown in Fig. 5.9. Notably, K-means-derived cluster numbers and REs of ANN-based log synthesis show a decent Pearson's correlation of 0.76. Unlike K-means, other clustering methods generate cluster numbers that exhibit negative or close to zero Pearson's correlation with REs of ANN-based log synthesis (Fig. 5.5). The K-means clustering results are shown in column 7 in Fig. 5.8, which demonstrates a strong

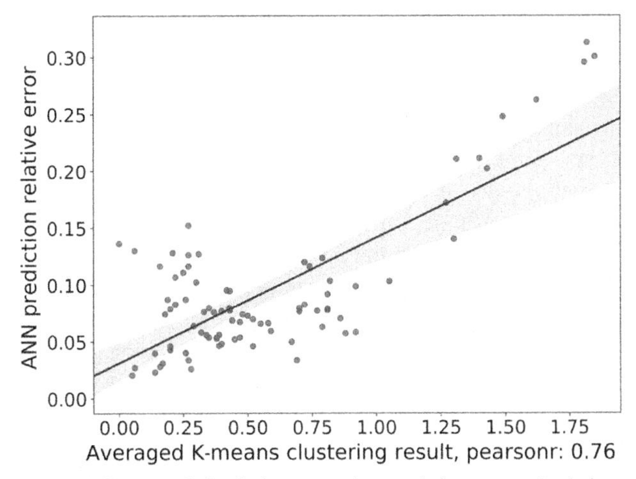

FIG. 5.9 Averaged K-means-derived cluster numbers and the averaged relative errors in the ANN-based synthesis of DTC and DTS logs, such that the averaging is done for each of the 50-ft depth intervals in the 4240-ft depth interval of Well 1. The *gray* region indicates 95% confidence interval.

correlation between the cluster numbers generated using the K-means clustering with the relative error of the log synthesis, especially those synthesized using the ANN model. According to Fig. 5.9, the reliability of ANN-based synthesis of DTC and DTS logs will be highest for the formation depths labeled as cluster number 1 by the K-means clustering. Formation depths assigned a cluster number 2 by the K-means clustering will have low reliability of ANN-based DTC and DTS log synthesis.

To better visualize the clustering results of each clustering method, we use t-distributed stochastic neighbor embedding (t-SNE) dimensionality reduction algorithm to project the 13-dimensional feature space on to a two-dimensional space. Following that, each sample (i.e., formation depth) projected on to the two-dimensional space is assigned a color based on the cluster number generated by the clustering method for the sample. Dimensionality reduction using t-SNE method enables us to plot each formation depth with 13 log responses as a point in a two-dimensional space while preserving the high-dimensional topological relationships of the formation depth with other neighboring depths and with other depths exhibiting similar log responses. This visualization technique can help us compare the characteristics of the various clustering methods implemented in our study.

t-SNE is one of the most effective algorithms for nonlinear dimensionality reduction. The basic principle is to quantify the similarities between all the possible pairs of data points in the original, high-dimensional feature space and construct a probability distribution of the topological similarity present in the dataset. When projecting the data points into low-dimensional space, the t-SNE algorithm arranges the data points to achieve a probability distribution of topological relationships similar to that in the original, high-dimensional feature space. This is accomplished by minimizing the difference between the two probability distributions, one for the original high-dimensional space and the other for the low-dimensional space. If a data point is similar to another in high-dimensional space, then it is very likely to be picked as a neighbor in low-dimensional space. To apply t-SNE, we need to define the hyperparameters, namely perplexity and training step. Perplexity defines the number of neighbors, and it usually ranges from 5 to 50, and it needs to be larger for a large dataset. In our study, we tested a range of values and selected the perplexity and the training steps to be 100 and 5000, respectively.

Fig. 5.10 presents the results of dimensionality reduction using the t-SNE method. Fig. 5.10 has four sub plots; each sub plot uses the same manifold from t-SNE but colored with different information, such as relative error in log synthesis, lithology, and cluster numbers obtained from various clustering methods. Mathematically, manifold is a continuous geometric structure. When dealing with high-dimensional data in machine learning, sometimes, we assume that data can be represented using a low-dimensional manifold. Each point on the plots represents one specific depth in the formation. The t-SNE algorithm projects all the input logs for each depth as a point on the t-SNE plot. Formation depths that are similar to each other

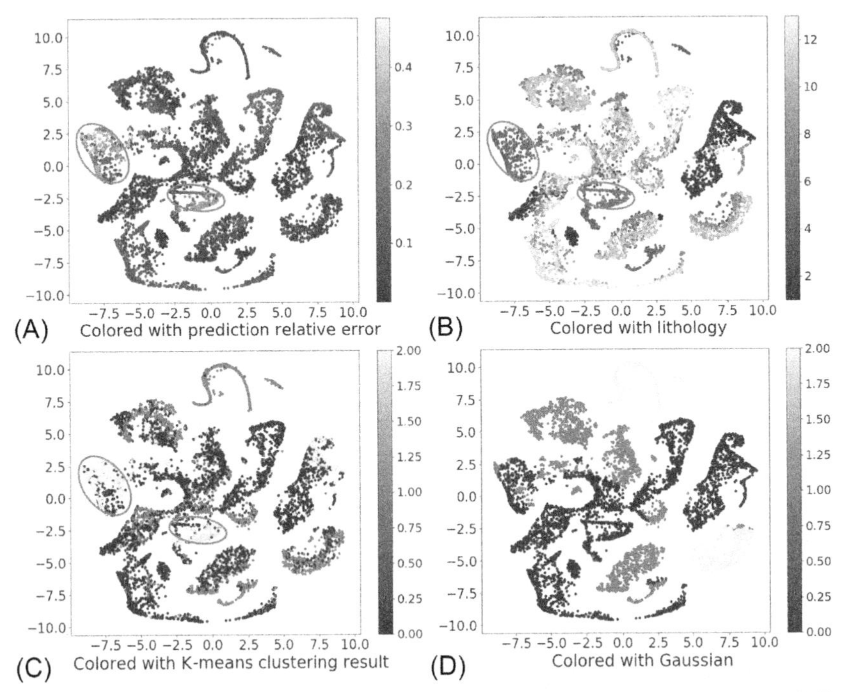

FIG. 5.10 Two-dimensional projection of formation depths generated using the t-SNE dimensionality reduction algorithm based on the topological relationships among the easy-to-acquire logs at each depth, which are colored according to (A) relative error in log synthesis, (B) lithology, (C) K-means-derived cluster numbers, and (D) Gaussian mixture model-derived cluster numbers. Each point represents a specific formation depth, and neighboring points in the two-dimensional projection have greater similarity. *Red circles* denote formation depths that exhibit high relative error in log synthesis.

based on the log responses will be projected as neighbors in the t-SNE manifold. t-SNE divides the formation depths into several blocks such that data points in one block are most similar. The shapes of blocks are random, and they change when the t-SNE algorithm is applied using different hyperparameters. In these sub plots, the value of the x-axis and y-axis does not have any physical meaning. Fig. 5.10A is colored with the relative error in ANN-based synthesis of DTS and DTC logs. In the figure, formations with higher relative errors are concentrated in the two blocks highlighted inside *red circles*. Fig. 5.10A indicates that only 5% of the formation depths have relative error higher than 0.4. Comparing Fig. 5.10A and B, the data points with low prediction accuracy (in *red circles*) are mostly from formation lithology 5, 6, 7, and 8. For other lithology, the relative errors of log synthesis are mostly lower than 0.2. Most of the formations 1 and 2 have relative errors lower than 0.1 and belong to K-means cluster of 0. Comparing Fig. 5.10A and B, the prediction relative error has a similar pattern with the lithology. Fig. 5.10C and D is colored with the cluster numbers computed using K-means and Gaussian mixture clustering. Compared with the other clustering techniques, both K-means and Gaussian mixture clustering exhibit good ability to differentiate formation depths. The Gaussian mixture clustering results are closely related to the t-SNE dimensionality reduction results. Different clusters in the Gaussian mixture model coincide with different blocks on the plots. However, the results from Gaussian mixture model do not have a similar pattern with the lithology and relative error. K-means clustering algorithm has the best correlation with the prediction relative error. This is also confirmed by analysis of Figs. 5.5 and 5.9. For the data points that have a high relative error in log synthesis (notated by the *red circle*), K-means clustering method labels them as cluster number 2. Moreover, the K-means clustering results have a very close pattern with the lithology plot in Fig. 5.10B.

4 Conclusions

Six shallow-learning regression models were used for synthesizing the compressional and shear travel-time logs in a shale reservoir. The regression models were trained using supervised learning to process 13 conventional easy-to-acquire logs. Artificial neural network (ANN) and multivariate adaptive regression spline (MARS) models achieve the best log-synthesis performance. ANN-based log synthesis is the best among the six regression models, and the ANN-based synthesis exhibits coefficient of determination (R-squared) of 0.85 on the test data from a single well. ANN-based log synthesis exhibits R-squared of 0.84 when deployed in the second well for blind testing, such that no information from the second well was used to train the model. In the entire first well, the six models show similar log synthesis performance, in terms of relative errors in synthesizing the compressional and shear travel-time logs.

For assessing the reliability of log synthesis when deploying the shallow-learning regression models in new wells/formations, where the compressional and shear wave travel-time logs are absent, we applied five clustering methods to group the formation depths by processing three specific easy-to-acquire conventional logs. Among the five clustering methods tested for their ability to assess the reliability of log synthesis, the centroid-based K-means clustering significantly outperforms other clustering methods. K-means-derived clusters exhibit strong correlation with the relative errors in log synthesis. Most formations that have prediction relative error higher than 0.3 are clustered into cluster number 2 by the K-means clustering. Formation depths assigned a cluster number 2 by the K-means clustering will have low reliability in ANN-based synthesis of compressional and shear travel time logs. By processing the 13 easy-to-acquire logs using the ANN model, we can synthesize the shear and compressional travel-time logs to facilitate geomechanical characterization under data constraint. At the same time, K-means clustering can be applied to evaluate the reliability of the log synthesis performed using the ANN model.

References

[1] Tariq Z, Elkatatny S, Mahmoud M, Abdulraheem A. A new artificial intelligence based empirical correlation to predict sonic travel time. In: International petroleum technology conference; Bangkok, Thailand. IPTC: International petroleum technology conference; 2016.
[2] Li H, Misra S. Prediction of subsurface NMR T2 distribution from formation-mineral composition using variational autoencoder. In: SEG technical program expanded abstracts; 2017. p. 3350–4.
[3] Li H, Misra S. Prediction of subsurface NMR T2 distributions in a shale petroleum system using variational autoencoder-based neural networks. IEEE Geosci Remote Sens Lett 2017;(99):1–3.
[4] Rezaee MR, Ilkhchi AK, Barabadi A. Prediction of shear wave velocity from petrophysical data utilizing intelligent systems: an example from a sandstone reservoir of Carnarvon Basin, Australia. J Petrol Sci Eng 2007;55(3–4):201–12.
[5] Asoodeh M, Bagheripour P. Prediction of compressional, shear, and stoneley wave velocities from conventional well log data using a committee machine with intelligent systems. Rock Mech Rock Eng 2012;45(1):45–63.
[6] Iverson WP, Walker JN. Shear and compressional logs derived from nuclear logs. In: SEG technical program expanded abstracts. Society of Exploration Geophysicists; 1988. p. 111–3.
[7] Greenberg M, Castagna J. Shear-wave velocity estimation in porous rocks: theoretical formulation, preliminary verification and applications. Geophys Prospect 1992;40(2):195–209.
[8] Maleki S, Moradzadeh A, Riabi RG, Gholami R, Sadeghzadeh F. Prediction of shear wave velocity using empirical correlations and artificial intelligence methods. NRIAG J Astron Geophys 2014;3(1):70–81.
[9] Keys RG, Xu S. An approximation for the Xu-White velocity model. Geophysics 2002;67(5): 1406–14.

Chapter 6

Index construction, dimensionality reduction, and clustering techniques for the identification of flow units in shale formations suitable for enhanced oil recovery using light-hydrocarbon injection

Hao Li* and Siddharth Misra†
*The University of Oklahoma, Norman, OK, United States, †Harold Vance Department of Petroleum Engineering, Texas A&M University, College Station, TX, United States

Chapter outline

Nomenclature

MMP Minimum miscibility pressure
$P_{i,j}$ Magnitude of formation parameter (j) in formation (i)
R_i Ranking characteristic parameter for formation (i)

Machine Learning for Subsurface Characterization. https://doi.org/10.1016/B978-0-12-817736-5.00002-8

V_o Volume fraction of miscible oil
V_w Volume fraction of movable water
V_b Volume fraction of bound fluid
V_k Volume fraction of kerogen

Subscripts

i Formation (*i*)
j Formation parameter (*j*)
o Oil, optimum
w Worst

1 Introduction

1.1 Geology of the shale formation

The shale formation under investigation is a 200-ft-thick formation divided into three distinct sections: upper, middle, and lower shale formations. The upper and lower sections are black shales, and the middle section is sandy siltstone. The upper and lower shale formations are hydrocarbon source rocks with total organic carbon (TOC) ranging from 12 to 36 wt%. The clay mineral content in these two sections is dominated by illite and quartz. The middle shale formation is the hydrocarbon-bearing reservoir and has a low TOC content ranging from 0.1 to 0.3 wt%. Below these three shale formations, there exists dolostone interbedded with clay-rich conglomeratic dolomudstone. The shale formations are characterized by ultralow matrix permeability that constrains oil mobilization [1]. Current recovery factor in such shale formations is around 3%–6% of the oil in place [2]. High oil-in-place estimates with low primary recovery mandates enhanced oil recovery (EOR) projects based on light, miscible hydrocarbon injection.

1.2 Literature survey

Due to the ultralow permeability and nanoscale pore sizes, the mechanisms of EOR in tight reservoir are different from those in conventional reservoir. In tight oil formation, EOR efficiency is controlled by the combined effects of miscibility, diffusion, sorption, dissolution, and capillary condensation, to name a few, out of which diffusion is the dominant mechanism [3]. Large permeability contrast between the organic-rich porous matrix and fracture can lead to fracture-dominated flow in fractured shales. In tight shale formations, the net displacement due to the injection of light hydrocarbon generally includes four steps: (1) injected light hydrocarbon flows through fractures, (2) injected hydrocarbon goes into the porous organic-rich shale matrix by diffusion and imbibition; following that, the injected hydrocarbon may interact with connate oil due to miscibility, (3) connate oil migrates into fractures via swelling and reduced viscosity, and (4) injected light hydrocarbon achieves equalization inside the organic-rich matrix [2]. Our focus in this study is to describe EOR efficiency in terms of above-mentioned processes 2 and 3.

Physical properties and engineering practices governing the EOR process can be assigned specific weights for purposes of ranking/screening reservoirs and hydrocarbon-bearing zones in terms of the efficacy of these reservoirs/zones for EOR using gas injection [4–7]. Rivas et al. [8] developed a ranking characteristic parameter to overcome the "binary characteristics" of conventional reservoir screening method and applied the method to rank reservoirs for EOR using gas injection, which was subsequently used and modified by Diaz et al. [9], Shaw and Bachu [10], and Zhang et al. [11]. Oil saturation index (OSI) has also been used to describe the potential producibility of shale formations. OSI is a simple geological normalization of oil content to TOC because kerogen has strong affinity to oil. For example, 80 mg of oil can be retained by 1 g of kerogen, which reduces the producibility of the formation [12]. OSI requires laboratory measurements to get the oil and TOC weight fractions. To overcome this requirement, Kausik et al. [13] introduced carbon saturation index (CSI) and reservoir producibility index (RPI) based entirely on the downhole logging tool measurements. CSI is the weight ratio of carbon in light oil to TOC. Unlike OSI, CSI only considers light oil. RPI is formulated by multiplying the light oil content and CSI. Compared to CSI, RPI accounts for organic richness that differentiates the reservoir qualities of organic-rich and organic-lean intervals.

1.3 Objectives

Hydrocarbon recovery potential of various flow units in the shale formation can be quantified from well logs to facilitate efficient reservoir development and management plans. We develop three log-based EOR-efficiency indices, namely, the ranking (R) index, microscopic displacement (MD) index, and K-means clustering (KC) index, to identify flow units suitable for light-hydrocarbon injection along the length of a well in shale formation. The R-index is a modification of Rivas et al.'s [8] reservoir ranking method and implements Jin et al.'s [14] findings from the laboratory investigation of miscible gas injection. On the other hand, MD-index is the ratio of positive to negative factors affecting miscible gas injection. MD-index involves a dimensionality reduction technique called factor analysis and a novel method to calculate the volume of miscible free oil in the presence of pore-confinement effect common in nanoporous shales. Finally the KC-index is obtained by K-means clustering of the available well logs.

Index is used to track variations in a phenomenon or process that cannot be captured in other ways. An index aggregates multiple features (physical properties, parameters, or attributes) to generate composite statistics that quantify the effects of changes in individual or group of features on the process or phenomenon of interest. Indices facilitate the summarization and ranking of the observations. Features implemented in an index can be differentially weighted to accurately reflect their effects on the process. To create an index, it is important to select relevant features, identify the nature of relationship between the features and the process to be tracked, and finally assign weights to the features. Features implemented in an index should be based on content validity, unidimensionality, amount of variance, and degree of specificity in which a dimension is to be measured.

2 Properties influencing EOR efficiency of light-hydrocarbon injection

In this chapter, EOR efficiency of light-hydrocarbon injection in shale formation is assumed to be dependent on the following properties:

a. Movable oil volume
b. Miscible oil volume
c. Water content
d. Kerogen and bitumen content
e. Minimum miscibility pressure (MMP)
f. Volume fraction, dip, aspect ratio, and orientation of fractures
g. Pore size distribution and dominant pore size
h. Compositions of the injected light hydrocarbon and connate hydrocarbon
i. Pore wettability

The oil displacement efficiency of miscible light-hydrocarbon injection depends on several parameters, such as MMP, pore structure, oil composition, gas composition, fractures, and formation dip. MMP is the lowest pressure required for connate oil and injected light hydrocarbon to achieve miscibility. When reservoir pressure is higher than MMP, the injected gas can achieve miscibility with reservoir oil resulting in viscosity reduction, oil swelling, interfacial tension reduction, and single-phase flow [15]. In this study, we are interested in the volume of miscible oil, defined as the portion of oil that can achieve miscibility with the injected light hydrocarbon. On the other hand, movable oil relates to the hydrocarbon residing in pores that do not trap the hydrocarbon due to high capillary pressure or due to pore isolation. Movable oil volume and miscible oil volume are both important parameters governing the EOR efficiency.

EOR efficiencies of gas injection in shale formations using CO_2, light-hydrocarbon (C1-C2) mixture, and N_2 have been investigated using laboratory core flooding and numerical modeling by Alharthy et al. [16]. Recovery factors of light hydrocarbons and CO_2 are comparable, whereas to achieve high EOR efficiency using N_2, a higher reservoir pressure is required to maintain the miscibility because N_2 has a relatively high MMP [17].

Miscible displacement in shales is different from that in conventional reservoirs because of the nanoscale pore sizes in shales. The injected light hydrocarbon mixes with the oil in the matrix by molecular diffusion and advection instead of direct displacement of oil in the matrix [16, 18]. Due to the drastic permeability contrast between fractures and matrix, the injected fluid prefers to flow through open fractures leaving behind large portion of oil in the matrix. Alharthy et al. [16] conclude that miscible mixing and solvent extraction near the fracture-matrix region is the primary oil mobilization mechanism of miscible light-hydrocarbon injection. Jin et al. [14] and Dadmohammadi et al. [19] showed that both viscous and molecular flow regimes can exist in organic-rich shales and ultra-tight formations; viscous flow exists in fractures, while molecular flow and viscous flow can simultaneously exist in matrix.

Shales have dominant pore size at nanoscale resulting in pore-confinement effects on the MMP. It is well known that the flow behavior in nanoscale is significantly different from that in larger scale [20, 21]. Fluid properties in nanopores differ greatly from bulk fluid, which means conventional calculation may not accurately describe the miscibility. Light-hydrocarbon injection in shales occurs at lower MMP because the critical temperature and pressure for the injected light hydrocarbon and reservoir oil decrease in nanopores [22]. Several methods have been proposed to calculate the MMP in nanopores [23, 24]. An alternation of MMP due to pore-confinement effect may result in miscibility variations in pores as a function of pore size, which is adequately accounted for in this study.

Volume content of bitumen, which is a nonproducible organic matter deposited in the pores, is another important factor governing the EOR efficiency. Due to large molecular size of bitumen and small pore throat size in the formation, the existence of bitumen clogs pores and inhibits the oil flow. Furthermore, bitumen increases kerogen swelling giving rise to smaller and more complicated pore structure [25]. Another factor influencing the EOR efficiency is the water content. At high water saturation, oil is surrounded by water in the pore; therefore the sweep efficiency of miscible displacement is reduced [26]. In this chapter, we take both bitumen and water as negative factors for the EOR using light-hydrocarbon injection.

Natural or induced fractures may pose two contrasting effects on the oil recovery with light-hydrocarbon injection. The presence of fracture enhances the mobilization of oil in matrix by molecular diffusion. Nonetheless, connected fracture system may result in low sweep efficiency. Natural or induced fractures increase the contact area between matrix oil and injected fluid and thus enhance the diffusion process near the fracture-matrix interface. The density, orientation, and size of fractures determine the producibility of shales [27], which has been ignored in this study.

Kerogen content and pore wettability also have significant impact on EOR efficiency when using light hydrocarbon. Experimental investigation showed that kerogen content is the most important factor affecting miscible gas injection performance [14]. Pores in shale formations can be classified as pores in organic matter and those in nonorganic matter. Kerogen is predominantly composed of micropores and mesopores that are oil wet and exhibit complex pore structure. Consequently, it is harder to displace hydrocarbon from pores in organic matter.

3 Methodology to generate the ranking (R) index

3.1 Description of the R-index

Rivas et al. [8] developed a ranking characteristic parameter to rank reservoirs/zones according to their suitability for miscible oil displacement when performing gas injection. Shaw and Bachu [10] and Zhang et al. [11] successfully applied Rivas et al.'s [8] method to rank reservoirs for purposes of EOR. Ranking characteristic parameters are obtained by comparing the actual reservoir parameters with fictive optimum and worst reservoir parameters using an exponentially varying function. The optimum and worst reservoir parameters are obtained through numerical simulation. We modified Rivas et al. [8] ranking method into the R-index, which is computed by processing well logs acquired in the shale formation of interest. In doing so, we demonstrate a procedure for index construction to understand and quantify the complex process of enhanced oil recovery due to light hydrocarbon injection.

3.2 Calculation of the R-index

First the formation of interest is partitioned into flow units. Following that, we need to decide the optimum and worst values for the parameters governing the EOR efficiency of the light-hydrocarbon injection into those flow units. Each property (j) of the flow unit (i) is normalized as follows:

$$X_{i,j} = \frac{|P_{i,j} - P_{o,j}|}{|P_{w,j} - P_{o,j}|} \tag{6.1}$$

where $P_{i,j}$ is the magnitude of property (j) in the flow unit (i) being ranked and $P_{o,j}$ and $P_{w,j}$ are the optimum and worst values of the property (j), respectively, such that $P_{o,j}$ and $P_{w,j}$ represent properties of two artificial flow units. One artificial flow unit has the best properties for miscible oil displacement using light-hydrocarbon injection, and the other has the worst properties. We use the maximum and minimum values of a property in the entire formation under investigation as the optimum and worst values to obtain the $P_{o,j}$ and $P_{w,j}$. The transformed variable $X_{i,j}$ ranges between 0 and 1. To better emphasize the properties that are not conducive for the proposed EOR, $X_{i,j}$ is transformed to $A_{i,j}$ using a heuristic exponential equation expressed as

$$A_{i,j} = 100e^{-4.6X_{i,j}^2} \tag{6.2}$$

where $A_{i,j}$ ranges from 1 to 100, such that $A_{i,j} = 100$ represents the best value of a specific property. Following that, $A_{i,j}$ is weighted by w_j, which represents the relative importance of each reservoir property in influencing the EOR efficiency of the light-hydrocarbon injection. We use the values of weight w_j as published by Jin et al. [14]. Weighted grade of property (j) in the flow unit (i) is expressed as

$$W_{i,j} = A_{i,j}w_j \tag{6.3}$$

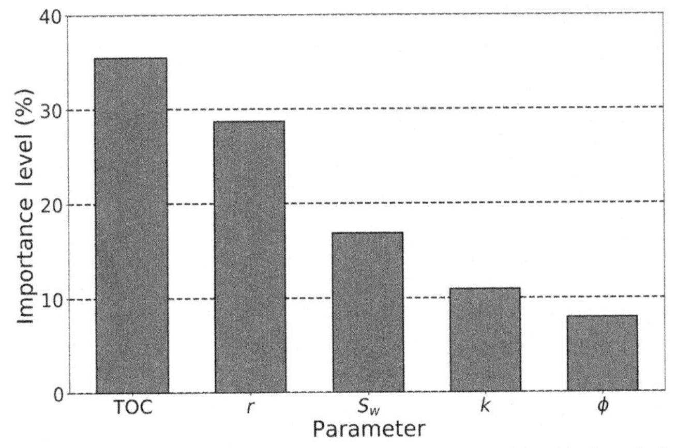

FIG. 6.1 Importance level of reservoir properties on the gas injection-based displacement efficiency. *(Modified from Jin L, Sorensen JA, Hawthorne SB, Smith SA, Bosshart NW, Burton-Kelly ME, et al., editors. Improving oil transportability using CO2 in the Bakken system—a laboratory investigation. SPE international conference and exhibition on formation damage control; 2016 24–26 February; Lafayette, Louisiana, USA: Society of Petroleum Engineers 2016.)*

The final score for each flow unit is the R-index expressed as

$$I_{R,i} = \frac{100\left[\sum_{1}^{j} W_{i,j}W_{j,i}\right]^{0.5}}{R_o} \tag{6.4}$$

where R_o is the ranking characteristic parameter for the formation with optimum properties conducive for the proposed light-hydrocarbon injection and $I_{R,\,i}$ is the R-index of flow unit i.

In the research conducted by Jin et al. [14], miscible CO_2 injection experiments were performed on 28 core samples. These cores are acquired from the shale formation under investigation in this study. The relative importance of various reservoir properties on the gas injection-based displacement efficiency was calculated by statistical analysis of the experimental results (Fig. 6.1). The key reservoir parameters identified by them include TOC, pore throat radius, water saturation, porosity, and permeability. Jin et al. [14] showed that TOC and pore throat radius are the most important properties governing the displacement efficiency. The contribution of TOC and pore throat radius is more than 60%.

EOR mechanisms due to CO_2 injection and light-hydrocarbon injection are relatively similar. Under miscible displacement, both CO_2 and light-hydrocarbon injection mobilize oil by a combination of oil swelling, reduced viscosity, and pressurization of reservoir [2, 15]. Some experiments on core samples showed similar oil recovery for CO_2 and light-hydrocarbon injection [16, 17]. Consequently, to generate the R-index for the shale formation under investigation in this study, we implement the aforementioned five properties along with their relative importance levels, which were identified for CO_2 injection experiments.

The proposed R-index calculation (Fig. 6.2) requires the following ranking characteristic parameters to be obtained from well logs: (1) Kerogen volume obtained from Schlumberger's Quanti-ELAN inversion, (2) apparent pore throat radius obtained from a Winland-type analysis of formation permeability and porosity derived from logs, (3) water saturation from the interpretation of induction resistivity measurements, (4) porosity from neutron-density logs, and (5) permeability from NMR T2 distribution log. The porosity and water porosity logs in Fig. 6.2 are derived using the Interactive Petrophysics NMR interpretation module. The oil porosity is derived from Techlog Quanti-ELAN mineralogy inversion module. As a result the water porosity log and oil porosity log do not add up to the total porosity log. The r35 log is an approximation of pore aperture radius corresponding to the 35th percentile mercury saturation.

The calculated R-index is shown in track 8 of Fig. 6.2. As per the R-index, light-hydrocarbon injection will have better EOR efficiency in the middle shale as compared with the upper and lower shales. Most parts of upper and lower shales have a low EOR efficiency. There are three disadvantages of the R-index: First, TOC has a relatively higher importance when calculating R-index that results in relatively lower R-index values in the upper and lower shales, which have high TOC. Second, the R-index is a relative indicator of EOR efficiency. Finally, the entire middle shale exhibits high R-index indicating the low vertical resolution of this index.

4 Methodology to generate the microscopic displacement (MD) index

4.1 Description of the MD-index

MD-index is a measure of pore-scale displacement efficiency of the injected light hydrocarbon in a hydrocarbon-bearing zone. Theoretically, microscopic displacement efficiency can be calculated by dividing the displaced oil volume by the initial oil volume [28]. This necessitates calculations of the volume fractions of free oil, free water, bound oil, and bound water in pores to generate the depth-specific MD-index. To have a good microscopic displacement efficiency during the miscible light-hydrocarbon injection, injection pressure and reservoir pressure should be above the MMP of the injected hydrocarbon and the pore-filling connate hydrocarbon. The MMP is determined based on the hydrocarbon composition and reservoir temperature. The fluid properties change in nanopores due to pore-confinement effect, which is a consequence of the large capillary pressures, electrostatic forces, and van der Waals forces giving rise to the changes in structural properties of the fluid [29]. MMP will be notably altered due to pore-confinement effect in nanoscale pores. Hydrocarbon residing in differently sized pores will have different MMP. Consequently, under similar reservoir temperature and pressure, oil in unconfined pores and nanopores may have different miscibility. Due to the pore-confinement effects, hydrocarbon in nanopores

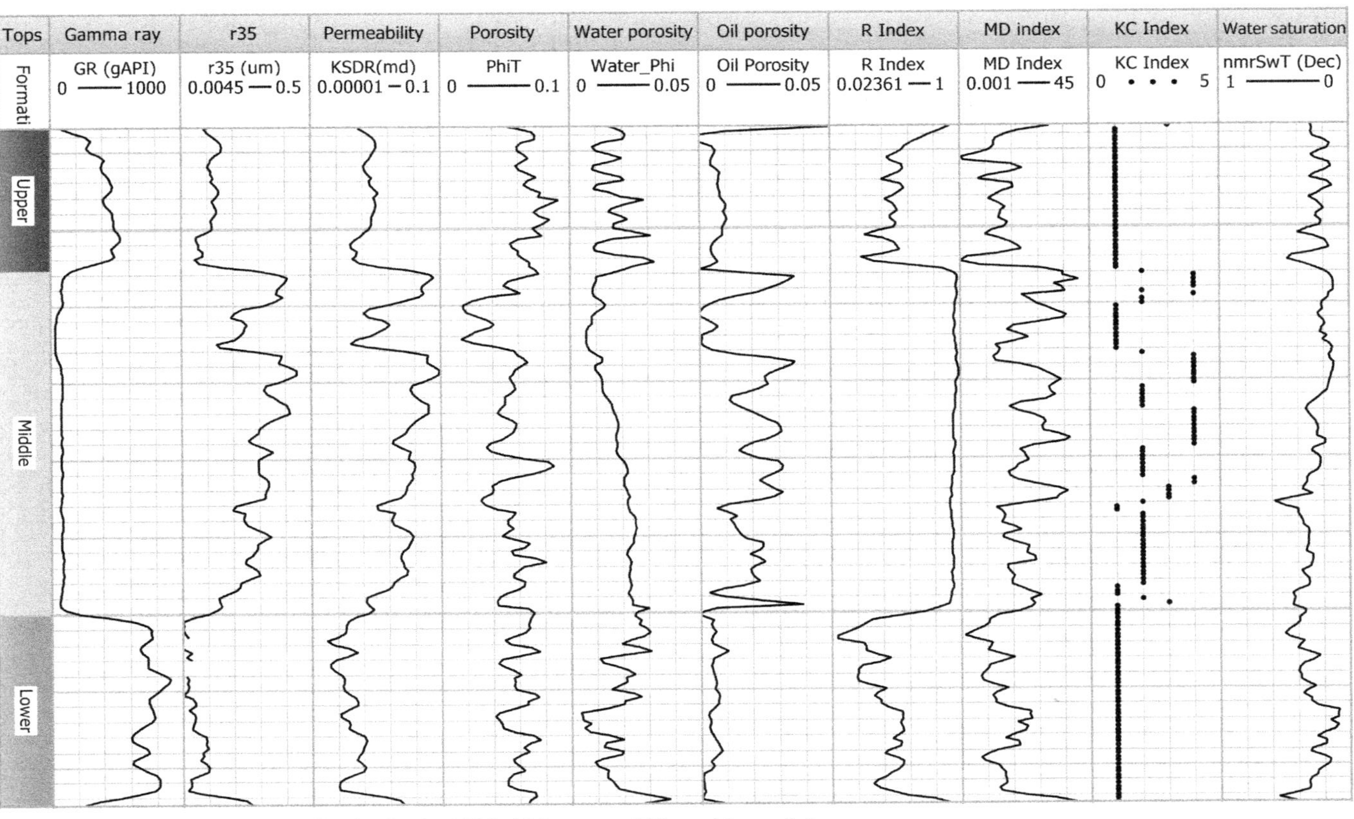

FIG. 6.2 R-index, MD-index, and KC-index for the 200-ft-thick upper, middle, and lower shales.

can achieve miscibility with the injected hydrocarbon, while the hydrocarbon residing in larger pores remains immiscible. The lack of miscibility results in interfacial tension and multiphase flow that reduces the displacement efficiency achieved by the injected light hydrocarbon.

In shales, the free-oil volume cannot be considered as a direct indicator of microscopic displacement efficiency. Instead the fraction of free-oil volume that can achieve miscibility is a better indicator of the displacement efficiency in shales. The presence of movable water in the pores will adversely affect the displacement efficiency. However, bound water has complex effect on light-hydrocarbon injection. Several studies have come up with contradictory results [30]. Here, we assume that bound fluid has a negative effect on the displacement efficiency. Furthermore, kerogen content increases pore complexity, and pores are preferentially oil wettability, thereby reducing the displacement efficiency. Another adverse factor is the presence of bitumen. Dominant pore throat size is 5 nm in the upper and lower shales and 25 nm in the middle shale [14]. Bitumen have diameters ranging from 5 to 200 nm [31]. Bitumen will tend to clog pores and inhibit flow in the shales. Such complex interactions and dependencies mandate the construction of an index to better quantify the complex process of oil recovery due to light hydrocarbon injection.

4.2 Calculation of the MD-index

The calculation process for the MD-index is shown in Fig. 6.3. MD-index incorporates volume fraction of miscible free oil, bound fluid, movable water, and kerogen. MD-index is formulated as a simple ratio of positively affecting parameters to the adversely influencing parameters expressed as

$$I_{MD} = \frac{V_o}{(V_w + V_b) \times V_k} \tag{6.5}$$

where V_o, V_b, and V_w is the pore volume fractions of miscible free oil, bound fluid, and movable water with respect to the bulk volume, V_k is the kerogen volume fraction with respect to bulk volume, and I_{MD} is the MD-index. The technical challenge is to accurately estimate the aforementioned properties.

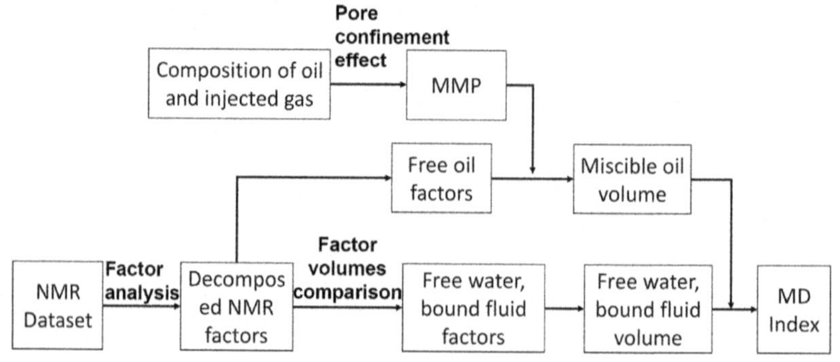

FIG. 6.3 Procedure for the MD-index calculation.

Though the formulation shown in Eq. (6.5) is simple, calculating the incorporated parameters from well logs is a challenge.

4.2.1 Step 1: NMR decomposition using factor analysis

Wireline NMR T2 distribution (Fig. 6.5, Track 1) is an overall response of all the fluid phases/components filling a range of pore sizes, such that the T2 signature of a fluid phase overlaps with those from other phases/components. The NMR T2 distribution used in this study are T2 amplitudes measured at 64 discrete T2 times in the range of 0.3 ms to 3000 ms. In other words, the T2 distribution has been discretized into 64 T2 bins. The mix of constituent signals masks the information related to the relative volumes and pore sets occupied by various fluid phases in the pore. Volume fractions of various fluid phases in Eq. (6.5) are obtained by decomposing the NMR T2 distribution into the contributions of individual fluid phases/components, which in our study are oil and water. The proposed decomposition assumes fluid phases possess distinct properties and occupy different sets of pore spaces. We implement Jain et al.'s [32] factor analysis to decompose the subsurface NMR-T2 distribution responses acquired in the shale formation. Factor analysis is a statistical method to extract unobserved latent variables, so it can only be applied to formations that share similar formation characteristics. Factor analysis describes variability among several observed, correlated variables in terms of a potentially lower number of unobserved variables called factors. Hence, it serves as a dimensionality reduction technique. For example, the variations in the 64 bins of NMR T2 distribution mainly reflect the variations in unobserved (underlying) variables/factors, such as bound water, bound oil, free water, and free oil.

Much like cluster analysis that involves grouping similar samples into clusters, factor analysis involves grouping similar features into fewer constructed features. The purpose of factor analysis is to explain observed features/variables (in our study, the measurements for the 64 bins of NMR T2 distribution) in terms of a much smaller number of variables called factors. These factors represent underlying concepts that cannot be adequately measured by the original variables. Factor analysis simplifies data. We are implementing an exploratory factor analysis to compute the underlying factors that influence the NMR T2 distribution response measured over 64 discrete bins without setting any predefined structure to the outcome. Factor analysis involves factor extraction followed by factor rotation. Factor extraction involves making a choice about the type of model as well as the number of factors to extract. Factor rotation comes after the factors are extracted, with the goal of achieving simple structure in order to improve interpretability. Factor analysis is applied to units with relatively similar pore structures. In our study the upper and lower shales are remarkably similar. The middle shale is associated with lower TOC and larger pore sizes. Consequently, factor analysis is jointly applied to the upper and lower shales and separately on the middle shale. Five factors were identified in the upper and lower shales, and eight factors were identified in the middle sections (Fig. 6.4). Each factor represents a fluid phase occupying a set of pores.

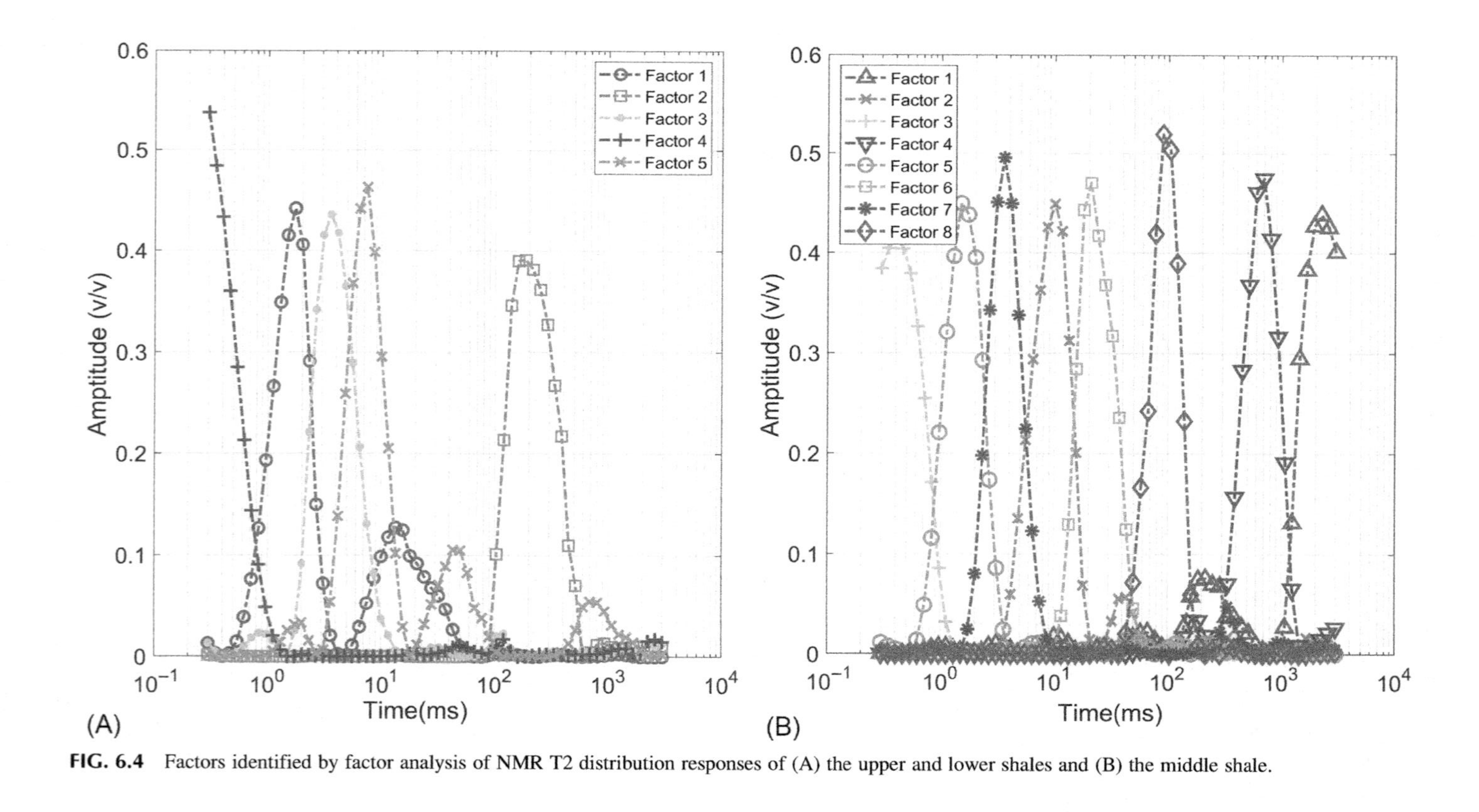

FIG. 6.4 Factors identified by factor analysis of NMR T2 distribution responses of (A) the upper and lower shales and (B) the middle shale.

4.2.2 Step 2: Determination of pore volumes of various fluid phases represented by the decomposed factors

The decomposed factors are converted to factor volumes using Interactive Petrophysics to associate each factor with a fluid phase. Various combinations of factor volumes are then compared with the bound- and free-fluid volumes estimated from NMR T2 distribution using a specific T2 cutoff [33]. For the formation of interest, the T2 cutoffs for clay-bound water and free fluid are reported to be around 0.411 and 0.958 ms, respectively. With these cutoffs, fluid volume for T2 $\leq$ 0.411 ms represents clay-bound fluid, fluid volume for 0.411 ms $\leq$ T2 $\leq$ 0.958 ms represents capillary-bound fluid, and fluid volume for T2 $\geq$ 0.958 ms represents free fluid. As shown in Fig. 6.5A, track 1 is the original T2 signal for the middle shale. The next set of tracks is the decomposed factors of the original T2 signal computed using factor analysis. Total bound fluid-filled porosity, free fluid-filled porosity, total water-filled porosity, and hydrocarbon-filled porosity were obtained using petrophysical interpretation. Bound- and free-fluid porosities can be directly calculated from NMR T2 and cutoffs. Total water saturation is calculated using dual-water equation with resistivity log and then can be transformed into total water porosity. The process is built in the NMR interpretation module in Interactive Petrophysics software. Following that, porosities obtained using various T2 cutoffs are compared with the porosities of factor volumes in Fig. 6.5A. Factor 4 is identified as the signature of bound fluid (both clay-bound and capillary-bound) for the upper and lower shales (Fig. 6.5B and C). For that case, the factors 1, 2, 3, and 5 are identified as the signatures of free fluid. By further comparing the porosity of free-fluid factor volume with free water and free hydrocarbon pore volumes, factors 1, 2, and 3 are identified as the signatures of free hydrocarbon, and factor 5 is identified as the signature of free water (Fig. 6.5B and C). The sum of factors shows good agreement with different fluid porosities acquired from T2 cutoffs, as shown in the last four tracks of Fig. 6.5A–C, which is the best for upper and lower shales.

4.2.3 Step 3: Correction of miscible, free-oil volume for pore confinement effect

MD-index is the relative volume of the positive and the negative petrophysical components influencing oil displacement by a light hydrocarbon (Eq. 6.5). Calculation of MD-index requires fluid type porosities, which are derived from the decomposed factors and corresponding fluid types. However, the free oil-filled porosity calculated using factor analysis cannot be directly used in Eq. (6.5), because not all free oil can achieve miscibility. Under certain conditions, only the free oil in smaller pores can achieve miscibility due to the pore-confinement effects. MMP of the injected light hydrocarbon and the pore-filling in situ hydrocarbon mixture is reduced in nanopores.

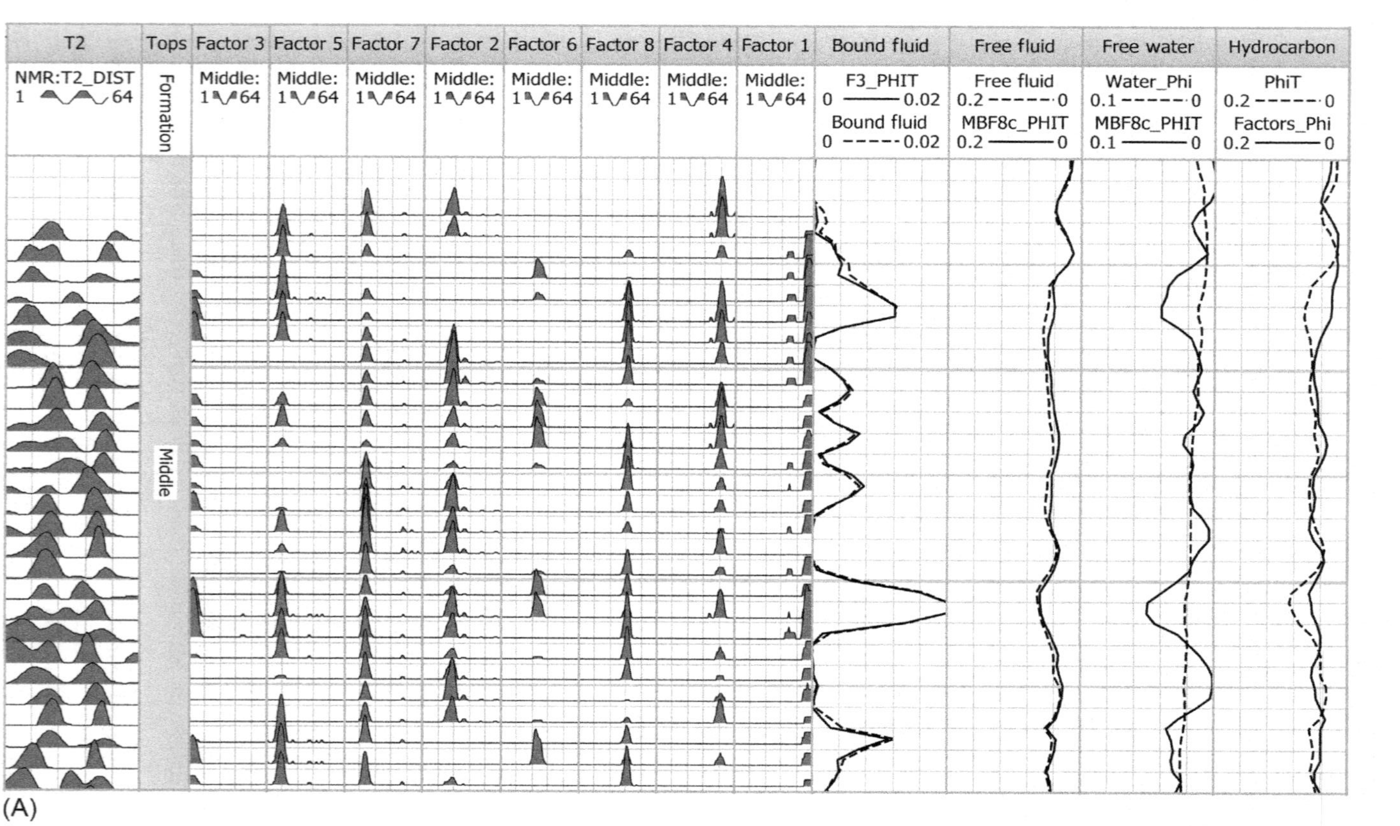

FIG. 6.5 (A) Decomposed factors (tracks 3–10) and comparison of the factor volumes with the fluid volumes obtained from T2 cutoff (tracks 11–14) for the middle shale.

(Continued)

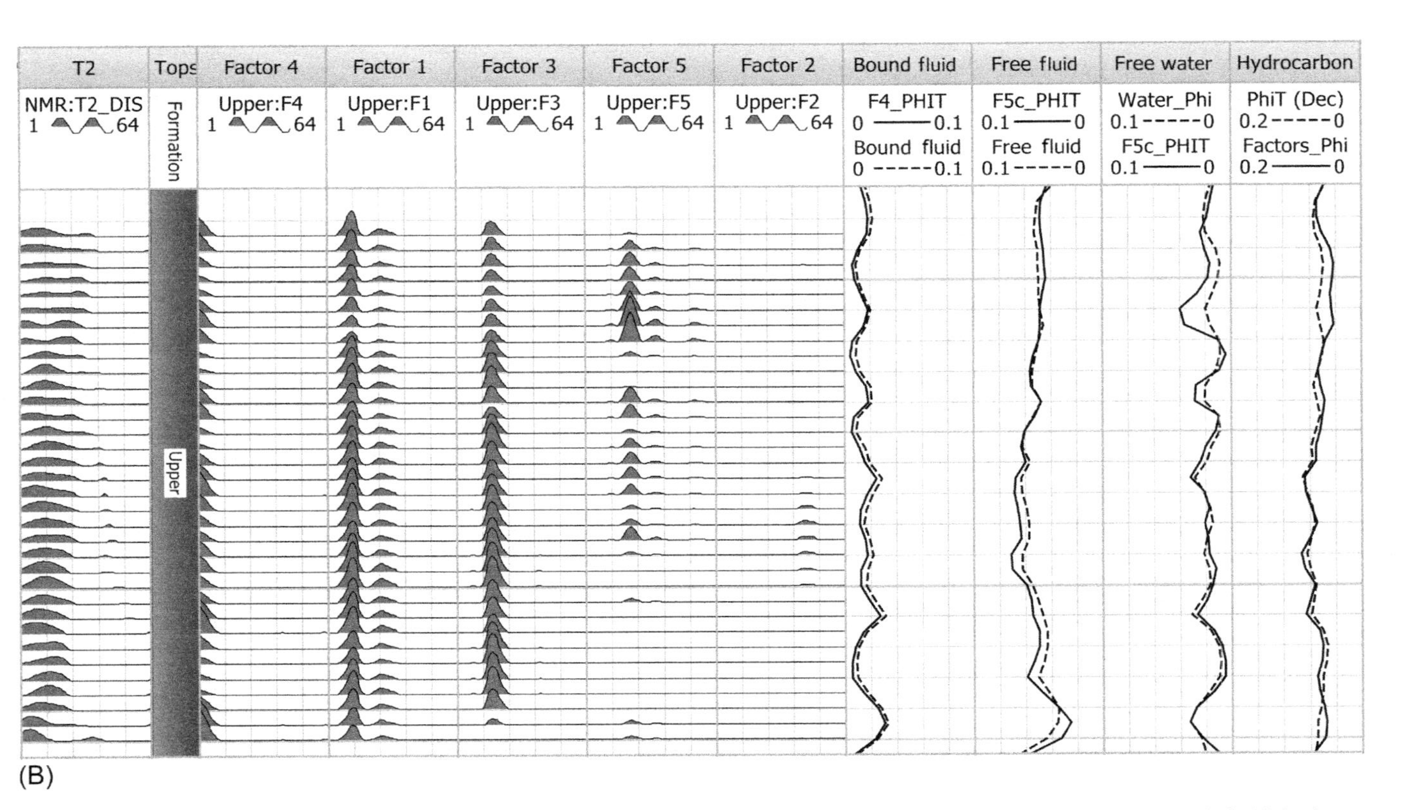

FIG. 6.5, cont'd (B) Decomposed factors (tracks 3–7) and comparison of factor volumes with the fluid volumes obtained from T2 cutoff (tracks 8–11) for the upper shale.

(Continued)

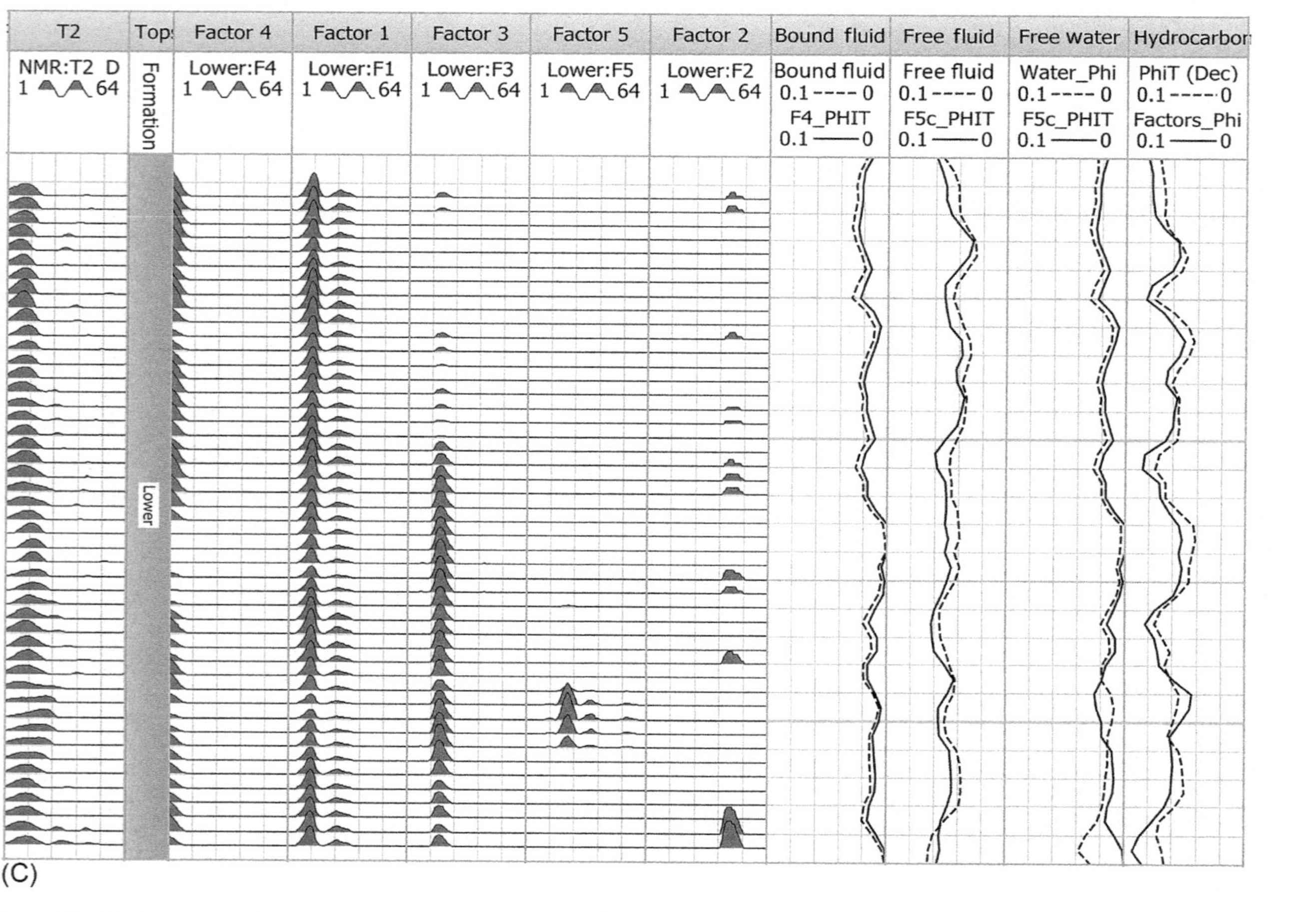

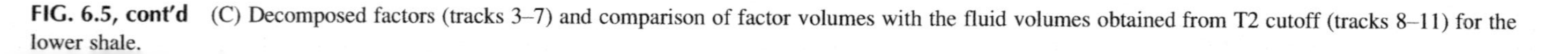

FIG. 6.5, cont'd (C) Decomposed factors (tracks 3–7) and comparison of factor volumes with the fluid volumes obtained from T2 cutoff (tracks 8–11) for the lower shale.

When the MMP is higher than the reservoir pressure, a portion of free oil cannot achieve miscibility during the injection, which is possible for free oil residing in larger pores. High MMP reduces the displacement efficiency of light-hydrocarbon injection. The amount of oil that can achieve miscibility under reservoir condition is a direct measure of displacement efficiency.

To estimate the volume fraction of free oil that can achieve miscibility in the presence of pore-confinement effect, the pore size distribution of a formation is first calculated from the measured NMR T2 distribution. The T2 relaxation captured in the NMR measurement (Fig. 6.5, Track 1) is a sum of bulk fluid relaxation, molecular relaxation, and surface relaxation. T2 distribution represents pore size distribution because the surface relaxation is proportional to the surface-to-volume ratio and surface relaxivity. In tight formation, where pore size is small and the pore is filled with light oil, the bulk relaxation and molecular diffusion effect can be neglected [34, 35]. Based on this assumption the T2 response is associated with pore diameter by using the surface relaxivity expressed as

$$\frac{1}{T_{2,Surface}} = \rho \frac{6}{d} \tag{6.6}$$

where $T_{2,\ surface}$ is the surface relaxation in ms, ρ is surface relaxivity in nm/ms, and d is pore body diameter in nanometers.

The middle shale has larger pore throat and pore diameter than the upper and lower shales. In the absence of SEM image, we refer to the literature and the maximum throat size in the middle section is assumed to be around 100 nm [36], and the maximum pore body size in the middle section is assumed to be around 1000 nm [37]. Similarly, the pore throat size in upper and lower sections is assumed to be around 25 nm [14]. We assume that the body size-to-throat size ratio in the upper and lower shales is equal to that in the middle section. From the NMR T2 distribution log, the maximum T2 value of upper and lower shale sections is 30 ms, and that in the middle section is 300 ms. Jiang et al.'s [34] method was then used to estimate the surface relaxivity using Eq. (6.6) with parameters mentioned in Table 6.1. Using the

TABLE 6.1 Surface relaxivity in the various sections of the shale formation under investigation.

	Maximum T2 (ms)	Maximum pore body size (nm)	Surface relaxivity (nm/ms)
Upper and lower shales	30	250	1.38
Middle shale	300	1000	0.56

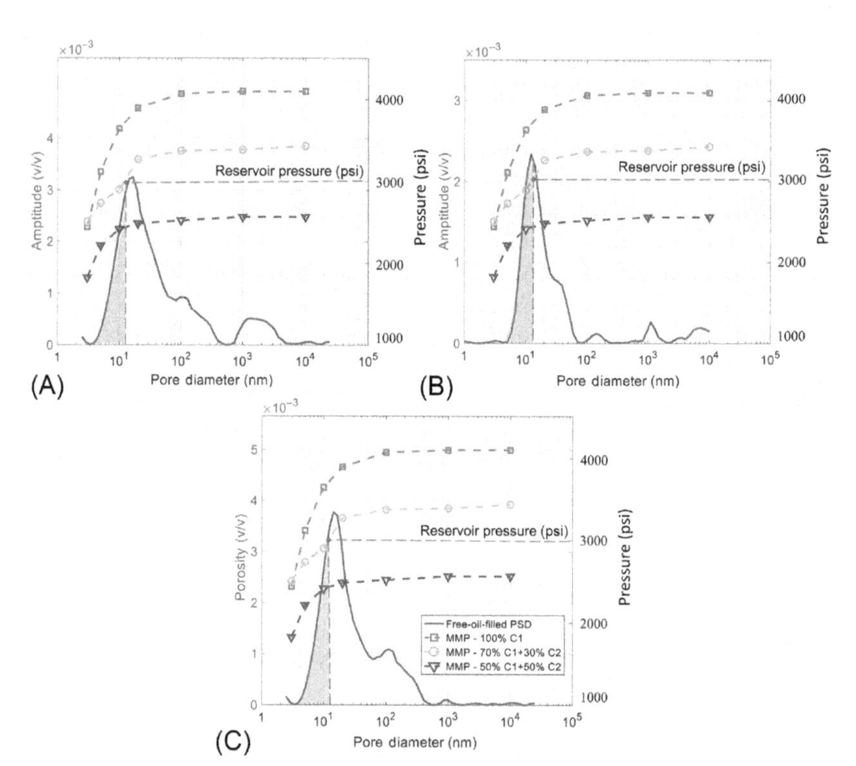

FIG. 6.6 Free oil-filled pore size distribution of (A) upper, (B) middle, and (C) lower shales along with MMP of injected hydrocarbon and in situ oil mixture in the presence of pore-confinement effect.

calculated surface relaxivity, the measured T2 distribution can be transformed into pore size distribution at each depth using Eq. (6.6).

The calculated free oil-filled pore size distributions for the shale formation are shown in Fig. 6.6. The MMP of the formation considering the pore-confinement effect is also calculated and presented in Fig. 6.6. The MMP is calculated using the WinProp software with the method presented by Teklu et al. [29] for various compositions of injected gas and a specific composition of oil present in the shale formation. The in situ oil composition is obtained from Nojabaei et al.'s [38] research. The MMP line as a function of pore diameter can be compared with the free-oil distribution data obtained using the Jain et al.'s [32] NMR factor analysis to determine the volume fraction of oil that can achieve miscibility under certain reservoir pressure. For instance, as shown in Fig. 6.6B, when the injected hydrocarbon contains 70% C1 and 30% C2 under a reservoir pressure of 3000 psi, the critical pore diameter to achieve miscibility in the middle shale is 10 nm. Above this critical pore diameter, MMP will be higher than reservoir pressure. Only the oil that resides in pores with a diameter smaller than 10 nm (shadowed part) can achieve miscibility under this circumstance. Therefore, using this method, we can now calculate the portion of in situ oil, residing in a certain pore size distribution, that can achieve miscibility with a specific injected fluid under a certain reservoir pressure and temperature.

Fig. 6.6 shows that the MMP is around 4000 psi in unconfined pores when the injected hydrocarbon is 100% C1. MMP decreases with an increase in the longer-chain carbon content. In the shale formation under investigation, the average reservoir pressure is above 4000 psi, which means miscibility is not a problem for light-hydrocarbon injection in the shale formation. All free oil in all the pore sizes of the shale formation under investigation can achieve miscibility when the reservoir pressure is maintained above 4000 psi during the miscible displacement.

4.2.4 Step 4: Compute the MD-index

MD-index is calculated using Eq. (6.5). This calculation requires estimates of miscible free oil-filled porosity corrected for pore-confinement effect (described in Step 3) and those of free water-filled and bound fluid-filled porosities calculated using the factor analysis (described in Step 2). The kerogen volume fraction needed in Eq. (6.5) is acquired from Quanti-ELAN inversion provided by Schlumberger. All these data are normalized and transformed using the heuristic Eq. (6.2). Such a transformation eliminated situations where the denominator in Eq. (6.5) becomes small making the MD-index unreasonably large.

The computed MD-index across the 200-ft depth interval of the upper, middle, and lower shales is shown in the second last track of Fig. 6.2. Based on the MD-index, the middle section has relatively better EOR potential when using light-hydrocarbon injection. Furthermore, the upper and lower shales of the formation exhibit low EOR potential for light-hydrocarbon injection. Around 50% and 20% of the middle shale exhibit good and intermediate EOR potential, respectively. Compared with R-index, the MD-index shows a similar trend; however, the MD-index better differentiates layers of low EOR potential in the middle shale.

5 Methodology to generate the K-means clustering (KC) index

5.1 Description of the KC-index

This index categorizes all the depths into one of the four groups, which indicate low, low-intermediate, high-intermediate, or high EOR potential for light-hydrocarbon injection. This index is computed using K-means clustering technique popular in various machine learning tasks for unsupervised learning. Input logs (features) from a single depth are arranged as n coordinates representing a point in an n-dimensional space. The K-means algorithm processes all input logs from all the depths and finds k cluster centers that minimizes the total Euclidean distances between cluster centers and the datapoints belonging to the corresponding clusters. The number of clusters k is specified prior to clustering the dataset; consequently, K-means benefits from using elbow method to determine the optimum number of clusters. With k equal to 4, we assign an integer-valued cluster number ranging from 1 to 4 to each depth in the shale formation under investigation. KC-index is a compilation of the cluster numbers. Each of the four clusters has an associated cluster center that determines the EOR potential of the cluster, which range from low, low-intermediate, high-intermediate to high EOR potentials.

5.2 Calculation of the KC-index

NMR-derived permeability, porosity, water saturation, oil saturation, NMR-derived bound-fluid porosity, and an approximation of pore aperture radius corresponding to the 35th percentile mercury saturation (r35) are used as inputs (features) to generate the KC-index. r35 characterizes the flow capacity of the rock and is generated using a Winland-type equation:

$$\log r_{35} = 0.732 + 0.588 \log K_a - 0.864 \log \Phi \tag{6.7}$$

For a low-dimensional feature space, correlations among the input logs do not significantly affect the clustering results because the clustering is based on the similarity between various samples in the dataset. However, increase in number of input logs (features) leads to the curse of dimensionality that adversely affects the K-means clustering because, in high-dimensional feature space, several data points have similar Euclidean distances from the cluster centers. As a result, the use of Euclidean distance or any other formulation of distance to quantify similarity of samples to the cluster centers for grouping the data into clusters becomes ambiguous and breaks down in a high-dimensional feature space. The center point of each cluster generated by K-means method represents the average of the log responses (feature values) of the samples in that cluster. The cluster centers of the four clusters (shown in Table 6.2) are physically consistent with the EOR potential represented by the corresponding clusters, that is, high, low, high-intermediate, and low-intermediate recovery potentials. Clustering assigns a cluster number to each depth that can be then used to identify the oil-recovery potential of light-hydrocarbon injection for the given depth.

Cluster 4 represents depths where light-hydrocarbon injection will yield the best displacement efficiency because of the high permeability and high oil saturation, whereas Cluster 1 represents depths of low EOR potential because of low permeability and low oil saturation. The water saturation and oil saturation do not sum to 1 because the cluster centers do not represent the real data. As listed in Table 6.2, Cluster 4 has relatively higher oil porosity, permeability, r35, lower water porosity, and lower bound-fluid porosity compared with other clusters. According to Fig. 6.2, there are very

TABLE 6.2 Centers of the four clusters required to compute the KC-index.

	Porosity (frac.)	Permeability (md)	Oil saturation (frac.)	Water saturation (frac.)	Bound fluid porosity (frac.)	r35 (μm)
Cluster 1	0.0593	0.0007	0.0804	0.2794	0.0145	0.0149
Cluster 2	0.0570	0.0127	0.3236	0.2886	0.0063	0.0870
Cluster 3	0.0380	0.0034	0.8750	0.3186	0.0042	0.0566
Cluster 4	0.0519	0.0296	0.5149	0.2043	0.0023	0.1484

few depths in the middle shale that exhibits KC-index of 3 and the upper and lower shales exhibit KC-index of 1 for the entire depth.

The KC-index, R-index, and MD-index are shown in Fig. 6.2 for purposes of comparison. The KC-index shows good agreement with the MD-index and R-index. Data points of upper and lower shales, where MD-index indicates low miscible-recovery potential, exhibit KC-index of 1, which is the least suitable for the light-hydrocarbon injection. KC-index in the middle shale shows a trend consistent with MD-index, as shown in Fig. 6.2. Layers in the middle shale with high values of MD-index and R-index have KC-index of 3 and 4, whereas layers with lower MD-index value have KC-index of 2. According to KC-index, around 40% and 50% of middle shale will demonstrate high and intermediate oil-recovery potentials when using light-hydrocarbon injection, respectively.

6 Limitations

There are few limitations of the proposed approach, especially those related to the assumptions of fluid properties, reservoir properties, and derivation of mean pore throat diameter in the absence of relevant laboratory data. Laboratory experiments and field studies are needed to further validate the predictions and input data related to the three indices. These indices are not based on actual measurements on fluid mobility, fluid displacement, and fluid trapping. MD-index is designed for formations where pore sizes are predominantly in the nanoscale range. For formations where pore sizes are large, pore-confinement effect will not produce a significant effect on the miscibility. MD-index is based on the analysis of microscopic displacement efficiency. However, parameters such as formation geometry, injector-producer orientations, API gravity, and fracture orientations that significantly influence miscible injection mechanism cannot be adequately addressed in the proposed method. The indices do not include the effects of (1) natural and induced fracture properties, (2) formation damage due to fracturing fluid, (3) gravity and capillary effects, and (4) temporal alterations in the spatial distribution of reservoir properties during production on the recovery potential. Few limitations of the R-index are as follows: (1) It is difficult to design experiments or develop theoretical formulations to determine the weights to be assigned to the ranking parameters, and (2) the interdependence of the ranking parameters is ignored. For KC-index, one limitation is that the cluster centers do not represent a real sample; as a result, the properties corresponding to the cluster center are unreliable in terms of their absolute value. Nonetheless, the relative values of cluster centers are indicative of EOR potential along the well length of the shale formation.

7 Conclusions

Wireline log-derived EOR-efficiency indices can be implemented for the identification of flow units in shale formations that are suitable for EOR using miscible light-hydrocarbon injection. The three EOR-efficiency indices, developed in this study, exhibit consistent predictions of the recovery potential in the 200-ft-thick upper, middle, and lower shale formations. Among these indices, the microscopic displacement (MD) index generates predictions at the highest resolution. MD-index calculation requires accurate estimations of miscible free-oil volume, free water-filled porosity, and bound fluid-filled porosity. Miscible free-oil volume estimate accounts for the alteration in the minimum miscibility pressure due to the pore-confinement effects. Free water-filled and bound fluid-filled porosities are estimated using factor analysis (a dimensionality reduction technique) of the NMR T2 distribution response measured over 64 discrete T2 bins. KC-index is computed using K-means clustering, which is consistent with the MD-index. KC-index has the lowest resolution in the low-productivity upper and lower shales, whereas the R-index has the lowest resolution in the productive middle shale. These indices predict that 50% and 20% of the middle shale should exhibit high and intermediate recovery potentials, respectively. Indices also indicate that the recovery potentials of the upper and lower shales are drastically lower than that of the middle shale formation.

References

[1] Han Y, Misra S, Simpson G, editors. Dielectric dispersion log interpretation in Bakken petroleum system. SPWLA 58th annual logging symposium. Society of Petrophysicists and Well-Log Analysts; 2017.

[2] Hawthorne SB, Gorecki CD, Sorensen JA, Steadman EN, Harju JA, Melzer S, editors. Hydrocarbon mobilization mechanisms from upper, middle, and lower Bakken reservoir rocks exposed to CO. SPE unconventional resources conference Canada; 2013 5–7 November. Calgary, Alberta, Canada: Society of Petroleum Engineers; 2013.

[3] Sorensen JA, Braunberger JR, Liu G, Smith SA, Hawthorne SA, Steadman EN, et al. Characterization and evaluation of the Bakken petroleum system for CO_2 enhanced oil recovery. In: Unconventional resources technology conference; 20–22 July; San antonio. Texas, USA. URTEC: Unconventional Resources Technology Conference; 2015. p. SPE-178659-MS.

[4] Taber JJ, Martin FD, Seright RS. EOR screening criteria revisited—part 1: introduction to screening criteria and enhanced recovery field projects. SPE Reserv Eng 1997;12(03):189–98.

[5] Thomas B. Proposed screening criteria for gas injection evaluation. J Can Pet Technol 1998; 37(11):14–20.

[6] Al-Adasani A, Bai B. Recent developments and updated screening criteria of enhanced oil recovery techniques. In: International oil and gas conference and exhibition in China; 8–10 June; Beijing, China. SPE: Society of Petroleum Engineers; 2010. p. SPE-130726-MS.

[7] Bourdarot G, Ghedan SG. Modified EOR screening criteria as applied to a group of offshore carbonate oil reservoirs. In: SPE reservoir characterisation and simulation conference and exhibition; 9–11 October; Abu Dhabi, UAE. SPE: Society of Petroleum Engineers; 2011.

[8] Rivas O, Embid S, Bolivar F. Ranking reservoirs for carbon dioxide flooding processes. SPE Adv Technol Ser 1994;2(01):95–103.

[9] Diaz D, Bassiouni Z, Kimbrell W, Wolcott J. Screening criteria for application of carbon dioxide miscible displacement in waterflooded reservoirs containing light oil. In: SPE/DOE

improved oil recovery symposium; 21-24 April; Tulsa, Oklahoma. Tulsa, Oklahoma: SPE: Society of Petroleum Engineers; 1996.

[10] Shaw J, Bachu S. Screening, evaluation, and ranking of oil reservoirs suitable for CO_2-flood EOR and carbon dioxide sequestration. J Can Pet Technol 2002;41(09):51–61.

[11] Zhang K, Qin T, Wu K, Jing G, Han J, Hong A, et al. Integrated method to screen tight oil reservoirs for CO_2 flooding. In: SPE/CSUR unconventional resources conference; 20–22 October; Calgary, Alberta, Canada. SPE: Society of Petroleum Engineers; 2015 [p. SPE-175969-MS].

[12] Jarvie DM. Shale reservoirs—giant resources for the 21st century. In: Breyer JA, editor. Shale reservoirs—giant resources for the 21st century. Shale resource systems for oil and gas: part 2—shale-oil resource systems. vol. 97. American Association of Petroleum Geologists; 2012. p. 89–119.

[13] Kausik R, Craddock PR, Reeder SL, Kleinberg RL, Pomerantz AE, Shray F, et al. Novel reservoir quality indices for tight oil. In: Unconventional resources technology conference; 20–22 July; San Antonio, Texas, USA. SPE: Society of Petroleum Engineers; 2015. p. SPE-178622-MS.

[14] Jin L, Sorensen JA, Hawthorne SB, Smith SA, Bosshart NW, Burton-Kelly ME, et al., Improving oil transportability using CO_2 in the Bakken system—a laboratory investigation. SPE international conference and exhibition on formation damage control; 2016 24–26 February; Lafayette, Louisiana, USA. Society of Petroleum Engineers; 2016.

[15] Ling K, Shen Z, Han G, He J, Peng P. A review of enhanced oil recovery methods applied in Williston Basin. In: Unconventional resources technology conference; 25–27 August; Denver, Colorado, USA. URTEC: Unconventional resources technology conference; 2014 [p. URTEC-1891560-MS].

[16] Alharthy N, Teklu T, Kazemi H, Graves R, Hawthorne S, Braunberger J, et al. Enhanced oil recovery in liquid-rich shale reservoirs: laboratory to field. In: SPE annual technical conference and exhibition; 28–30 September; Houston, Texas, USA. SPE: Society of Petroleum Engineers; 2015. p. 28–30 September.

[17] Jin L, Hawthorne S, Sorensen J, Kurz B, Pekot L, Smith S, et al. A systematic investigation of gas-based improved oil recovery technologies for the Bakken tight oil formation. In: Unconventional resources technology conference; 1–3 August; San Antonio, Texas, USA. URTEC: Unconventional Resources Technology Conference; 2016. p. URTEC-2433692-MS.

[18] Fai-Yengo V, Rahnema H, Alfi M. Impact of light component stripping during CO_2 injection in Bakken formation. In: Unconventional resources technology conference; 25–27 August; Denver, Colorado, USA. URTEC: Unconventional Resources Technology Conference; 2014 [p. URTEC-1922932-MS].

[19] Dadmohammadi Y, Misra S, Sondergeld C, Rai C, editors. Simultaneous estimation of intrinsic permeability, effective porosity, pore volume compressibility, and klinkenberg-slip factor of ultra-tight rock samples based on laboratory pressure-step-decay method. SPE low perm symposium. Society of Petroleum Engineers; 2016.

[20] Wu K, Chen Z, Li X, Dong X. Methane storage in nanoporous material at supercritical temperature over a wide range of pressures. Sci Rep 2016;6.

[21] Wu K, Chen Z, Li X, Xu J, Li J, Wang K, et al. Flow behavior of gas confined in nanoporous shale at high pressure: real gas effect. Fuel 2017;205:173–83.

[22] Adekunle OO, Hoffman BT. Minimum miscibility pressure studies in the Bakken. In: SPE improved oil recovery symposium; 12–16 April; Tulsa, Oklahoma, USA. SPE: Society of Petroleum Engineers; 2014. p. SPE-169077-MS.

[23] Ahmadi K, Johns RT. Multiple-mixing-cell method for MMP calculations. SPE J 2011;16 (04):733–42.

[24] Teklu TW, Alharthy N, Kazemi H, Yin X, Graves RM. Hydrocarbon and non-hydrocarbon gas miscibility with light oil in shale reservoirs. In: SPE improved oil recovery symposium; 12–16 April; Tulsa, Oklahoma, USA. SPE: Society of Petroleum Engineers; 2014. p. SPE-169123-MS.

[25] Reeder SL, Craddock PR, Rylander E, Pirie I, Lewis RE, Kausik R, et al. The reservoir producibility index: a metric to assess reservoir quality in tight-oil plays from logs. Petrophysics 2016;57(02):83–95.

[26] Rampersad PR, Ogbe DO, Kamath VA, Islam R. Impact of trapping of residual oil by mobile water on recovery performance in miscible enhanced oil recovery processes. In: Low permeability reservoirs symposium; 19–22 march; Denver, Colorado. SPE: Society of Petroleum Engineers; 1995 [p. SPE-29563-MS].

[27] Kurtoglu B, Sorensen JA, Braunberger J, Smith S, Kazemi H. Geologic characterization of a Bakken reservoir for potential CO_2 EOR. In: Unconventional resources technology conference; 12–14 August; Denver, Colorado, USA. SPE: Society of Petroleum Engineers; 2013. p. 12–4 August.

[28] Sehbi BS, Frailey SM, Lawal AS. Analysis of factors affecting microscopic displacement efficiency in CO_2 floods. In: SPE Permian Basin oil and gas recovery conference; 15–17 May; Midland, Texas. SPE: Society of Petroleum Engineers; 2001 [p. SPE-70022-MS].

[29] Teklu TW, Alharthy N, Kazemi H, Yin X, Graves RM, AlSumaiti AM. Phase behavior and minimum miscibility pressure in nanopores. SPE Reserv Eval Eng 2014;17(03):396–403.

[30] Kulkarni MM. Multiphase mechanisms and fluid dynamics in gas injection enhanced oil recovery processes. [Ph.D. General Exam Report] Baton Rouge, LA: The Craft and Hawkins Department of Petroleum Engineering, Louisiana State University and A&M College; 2004 (Jul).

[31] Zhao B, Shaw JM. Composition and size distribution of coherent nanostructures in Athabasca bitumen and Maya crude oil. Energy Fuel 2007;21(5):2795–804.

[32] Jain V, Minh CC, Heaton N, Ferraris P, Ortenzi L, Ribeiro MT. Characterization of underlying pore and fluid structure using factor analysis on NMR data. In: SPWLA 54th annual logging symposium; 22–26 June; New Orleans, Louisiana. SPWLA: Society of Petrophysicists and Well-Log Analysts; 2013 [p. SPWLA-2013-TT].

[33] Jiang T, Jain V, Belotserkovskaya A, Nwosu NK, Ahmad S. Evaluating producible hydrocarbons and reservoir quality in organic shale reservoirs using nuclear magnetic resonance (NMR) factor analysis. 2015/10/20/, SPE: Society of Petroleum Engineers; 2015.

[34] Jiang T, Rylander E, Singer PM, Lewis RE, Sinclair SM, editors. Integrated petrophysical interpretation of Eagle Ford Shale with 1-D and 2-D nuclear magnetic resonance (NMR). SPWLA 54th annual logging symposium; 2013 22–26 June; New Orleans, Louisiana. Society of Petrophysicists and Well-Log Analysts; 2013.

[35] Sun B, Yang E, Wang H, Seltzer SJ, Montoya V, Crowe J, et al. Using NMR to characterize fluids in tight rock unconventional and shale formations. In: SPWLA 57th annual logging symposium; 25–29 June; Reykjavik, Iceland. SPWLA: Society of Petrophysicists and Well-Log Analysts; 2016 [p. SPWLA-2016-PP].

[36] Ramakrishna S, Balliet R, Miller D, Sarvotham S, Merkel D. Formation evaluation In the Bakken complex using laboratory core data and advanced logging technologies. In: SPWLA 51st annual logging symposium; 19–23 June; Perth, Australia. SPWLA: Society of Petrophysicists and Well-Log Analysts; 2010. p. SPWLA-2010–74900.

[37] Saidian M, Prasad M. Effect of mineralogy on nuclear magnetic resonance surface relaxivity: a case study of middle Bakken and three forks formations. Fuel 2015;161:197–206.

[38] Nojabaei B, Johns RT, Chu L. Effect of capillary pressure on phase behavior in tight rocks and shales. SPE Reserv Eval Eng 2013;16(3):281–9.

Chapter 7

Deep neural network architectures to approximate the fluid-filled pore size distributions of subsurface geological formations

Siddharth Misra* and Hao Li†

*Harold Vance Department of Petroleum Engineering, Texas A&M University, College Station, TX, United States, †The University of Oklahoma, Norman, OK, United States

Chapter outline

Machine Learning for Subsurface Characterization. https://doi.org/10.1016/B978-0-12-817736-5.00007-7

1 Introduction

1.1 Log-based subsurface characterization

Subsurface characterization involves estimation, computation, and measurement of the physical properties of the subsurface geological formations. Surface-based deep sensing measurements, borehole-based near-wellbore measurements (logs), and laboratory measurements of geological core samples extracted from wellbores are interpreted using empirical, numerical, and mechanistic models to quantify the physical properties of the subsurface formations. Subsurface measurements (referred as logs), acquired using downhole logging tools, sense the near-wellbore subsurface formation volume by inducing/monitoring various physical/chemical processes. Subsequently, relevant tool physics modeling and geophysical interpretation models are used to process the logs for purposes of subsurface characterization. For example, multifrequency electromagnetic logs are processed using stochastic inversion for fluid saturation estimation [1], various petrophysical models are used to process dielectric dispersion logs to characterize hydrocarbon pore volume and salinity in shales [2], electromagnetic short pulse borehole imaging method is used to characterize cracks and rugosity [3], poroelastic inversion of sonic velocity logs improves permeability characterization [4], and triaxial electromagnetic induction measurement facilitates the estimation of dip and anisotropy of the formation [5].

Use of logging tools, geophysical models, and inversion- and machine learning-based data interpretation techniques for purposes of subsurface characterization has been evolving with the advancements in sensor physics and computational methods. For example, Wong et al. [6] classified well log data into different lithofacies followed by the estimation of porosity and permeability using genetic neural networks. Similarly, lithology determination from well logs was performed by Chang et al. [7] in Ordovician rock units in northern Kansas using fuzzy associative memory neural network. Xu et al. [8] listed the recent advances in machine learning applications on well logs for purposes of improved subsurface characterization. He and Misra [9] used various architectures of shallow neural networks to synthesize dielectric dispersion logs in shales. In another application of neural networks and shallow learning methods, Li et al. [10] generated compressional and shear wave travel times in shale oil reservoir for improved geomechanical characterization. Other than simple machine learning methods, there have been very limited public demonstrations of development of deep learning methods and their applications for formation evaluation and well log analysis.

1.2 Deep learning

Deep learning is a specialized form of machine learning that uses several processing layers to learn complex abstractions (concepts) in data by building a hierarchy/levels of abstractions, wherein each level of abstraction is created using the lower-level abstractions learnt by the preceding layer in the hierarchy. In other words, a model developed using deep learning techniques learns complicated concepts using simpler ones. There are many computational layers between the input and output resulting in multiple linear and nonlinear transformations at each layer. Deep learning uses multiple layers of hierarchical, sequential, and/or recurring computational units to extract features from raw data at multiple levels, such that the collection of extracted features at a certain level forms a specific level of abstraction. In simple terms, a deep learning model is a chain of simple, continuous transformations that map the input vector space into output vector space by learning from a dense sampling of available inputs and outputs (referred as training data). Learning each suitable transformation in the chain of transformations requires computation of certain parameters, which are iteratively updated using various optimization algorithms based on the model performance.

Unlike traditional machine learning, deep learning does not require manual feature engineering prior to the model development because of their capability to perform hierarchical feature learning, where higher-level features are defined in terms of lower-level features. Learning of features at multiple levels of abstraction allows deep learning methods to learn complex functions that map the input to the output directly from data, without depending on human-engineered features. Consequently, deep learning is popular method when dealing with unstructured data, such as images, video, audio, speech, text, language, analog data, health records, metadata and game play. Deep learning model is randomly initiated and then generally gradient-based optimization is used to converge the model parameters (weights and biases) to an optimal solution, which might not necessarily be the global optimum. This optimization process for multiple layers containing multiple computational units requires a lot of data.

Deep learning came into prominence due to the following reasons:

1. Availability of large data volumes being generated at a fast rate by multiple sources.
2. Access to computational infrastructure and advanced hardware for data processing and model training.
3. Easily accessible, robust optimization and learning platforms.
4. Opportunity for large-scale training and deployment of the data-driven models.
5. Need for real-time decision-making and precision engineering/operations.

Few disadvantages and limitations of deep learning are as follows:

1. Not suited when the data size is in the order of thousand samples or smaller.
2. Not easy to interpret and explain due to the multiple levels of abstractions without data augmentation techniques.
3. Slow and expensive to setup and train the model, which requires substantial computation power.
4. Data preprocessing is required to avoid garbage-in/garbage-out scenarios because deep learning methods can easily pick up spurious correlations and biases.
5. Not suitable for implementations in limited/constrained memory devices.
6. For structured data, deep learning performance should be compared with ensemble and nonlinear kernel-based machine learning methods for structured data problems.
7. Models do not generalize well due to the tendency to overfit for simple tasks.

Few advantages of deep learning are as follows:

1. Suited for complex problems based on unstructured data.
2. Best-in-class performance that beats traditional methods by a large margin.
3. Deep learning algorithms scale with data, that is, they continue to improve as the size of data increases, while the performance of the traditional method flattens out.
4. Scalable across various domains and easily adapted to new problems, unlike domain specific methods such as those traditional ones specialized for natural language processing and image analysis.
5. No need for feature engineering. Deep learning does not require domain expertise because it does not need feature engineering. The focus is more on data engineering.

1.3 NMR logging

Nuclear magnetic resonance (NMR) logging tool was introduced for well logging applications in the 1980s. These instruments are typically low-field NMR spectrometers. NMR tool is deployed in an open borehole to primarily acquire the depth-wise T2 distribution response of the subsurface formation volume intersected by a borehole. T2 distribution response is produced because of the T2 relaxations of the excited hydrogen nuclei as they move through the pore-filling fluid while interacting with the grain surfaces. T2 distribution is further processed to obtain the physical properties of the formations, such as pore size distribution, fluid-filled porosity, bound fluid saturations, and permeability, critical for developing and producing from hydrocarbon- and water-bearing reservoirs. However, the acquisition of subsurface NMR log is limited due to the financial expense and operational challenges in its subsurface deployment. In the absence of a downhole NMR logging tool, we propose four deep neural networks (DNNs) architectures that process the conventional "easy-to-acquire" subsurface logs to generate the NMR T2 distributions, which approximate the fluid-filled pore size distributions. This chapter implements the variational autoencoder (VAE) assisted neural network (VAE-NN), generative adversarial network (GAN) assisted neural network (GAN-NN), long short-term memory (LSTM) network, and variational autoencoder with a convolutional layer (VAEc) assisted neural network (VAEc-NN) to synthesize the NMR T2 distribution response of fluid-filled porous subsurface formations around the wellbore.

NMR log contains valuable pore- and fluid-related information that cannot be directly estimated from other conventional logs, such as porosity, resistivity, mineralogy, and saturation logs. Acquisition of NMR log is more expensive and requires better well conditions as compared with other conventional logs. During data acquisition, NMR logging tool is run at a relatively slower speed of 2000 ft/h. Moreover, the NMR logging tool is usually run in bigger-diameter uniform boreholes around 6.5-in. in diameter. NMR signals due to the inherent physics have a poor signal-to-noise ratio [11]. Furthermore, the acquisition of high-quality NMR log in subsurface borehole environment requires highly trained wireline field engineers. Consequently, the oil and gas companies do not deploy NMR logging tool in each well in a reservoir due to the above mentioned financial and operational challenges involved in running the NMR tool. In the absence of NMR logging tool, neural networks can synthesize the entire NMR T2 distribution spanning 0.3–3000 ms along the length of a well by processing the "easy-to-acquire" conventional logs. An accurate synthesis of NMR T2 distribution will assist the geoscientists and engineers to quantify the pore size distribution, permeability, and bound fluid saturation in water-bearing and hydrocarbon-bearing reservoirs; thereby, improving project economics and subsurface characterization capabilities.

2 Introduction to nuclear magnetic resonance (NMR) measurements

2.1 NMR relaxation measurements

In the absence of an external magnetic field, the directions of nuclear magnetic moment associated with the spin of hydrogen nuclei in the pore-filling fluid are randomly oriented. In an external magnetic field, the nuclear magnetic moments associated with spins will get aligned in specific orientations. Such ordered nuclei, when subjected to EM radiation of the proper frequency, will absorb energy and "spin-flip" to align themselves against the field at a higher energy state. The energy transfer takes place at a wavelength that corresponds to radio frequencies, and when the spin returns to its base level, energy is emitted at the same frequency. When this spin-flip occurs, the nuclei are said to be in "resonance" with the field; hence, the name for the technique to excite the nuclei to higher energy states is referred as the nuclear magnetic resonance (NMR).

When the EM radiation is switched off, the excited nuclei release their excess energy (NMR relaxation/decay signal) and return to low energy levels by two relaxation processes. Relaxation is a measure of deterioration of NMR signal with time during the conversion of the excited nonequilibrium population to a normal population at thermal equilibrium. For material characterization the NMR relaxation/decay measurement is analyzed in terms of two processes, namely, T1 (spin-lattice) and T2 (spin-spin) relaxations. T1 relaxation is the loss of signal intensity due to the interaction with environment resulting in the relaxation of the components of the nuclear spin magnetization vector parallel to the external magnetic field, whereas T2 is the broadening of the signal due to the exchange of energy with neighboring nuclei at lower energy levels resulting in the relaxation of the components of the nuclear spin magnetization vector perpendicular to the external magnetic field. T1 relaxation indicates the inter- and intramolecular dynamics. T2 relaxation involves energy transfer between interacting spins via dipole and exchange interactions.

For fluids contained within pores, both T1 and T2 relaxations depend on surface relaxation in addition to primary controls of viscosity, composition, temperature, and pressure. Surface relaxation occurs at the fluid-solid interface between a wetting-pore fluid and the rock-pore walls, which is very different from the individual bulk relaxations of pure fluid and solid. Surface relaxation dramatically reduces T1 and T2, such that surface relaxation is the dominant contributor to T2 relaxation. Surface relaxation depends on the surface-to-volume ratio, which is determined by the mineralogy and is a measure of permeability. T2 relaxation of a pore-filling fluid is considered a combination of bulk, surface, and diffusion relaxation, wherein the surface relaxation dominates. Surface relaxation occurs at the fluid-solid interface and is affected by pore shape, pore network characteristics, and mineralogy. Bulk relaxation is affected by the fluid type, the hydrogen content, and its mobility. Diffusion relaxation occurs due to the nonzero gradient of the magnetic field exciting the sample/reservoir under investigation.

The rate of NMR T2 relaxation of fluids in pores most significantly depends on the frequency of collision of hydrogen nuclei with the grain surface, and this is controlled by the surface-to-volume ratio of the pore in which the hydrogen nuclei are located. Collisions are less frequent in larger pores, resulting in a slower decay of the NMR signal amplitude and allowing the characterization of fluid-filled pore size distribution. When all pores are assumed to have a similar geometric shape, the largest pores have the lowest surface to volume and, thus, the longest T2. However, subsurface rocks have pores of varying sizes containing more than one fluid type, so the T2 relaxation consists of several decays, such that T2 is directly proportional to pore size. Consequently, T2 distribution, which is a spectrum of T2 amplitudes as a function of T2 times discretized into 64 T2 bins in the range of 0.3–3000 ms, can be obtained by processing the T2 decay/relaxation measurements. NMR logging tool is generally deployed in the boreholes to obtain the T2 distribution response of the fluid-filled geological formation intersected by the wellbore.

2.2 Relationships between NMR T2 distribution and conventional logs

NMR T2 relaxation response is used as a diagnostic technique to map the fluid-filled pores in a geomaterial for purposes of reservoir characterization. Pore-filling fluid, such as water and hydrocarbon, contain hydrogen nuclei, which are sensitive to the NMR excitation. In fluid-bearing reservoirs, NMR response is primarily generated by the relaxation of hydrogen nuclei originally excited by the electromagnetic field exerted by NMR tool. NMR relaxation depends on pore structure, pore size distribution, grain texture, and surface relaxivity that governs the relaxation of the hydrogen nuclei. NMR response is quantified as a T2 distribution obtained from the NMR T2 relaxation/decay signal. T2 distribution is a spectrum of T2 amplitudes as a function of T2 times discretized into 64 T2 bins in the range of 0.3–3000 ms (Fig. 7.1, Track 6).

Generation of NMR T2 distribution log using conventional "easy-to-acquire" logs is feasible because these logs are sensitive to fluid saturations, pore size distribution, and mineralogy. Resistivity measurement and the invasion profile seen in resistivity logs are influenced by the pore sizes, pore throat sizes, tortuosity, and Maxwell-Wagner polarization, which depend on pore size distribution. Neutron log is sensitive to hydrogen index of fluids in the formation, which correlates with the average porosity of the formation. GR is sensitive to clay volume that determines the clay-bound water and capillary-bound water, which is sensed by the NMR T2 distribution. Density measurement is sensitive to mineral densities and fluid densities in the formation. Mineral density is controlled by mineralogy that controls the lithology and pore size distribution of the formation. For example, sandstone formation with predominantly quartz mineralogy will have a distinct pore size distribution compared with carbonate formation with predominantly calcite mineralogy, which will be distinct from shale formation with predominantly clay mineralogy. Dielectric dispersion measurement is sensitive to the water saturation and pore size distribution due to Maxwell-Wagner polarization dependent on the pore size distribution. Notably, these logs are influenced by water and hydrocarbon saturation and the vertical distribution of fluid saturations along the depth that is controlled by capillary pressure, which is governed by the pore size distribution.

NMR T2 distribution (Fig. 7.1, Track 6) depends on fluid saturations, fluid mobility, pore size distribution, and surface relaxation, which depends on the mineralogy of the formation. These physical properties that govern NMR T2 also influence the conventional "easy-to-acquire" logs, namely, neutron, density, resistivity, dielectric, and GR (Fig. 7.1, Tracks 1–4). These conventional logs are sensitive to pore size distribution, fluid saturation, and mineralogy. We use deep neural networks to identify and extract the complex relationships between these conventional logs and the NMR T2 distribution that cannot be easily identified/quantified by any other mechanistic models. Deep neural networks cannot find the mechanistic model but can find a nonlinear higher-order functional mapping that describes the physical relationships between the conventional logs and NMR T2 distribution.

3 Data acquisition and preprocessing

3.1 Data used in this chapter

A set of well log data were retrieved from a shale petroleum system comprising seven formations as shown in Fig. 7.1. The top three formations (F1–F3) constitute a shale formation, and the bottom four (F4–F7) are dolostone interbedded with clay-rich conglomeratic dolo-mudstone of Devonian age. F1 is an upper black shale, F2 is a middle sandy siltstone, and F3 is a lower black shale. Formations F1 and F3 are hydrocarbon source rocks are organic-rich with total organic carbon ranging from 12 to 36 wt%. The clay mineral content is dominated by illite and quartz. Formation F2 is the hydrocarbon-bearing reservoir and has a low total organic carbon (TOC) content ranging from 0.1 to 0.3 wt%. Variation in formation mineral compositions leads to changes in the pore structure, grain texture, and surface relaxivity. These characteristics along with fluid saturations and their distribution in the pore network govern the NMR T2 distribution. Mineral compositions and fluid saturations (Fig. 7.1, Track 5) in the seven formations were obtained by numerical inversion of resistivity, neutron, density, gamma ray (GR), and dielectric logs using the mineral inversion module in TECHLOG. Our goal is to apply the four neural networks to process 10 inversion-derived logs, including 7 mineral content logs and 3 fluid saturation logs, namely, bound water, free water, and oil, to synthesize the NMR T2 distributions discretized into 64 T2 bins, which approximate the fluid-filled pore size distribution of the shale petroleum system. The 10 inversion-derived logs were obtained from conventional logs and will be referred as the inversion-derived logs. Therefore, the deep neural networks are trained to relate 10 features (inversion-derived logs) with 64 targets (NMR T2 distribution discretized into 64 T2 bins).

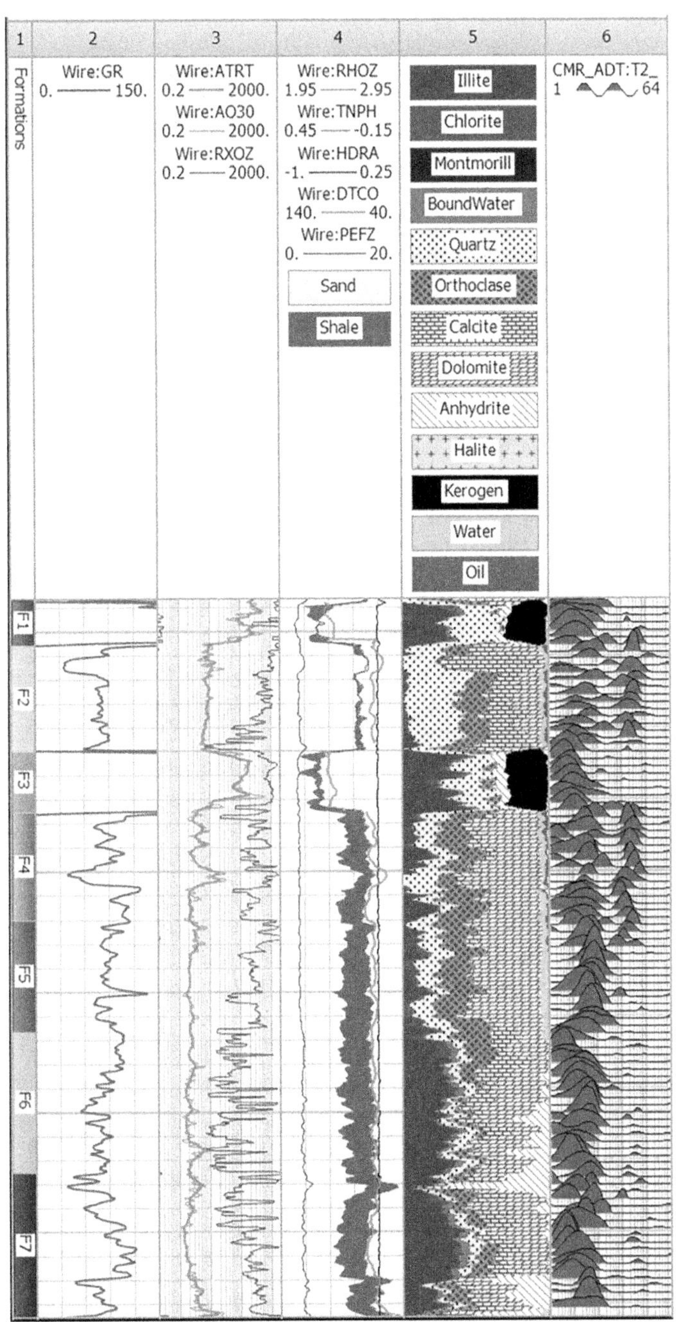

FIG. 7.1 Well logs acquired in the shale petroleum system, where GR is gamma ray, ATRT is 2-feet induction resistivity, AO30 is 1-feet induction resistivity, RXOZ is shallow induction resistivity, RHOZ is electronic density, TNPH is neutron porosity, HDRA is density standoff correction, DTCO is compressional sonic travel time, and PEFZ is photoelectric factor. Track 1 contains the

(See figure legend on next page)

3.2 Data preparation

Data preparation, or data preprocessing, affects the convergence time of neural networks during the training phase. Each input has a different range; for example, illite volume fraction ranges from 0.1 to 0.5 m^3/m^3, whereas water and oil volume fraction is below 0.1 m^3/m^3. Each input data is scaled to have zero mean and unit variance, which is also referred as the standard scaler. Such scaling is suitable when there are only a few outliers in the data. As the number of outliers increase, it is advisable to use robust scaling. Output/target data generally do not require either scaling or normalization. Outliers in targets were screened and removed by fitting NMR T2 distribution, discretized into 64 T2 bins, with a cubic spline and then removing those with three peaks and unusual characteristics when fitting the NMR T2. Depth shift was performed to ensure that the 10 inversion-derived logs and the corresponding NMR T2 distribution correspond to the same depth. This ensures that the deep neural networks are well trained to relate 10 features (inversion-derived logs) with 64 targets (NMR T2 distribution discretized into 64 T2 bins).

4 Neural network architectures for the NMR T2 synthesis

4.1 Introduction to NMR T2 synthesis

We implement four deep neural network architectures that can learn to process the seven mineral volume content logs, namely, illite, chlorite, quartz, calcite, dolomite, anhydrite, and kerogen, and three fluid saturation logs, namely, bound water, free water, and oil, to synthesize NMR T2-distribution responses along a well length. These 10 logs will be referred as the inversion-derived logs. The first half of the VAE-NN, GAN-NN, and VAEc-NN architectures learn to abstract and understand the concept of NMR T2 distribution. These three architectures first train deep neural networks to accurately reproduce the NMR T2 distribution in the training dataset by extracting certain representative features of the NMR T2 distribution; following that, a simple multilayer neural network is connected with the pretrained deep neural network to reliably synthesize the NMR T2 distribution from the 10 inversion-derived logs. Instead of individually predicting T2 amplitude for each of the 64 T2 bins of the NMR T2 distribution, we implement deep networks to simultaneously predict T2 amplitudes for the 64 T2 bins to generate the complete NMR T2 spectra, thereby providing greater constraint

FIG. 7.1, CONT'D formation tops or lithology indicators provided by the geology expert. Tracks 1, 2, 3, and 4 are the conventional logs on which inversion is performed to obtain the inversion-derived logs in Track 5. CMR_ADT in Track 6 is NMR T2 distribution response that is the target for the data-driven modeling. NMR T2 distribution response at a specific depth comprises 64 T2 amplitudes as a function of T2 times. Deep neural networks were trained to relate the 10 inversion-derived logs to the entire NMR T2 spectra, comprising 64 T2 amplitudes as a function of T2 times.

to the neural network-based synthesis that mitigates overfitting. Unlike the VAE-NN, GAN-NN, and VAEc-NN architectures, the LSTM architecture considers the NMR T2 synthesis problem as a transformation task (similar to many-to-many language translation), wherein certain subsamples of the 10 inversion-derived logs are used for synthesizing amplitudes for certain subsamples of the 64 T2 bins. LSTM architecture learns to relate the T2 spectra to sequential variations between various combinations of inversion-derived logs. All layers in the four neural networks are fully connected except the convolution and max-pooling layers in the VAEc and the recurrent layer in the LSTM. Fully connected layers connect every neuron in one layer to every neuron in the previous layer.

4.2 VAE-NN architecture, training, and testing

VAE-NN stands for variational autoencoder (VAE) assisted neural network. An autoencoder is a type of deep neural network that is trained to reproduce its high-dimensional input (in our case, 64-dimensional NMR T2) by implementing an encoder network followed by a decoder network [12]. A variational autoencoder (VAE) provides a probabilistic manner for describing an observation in latent space, such that the encoder describes a probability distribution for each latent attribute. On the encoder side, a neural network learns to project the high-dimensional input on to a low-dimensional latent space (in our case, two-dimensional space). Following that a decoder neural network learns to decode a vector in the low-dimensional latent space to reproduce the high-dimensional input. With this bottleneck structure an autoencoder learns to extract the most important information when the input goes through the latent layers. Therefore, an autoencoder is an effective way to project data from a high dimension to a lower dimension by extracting the most dominant features and characteristics. A variational autoencoder is a specific form of autoencoder, wherein the encoding network is constrained to generate latent vectors that roughly follow a unit Gaussian distribution [13]. In doing so, a trained decoder can be later used to independently synthesize data (similar to the training data) by using a latent vector sampled from a unit Gaussian distribution. More details about the latent layer are provided in the subsequent description of the VAEc architecture in Section 4.4. VAE arranges the learned features with similar shapes close to each other in the projected latent space, thereby reducing the loss in the reproduction of input.

As mentioned earlier, the synthesis of NMR T2 distributions using VAE-NN requires a two-stage training process prior to the testing and deploying the neural network (Fig. 7.2). In the first stage of training the VAE-NN, the VAE is trained to reconstruct the NMR T2 in the training dataset by extracting the dominant features of the NMR T2 distribution. Encoder network has two fully connected layers, 64-dimensional input layer followed

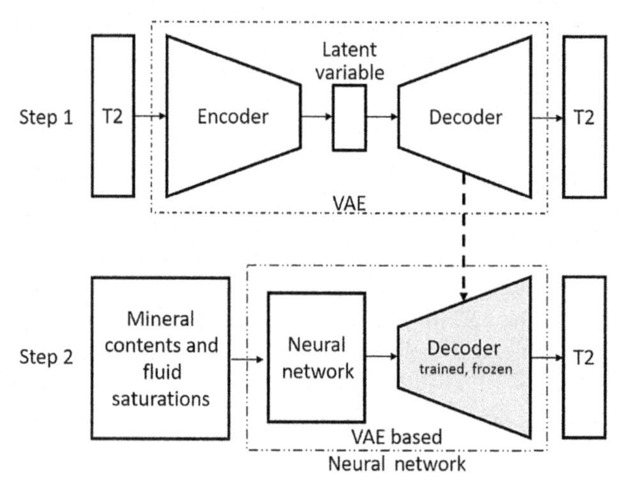

FIG. 7.2 Schematic for training the VAE-NN architecture. The number of hidden layers in each network and the number of neurons in each hidden layer are determined by hyperparameter optimization.

by a 16-dimensional hidden layer that project the 64-dimensional input to a 2-dimensional latent space. Decoder network has two fully connected layers, namely, 16-dimensional hidden layer followed by a 64-dimensional output layer that decode the projected vectors in the 2-dimensional latent space. When the shapes of T2 distributions in the training dataset are relatively simple and the size of the training dataset is limited, VAE should be designed to have a two-dimensional latent layer. An increase in the dimensionality of the latent layer without an increase in the complexity of T2 data will cause the VAE to extract unrepresentative features and to overfit the training dataset.

The primary objective of VAE training is that the encoder network of the VAE should learn to project the T2 data on to a two-dimensional latent space, such that the projected vectors in the latent space can then be processed by the decoder network to reproduce the input T2 data. VAE accomplishes this goal by minimizing a loss function, wherein the first component $||X - f(z)||^2$ represents the difference between the input T2 (denoted as X) and reproduced T2 (denoted as $f(z)$) and the second component of the loss function represents the KL divergence [14], expressed in Eq. (7.1), which forces the latent variable z to follow a unit Gaussian distribution:

$$KL[\mathcal{N}(\boldsymbol{\mu}(\boldsymbol{X}),\boldsymbol{\Sigma}(\boldsymbol{X}))\,||\mathcal{N}(0,I)] = \frac{1}{2}\left(tr(\boldsymbol{\Sigma}(\boldsymbol{X})) + (\boldsymbol{\mu}(\boldsymbol{X}))^2(\boldsymbol{\mu}(\boldsymbol{X})) - k - \log\det(\boldsymbol{\Sigma}(\boldsymbol{X}))\right) \tag{7.1}$$

where k is the dimensionality of the distribution and $\mathcal{N}(\boldsymbol{\mu}, \boldsymbol{\Sigma})$ represents Gaussian distribution with mean $\boldsymbol{\mu}$ and covariance $\boldsymbol{\Sigma}$.

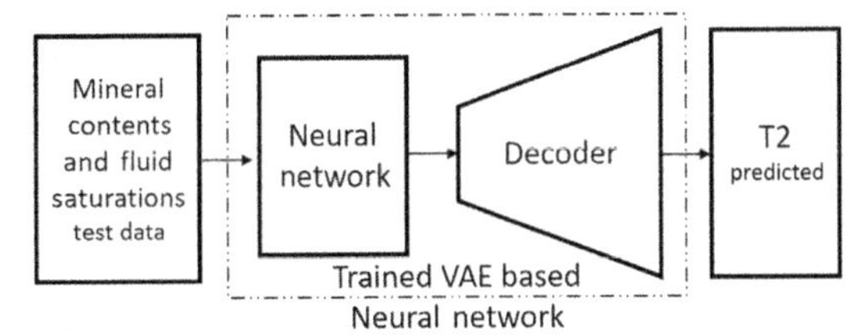

FIG. 7.3 Schematic for testing or deploying the VAE-NN model.

After the VAE is trained, in the second stage of training the VAE-NN, a four-layered fully connected neural network followed by the frozen pretrained decoder learns to relate the three formation fluid saturation logs and seven mineral content logs with the NMR T2 distribution. For the second stage of training, the trained decoder (the second half of the VAE described in the previous paragraph) is frozen, and a four-layered neural network is connected before the frozen decoder. Only the four-layered neural network undergoes weight updates in the second stage of training. The four-layered neural network comprises 10-dimensional input layer, two 30-dimensional hidden layers, and one final 6-dimensional hidden layer attached to the frozen decoder. In doing so the four-layered neural network learns to transform the 10 inversion-derived logs into dominant NMR T2 features extracted by the encoder network in the first stage of the training process. Overall the two-stage training process ensures that the VAE-NN will generate similar NMR T2 distributions for formations with similar fluid saturations and mineral contents. After the training process is complete, for purposes of testing and deployment, the trained four-layered neural network followed by the frozen decoder synthesizes T2 distributions of subsurface formations by processing the seven formation mineral content logs and three fluid saturation logs (Fig. 7.3).

4.3 GAN-NN architecture, training, and testing

GAN-NN stands for the generative adversarial network (GAN) assisted neural network. Like the VAE-NN, the GAN-NN also undergoes two-stage training, such that the first stage focuses on training the GAN and the second stage focuses on training a three-layered neural network followed by the frozen pretrained generator. GANs have been successfully applied to image generation [15] and text to image synthesis [16]. For our study, GAN is a type of deep neural network that learns to generate 64-dimensional NMR T2 distribution by having competition between a generator network (G) and a discriminator network (D). The generator network G learns to upscale (transform) random noise to generate synthetic T2 distribution that is very similar to the original 64-dimensional T2 distribution, whereas the discriminator network D learns to correctly distinguish between the synthetic and the original T2 distributions. T2 distributions generated by the generator is fed into the discriminator along with samples taken from the actual, ground-truth NMR T2 data. The discriminator takes in both real and synthetic T2 and returns probabilities, a number between 0 and 1, where 1 denotes authenticity of the T2 and 0 denotes fake/synthetic T2 produced by the generator network. Generator learns to generate T2 data that get labeled as 1 by the discriminator, whereas as the discriminator learns to label the T2 data generated by the generator as 0.

In the first stage of training the GAN-NN, the GAN is trained to synthesize the NMR T2 similar to those in the training dataset. Generator network G has two fully connected layers, 6-dimensional input layer followed by two 64-dimensional hidden layers that upscale (transform) the 6-dimensional noise input to 64-dimensional synthetic NMR T2. Discriminator network D has four fully connected layers, namely, 64-dimensional input layer, 64-dimensional and 16-dimensional hidden layers, and finally a 2-dimensional output layer for classifying each input data fed into the discriminator as either original or synthetic T2. The primary objective of GAN training is that the generator network of the GAN should learn to synthesize realistic, physically consistent 64-dimensional T2 data. To achieve the desired reconstruction [15], GAN learns to maximize the objective function $V(D,G)$ of the two competing networks represented as

$$\min_{G}\max_{D} V(D,G)=\mathbb{E}_{x\sim p_{\text{data}}}[\log D(x)]+\mathbb{E}_{z\sim p_z}[\log(1-D(G(z)))] \tag{7.2}$$

where x represents the entire real training data, z represents a random vector, $G(z)$ represents synthetic data generated by generator G from a random vector, $D(x)$ represents the binary classes predicted by the discriminator D for real data x, and $D(G(z))$ represents the binary classes predicted by the discriminator D for noise z. The objective function is composed of two parts, which represent the expected ($\mathbb{E}$) performances of discriminator when given real data and fake data generated from random vector, respectively. The generator and discriminator are alternatively trained to compete. Individually, both networks are trying to optimize a different and opposing objective function, or loss function, in a zero-sum game. GAN training ends when the generator can synthesize NMR T2 that fools the discriminator to label the new synthetic data as real data.

The training process of the GAN-NN is similar to the training process of the VAE-NN involving a two-stage process. In the first stage of training (Fig. 7.4), the GAN learns the dominant features in T2 distributions and to synthesize realistic T2. After the GAN is trained, the frozen/pretrained generator is connected to a three-layered fully connected neural network (Stage 2 in Fig. 7.4) to learn to associate the 10 mineral content logs and 3 fluid saturation logs with the NMR T2 distribution. For the second stage of training the GAN-NN, the trained generator (the first half of the GAN described in the previous paragraph) is frozen, and a three-layered neural

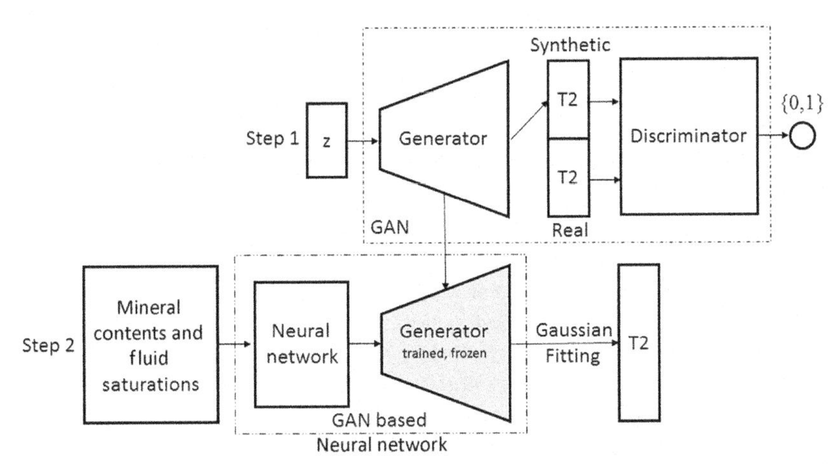

FIG. 7.4 Schematic for training the GAN-NN architecture. The number of hidden layers in each network and the number of neurons in each hidden layer are determined by hyperparameter optimization.

network is connected before the frozen generator. Only the three-layered neural network undergoes weight updates in the second stage of training. The three-layered neural network comprises 10-dimensional input layer, one 8-dimensional hidden layer, and one final 2-dimensional hidden layer attached to the frozen generator. In doing so the three-layered neural network learns to transform the 10 inversion-derived logs into dominant NMR T2 features extracted by the generator network in the first stage of the training process. Overall, the two-stage training process ensures that the GAN-NN will generate similar NMR T2 distributions for formations with similar fluid saturations and mineral contents. After the training process is complete, for purposes of testing and deployment, the trained three-layered neural network followed by the frozen generator synthesizes the T2 distributions of subsurface formations by processing the seven formation mineral content logs and three fluid saturation logs (similar to Fig. 7.3). Unlike the VAE-NN, the T2 distributions generated by GAN-NN are not smooth. Therefore, we add a Gaussian fitting process to smoothen the synthetic NMR T2 generated by the GAN-NN.

4.4 VAEc-NN architecture, training, and testing

VAEc-NN stands for variational autoencoder with convolution layer (VAEc) assisted neural network. As explained in Section 4.2, an autoencoder is a type of deep neural network that is trained to reproduce its high-dimensional input (in our case, 64-dimensional NMR T2) by implementing an encoder network followed by a decoder network [12]. VAEc-NN implements a convolutional layer in the encoder network of the VAEc to better extract spatial features of the NMR T2 distribution, which is processed as a 64-dimensional 1D vector. The training schematic for the VAEc-NN architecture is presented in Fig. 7.5.

In the first stage of training the VAEc-NN (Fig. 7.5), the VAEc learns to reproduce the measured NMR T2 distributions. The encoder network of the VAEc projects the 64-dimensional T2 data to 3-dimensional latent space. Subsequently the decoder upscales the latent vectors sampled from the latent space to reconstruct the measured T2 data, which was fed to the encoder network. VAEc comprises 8 layers, namely, one 64-dimensional input layer, one 64×16 convolutional layer, one 16×16 max-pooling layer, 16-dimensional fully connected hidden layer, one 3-dimensional probabilistic latent layer, and finally the 3-layered decoder network comprising fully connected 3-dimensional and 16-dimensional hidden layers and 64-dimensional output layer.

The original NMR T2 are first fed into the 64-dimensional input layer of the encoder. The 64×16 convolutional layer is then generated by filtering the 64-dimensional input layer using 16 filters (kernels) of size 3×1. VAEc training aims at finding the mathematical forms of these 16 filters most suited for extracting features from the NMR T2 that can later be used for the NMR T2 synthesis. Unlike a fully connected layer that learns features from all the combinations of the features of the previous layer, a convolutional layer relies on local spatial coherence within a small receptive field, as

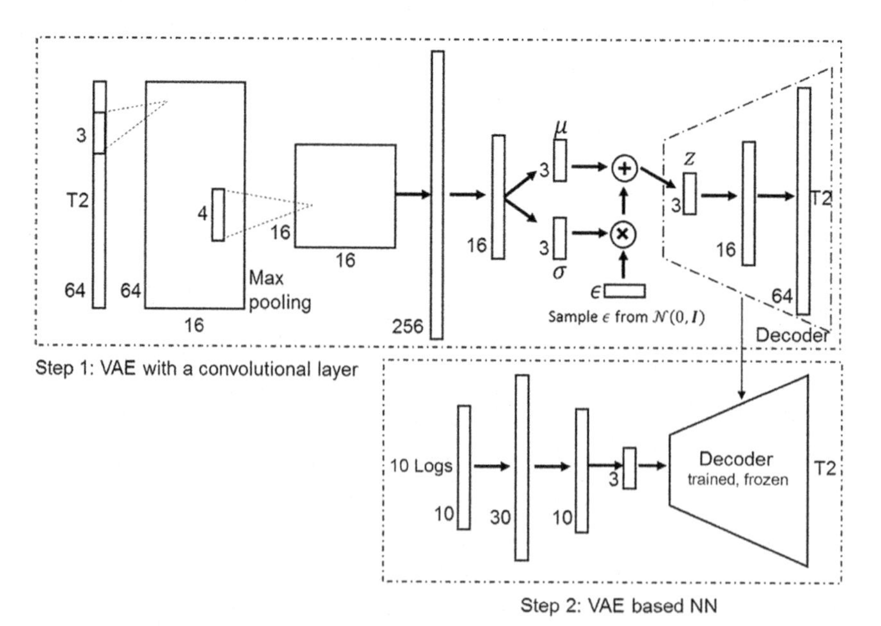

FIG. 7.5 Schematic for training the VAEc-NN architecture. The number of hidden layers in each network and the number of neurons in each hidden layer are determined by hyperparameter optimization.

defined by the filters. Convolution is beneficial when the data (such as audio and image) have local structures such that spatially proximal features exhibit strong correlations. For images and audio signals, multiple convolutional layers are used in series to extract features at multiple scales, such that with the addition of convolutional layer, the entire architecture adapts to the higher-level features.

After the convolution the max-pooling operation processes the 64×16 convolutional layer with a filter size of 4×1 to obtain a 16×16 layer. Filters for max pooling select maximum values among the four spatially neighboring features subsampled from the convolutional layer. Consequently, max pooling neglects a large portion of the data in the convolution layer and retains only one-fourth of data, especially the large-valued features. Max pooling reduces overfitting, reduces computational time, extracts rotation- and position-invariant features, and improves the generalizability of the lower-dimensional output features. While the convolution operation helps in obtaining the features maps, the pooling operations play an important role in reducing the dimensionality of the convolution-derived features. Sometimes, average pooling is used as an alternative to the max pooling.

The output of the max-pooling layer is then flattened and fed to a fully connected 16-dimensional layer. The fifth layer of the VAEc is a 3-dimensional latent layer, which samples from the output of the previous 16-dimensional layer to generate mean and variance vectors. The 64-dimensional NMR T2 data are compressed to 3 dimensions (3 means and 3 variances) through these 5 layers. Rather than directly outputting single discrete values for the latent attributes, the encoder model of a VAEc will output mean and variance, which serve as parameters describing a distribution for each dimension in the latent space. The three-dimensional latent layer contains the encoded Gaussian representations of the input, such that each dimension represents a learned attribute about the data. As shown in the Fig. 7.5, the latent layer is not merely a collection of single discrete values for each dimension (latent attributes). Instead, each latent attribute is represented as a range of possible values (probability distribution) by learning mean and variance for each latent attribute. Following that a random sample is selected from the probability distribution encoded into the latent space and is then fed into the decoder network for the desired NMR T2 synthesis. The mean and variance vectors in the latent layer and the random sampling of the latent vector from the latent space enforce a continuous, smooth latent space representation and target reconstruction. Similar latent layer is used for the VAE architecture discussed in Section 4.2.

After the encoding, the first three-dimensional layer of the decoder generates a latent vector by randomly sampling from the probability distribution encoded into the latent space. Random sampling process leverages "reparameterization trick" that samples from a unit Gaussian and then shifts the random sample by the mean vector in the latent layer and scales it by the variance vector in the latent layer. The first 3-dimensional layer of the decoder is followed by 16-dimensional hidden layer and the 64-dimensional output layer that upscale the latent vector to develop a reconstruction of the original input. For any sampling of the latent distributions, the decoder model learns to accurately reconstruct the input, such that the samples that are nearby in the latent space result in very similar reconstructions. Similar to the VAE, the loss function of the VAEc model has two components: reconstruction loss and the Kullback-Leibler (KL) divergence loss. The reconstruction loss emphasizes the difference between the original and synthesized NMR T2 distribution, and the KL divergence loss evaluates the difference between the distribution of the latent vector and a unit Gaussian distribution.

In the second stage of training the VAEc-NN (Fig. 7.5), a four-layered NN model is connected to the frozen pretrained three-layered decoder. The decoder trained in the first stage of VAEc-NN training is frozen to preserve the generalization of the T2 distributions. Only the four-layered neural network undergoes weight updates in the second stage of training. The 4-layered neural network comprises one 10-dimensional input layer, one 30-dimensional hidden layer, one 10-dimensional hidden layer, and one final 3-dimensional hidden layer attached to the frozen decoder network. In doing so, the 4-layered neural network processes the 7 mineral composition and 3 fluid saturation logs to produce a 3-dimensional vector, which can be decoded by the trained decoder to generate the 64-dimensional T2 distribution. After the training process is complete, for purposes of testing and deployment, the trained four-layered neural network followed by the frozen three-layered decoder synthesizes the T2 distributions of subsurface formations by processing the seven formation mineral content logs and three fluid saturation logs (similar to Fig 7.3).

4.5 LSTM architecture, training, and testing

Long short-term memory (LSTM) network is a type of recurrent neural network (RNN) that operates over sequences of input/output vectors. Unlike other neural networks that are stateless, LSTM and RNN use the internal state at one timestep as one of the inputs for calculating the internal state and output at the next timestep. The recurrence of internal state serves as a form of memory that allows contextual information to flow through the network so that relevant outputs or hidden states from previous timesteps can be used by the network at the current timestep. In our implementation the LSTM network processes the sequence of 10 inversion-derived logs to encode an intermediate vector that can be used to generate the sequence of 64 discrete values that constitute the 64-dimensional NMR T2 distribution (Fig. 7.6).

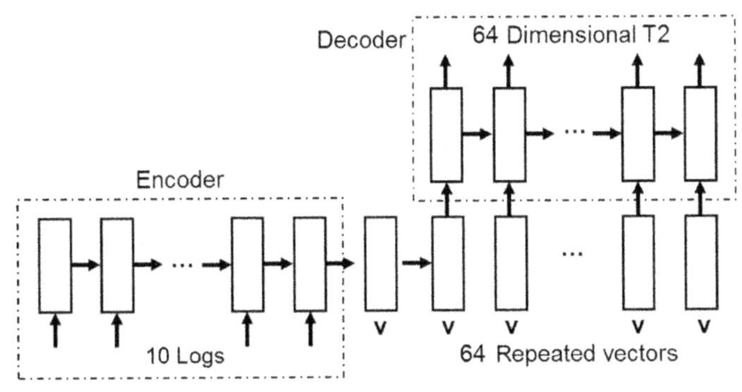

FIG. 7.6 Schematic for training the LSTM architecture. The number of hidden layers in each network and the number of neurons in each hidden layer are determined by hyperparameter optimization.

In this study, we use the LSTM network for sequence-to-sequence modeling (i.e., many-to-many mapping), wherein for each sequence of one feature vector, the LSTM network learns an intermediate vector that can be decoded to generate a distinct sequence of one target vector. LSTM first encodes the relationships between the various combinations of inversion-derived logs and the amplitudes of various combinations of T2 bins into an intermediate vector. Consequently the use of LSTM frees us from knowing any of the mechanistic rules that govern the multitude of physical relationships between the various logs and those between the logs and physical properties of the formation. LSTM-based sequence-to-sequence model generally contains three components, namely, encoder, intermediate vector, and decoder. The encoder is tasked with learning to generate a single embedding (intermediate vector) that effectively summarizes the input sequence, and the decoder is tasked with learning to generate the output sequence from that single embedding.

Unlike the three other neural network architectures discussed in the previous sections, LSTM training requires only one stage. LSTM excels other deep neural network architectures for data that have long-term dependencies and unfixed lengths of input/target vectors. LSTM sequence-to-sequence modeling has been successful in language translation. Language translation relies on the fact that sentences in different languages are distinct representations of one common context/theme. In a similar manner, different subsurface logs acquired at a specific depth are distinct responses to one common geomaterial having specific physical properties. We implement LSTM to capture the dependencies among the various T2 bins in the T2 distribution and also to capture the dependencies between various combinations of inversion-derived logs and the amplitudes for various combinations of T2 bins.

For each depth in the training dataset, the encoder network receives, one at a time, each element (i.e., one log) of the input sequence of 10 inversion-derived logs. Encoder updates the intermediate vector based on each element and propagates the updated intermediate vector for further updates based on the subsequent elements of the input sequence. Intermediate vector (also referred as encoder vector or context vector) is the final hidden state produced by encoder network. The intermediate vector contains information about the input sequence in an encoded format. Decoder learns to process the intermediate vector to compute an internal state for generating the first element of the output sequence constituting the 64-dimensional NMR T2. For generating each of the subsequent elements in the output sequence, the decoder learns to compute the corresponding internal states by processing the intermediate vector along with the internal state calculated when generating the previous element in the output sequence.

The encoder and decoder networks of the LSTM network architecture are collections of LSTM modules. The encoder and decoder networks comprise 10 and 64 chained LSTM modules, respectively. Each module has gates controlling the flow of data. Controlled by various gates the LSTM can choose to forget or update the information flowing through the modules. The encoder compresses the seven formation mineral logs and three fluid saturation logs into a single intermediate vector. Then the decoder sequentially decodes the intermediate vector to generate the 64 elements of the target NMR T2 sequence. The 10 inversion-derived logs are taken as sequence, and the encoder processes one of the 10 logs at each timestep to generate an internal state. After processing all the 10 input inversion-derived logs, the encoder generates the 15-dimensional intermediate vector **v** in the last step. The intermediate vector **v** is fed to each module in the decoder. The decoder modules sequentially generate each element of the output NMR T2 sequence. Each decoder module processes the 15-dimensional intermediate vector **v** along with the internal state from the previous module to construct a corresponding element of the output sequence. The full synthesis of NMR T2 for a single depth requires 64 timesteps. The loss function used for the LSTM model is mean squared error function, like the second training steps of the previous three neural network architectures. The optimizer used to train the LSTM model is RMSprop that updates the weights of neurons based on the loss function during the backpropagation.

4.6 Training and testing the four deep neural network models

NMR T2 distribution log and inversion-derived mineral content and fluid saturation logs are split randomly into testing and training datasets. Data from 460 depths were used as the training data, and other data from 100 depths were used as the testing data. In a more realistic application, the dataset should be of larger size for robust development of the deep neural networks. The four models were trained/tested on a desktop computer with 3.5 GHz CPU and 32 GB RAM. With the increase in number of training steps, the model performances on both the training and testing dataset improve, that is, the prediction accuracies for both dataset increase. When training a deep neural network model, the model performances on both the training and testing dataset need to be monitored to identify the optimum training steps. When the number of training steps is more than the optimal value, the performance on training loss progressively improves, but the performance on the testing dataset first stabilizes and then starts reducing. This indicates that the training of the model beyond the optimal training steps results in overfitting the training dataset and reduction in the model generalization capability. The training steps are selected by monitoring the mean squared error (MSE) of the learning on both training and testing datasets. In the initial training stage, MSEs of the model on the training and testing dataset decrease as the model learns from the training dataset. After a certain number of steps of training, the testing MSE starts increasing; this is the point where the model starts overfitting the training datasets. To avoid overfitting, we invoke an early stopping criterion based on the performances on the training and testing dataset.

For purposes of evaluating the model performances, we use the coefficient of determination, R^2, for each depth or averaged over all the depths of the testing dataset. R^2 as a performance metric has limitations and assumptions that need to considered prior to the implementation. R^2 for a specific depth is calculated using the predicted NMR T2 and measured NMR T2 distributions, such that

$$R^2(y,\hat{y}) = 1 - \frac{\sum_{i=1}^{n}(y_i - \hat{y}_i)^2}{\sum_{i=1}^{n}(y_i - \overline{y})^2} \tag{7.3}$$

where $\hat{y}_i$ and y_i are predicted and measured NMR T2 amplitude for bin i, $\overline{y}$ is mean of measured NMR T2 amplitude for bin i, and n is the total number of bins in the T2 distribution (i.e., 64).

Each model needs to compute a different number of parameters based on the model architecture. For example, 4608 and 5234 parameters are computed when training the generator and discriminator networks, respectively, of the GAN. Each model requires a different number of training steps and has different training times depending on the model architecture (see Table 7.1). For example, it takes 41.25 s for the two-stage training of VAEc-NN to train 1000 times on the training data. LSTM requires more time to train than the VAEc-NN. It took 566.33 s to train the LSTM model with 1000 training steps. VAEc-NN model training computes 1825 parameters, whereas the LSTM model training computes 2071 parameters. Even though the two models have a similar number of parameters and LSTM has one training stage as compared with the VAEc-NN, the large difference in the training times of LSTM and VAEc-NN is due to the complexity of calculations in

TABLE 7.1 Summary of few parameters of the deep neural networks.

	GAN-NN[a]	VAEc-NN	LSTM
Number of parameters	9842 (for GAN)	1825	2071
Training time	–	41.25 s/1000 steps	566.33 s/1000 steps

[a]*Different training schedule is used in GAN-NN. Two models were trained alternatively.*

the LSTM network, which relies on the recurrence of the internal state. The LSTM model takes one of the 10 inversion-derived logs at each timestep to generate the final form of the intermediate vector. Following that, when predicting the NMR T2 distribution, the LSTM model predicts one element of the 64-dimensional NMR T2 sequence at each timestep, which requires 64 updates of all the decoder parameters for the sequential generation of the entire NMR T2.

5 Application of the VAE-NN model

In the first stage of training the VAE-NN, VAE is trained to memorize and generalize the sequential and shape-related features in the T2 distribution. The sparsely distributed 64-dimensional T2 distribution is projected to a 2-dimensional latent. VAE learns the manifold of T2 in the latent space by minimizing the loss functions. There are infinite points in a 2-dimensional latent space on which infinite projections of 64-dimensional T2 can be generated. One hundred samples from the two-dimensional latent space are shown in Fig. 7.7. In Fig. 7.7, the learned T2 distribution changes gradually in the 2D latent space with slight successive variations in the number of peaks, smoothness of the distribution, the positions of the peaks, and heights and widths of the peaks. VAE learning process ensures that the predictive model will have robust predictions in the presence of noise in input signals. Once the VAE is trained to memorize and generalize the dominant features in the T2 distributions, any new input of T2 distribution is projected to a corresponding location in the 2D latent space where there already exists a projection of a similar T2 distribution, which was fed during the prior training phase. The decoder learns to randomly sample from this latent space to synthesize realistic NMR T2 distributions.

The trained VAE-NN was applied on the 7 formation mineral content logs and 3 formation fluid saturation logs from 100 testing depths to synthesize the NMR T2. VAE-NN-based synthesis of NMR T2 shows good agreement with the original T2 distributions (Fig. 7.8). Histogram of the coefficient of

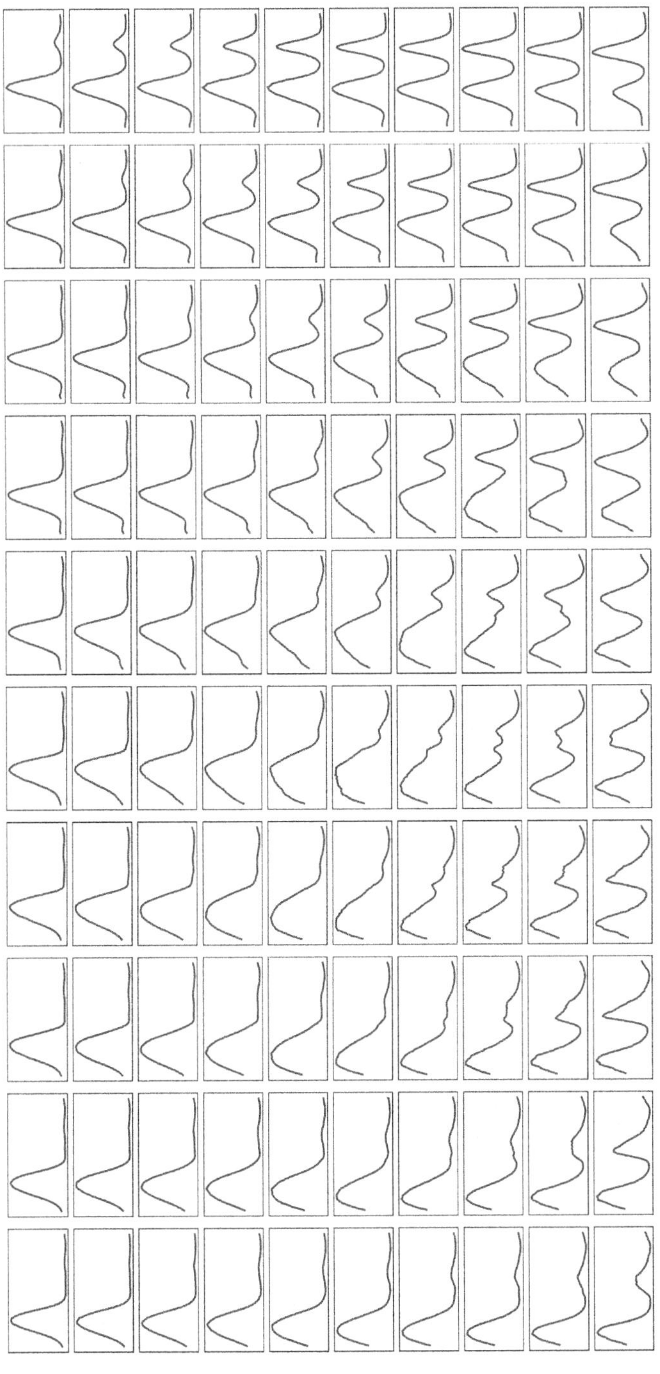

FIG. 7.7 Manifold of T2 in the two-dimensional latent space as learnt by VAE from the 460 depths in the training dataset. x-axis represents T2 relaxation time from 0.3 to 3000 ms; y-axis is normalized NMR signal amplitude ranging from 0 to 0.25 (unitless).

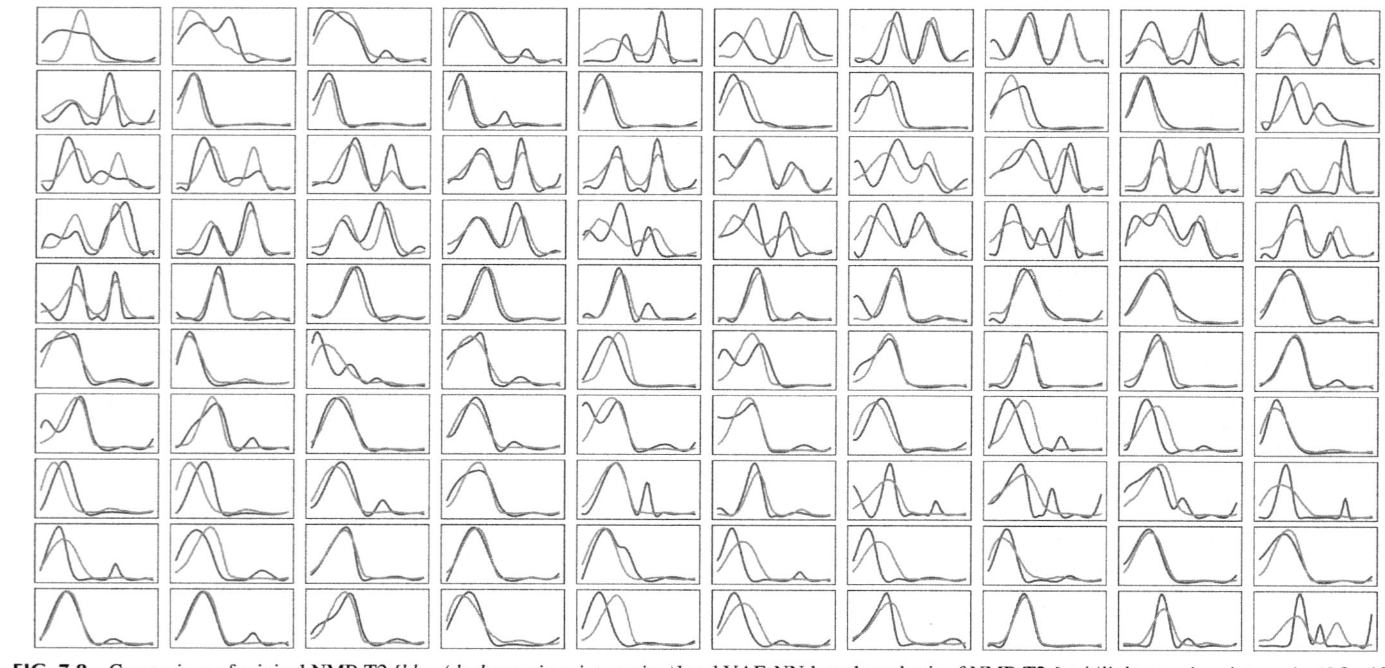

FIG. 7.8 Comparison of original NMR T2 [*blue (dark gray in print version)*] and VAE-NN-based synthesis of NMR T2 [*red (light gray in print version)*] for 100 depths from the testing dataset. *x*-axis represents the T2 relaxation time from 0.3 to 3000 ms; *y*-axis is normalized NMR signal amplitude ranging from 0 to 0.25 (unitless).

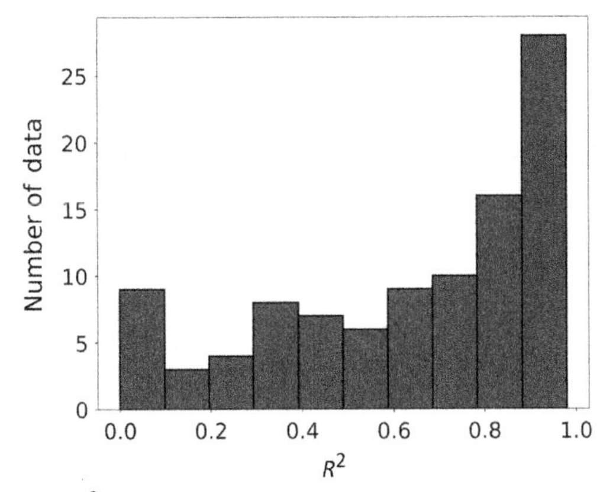

FIG. 7.9 Histogram of R^2 for the T2 distributions synthesized for the 100 discrete depths of the testing dataset when using the VAE-NN model having a two-dimensional latent layer.

determination, R^2, is suitable for evaluating the performance of VAE-NN-based synthesis of NMR T2. Another metric to evaluate the T2-synthesis performance is the normalized root-mean-square deviation (NRMSD), which is the percentage of the residual variance to the range of the NMR T2. Lower R^2 and higher NRMSD indicate poorer performance. During the testing phase, VAE-NN performs at R^2 of 0.8 and NRMSD of 14% when synthesizing NMR T2 with single peak. For NMR T2 with two peaks, the synthesis performance is poorer than the single-peak cases. Overall, VAE-NN performs at R^2 of 0.75 (Fig. 7.9) and NRMSD of 15% when using VAE with two-dimensional latent space. VAE-NN performance is better than R^2 of 0.7 for 50 out of the 100 depths. VAE-NN had a limited exposure to NMR T2 distributions with two peaks during the training phase leading to the poorer performance for T2 distributions with two peaks. Less than 1/3 of the T2 data in the training dataset have two peaks. The performance of a trained deep neural network relies on the quality and quantity of the training dataset. Wang et al. indicate that imbalanced data exist widely in real word that lowers the quality of training a deep neural network [17]. This is especially true in well log dataset when the reservoir/formation exhibits a dominant petrophysical feature. In our dataset, there is an imbalance of T2 distributions with one peak versus those with two peaks that adversely affects the NMR T2 synthesis.

6 Application of the GAN-NN model

GAN-NN training is a two-stage process similar to VAE-NN training. The generator is a three-layer neural network that takes six-dimensional noise vector as input. The generator learns to generate NMR T2 with different shapes with the aid of discriminator. Fig. 7.10 presents the T2 distribution

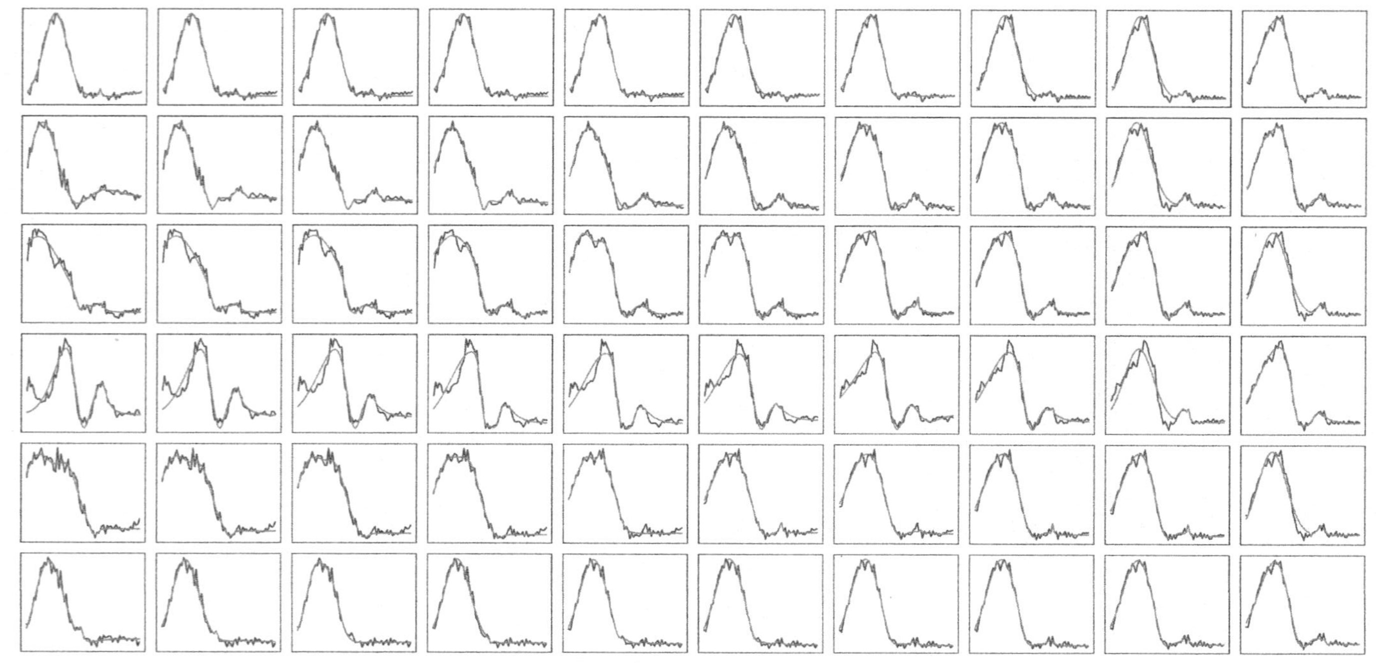

FIG. 7.10 Comparison of GAN-based synthesis of NMR T2 *[blue (dark gray in print version)]* and corresponding Gaussian smoothening *[red (light gray in print version)]* for 100 depths from the training dataset. x-axis represents the T2 relaxation time from 0.3 to 3000 ms; y-axis is normalized NMR signal amplitude ranging from 0 to 0.25 (unitless).

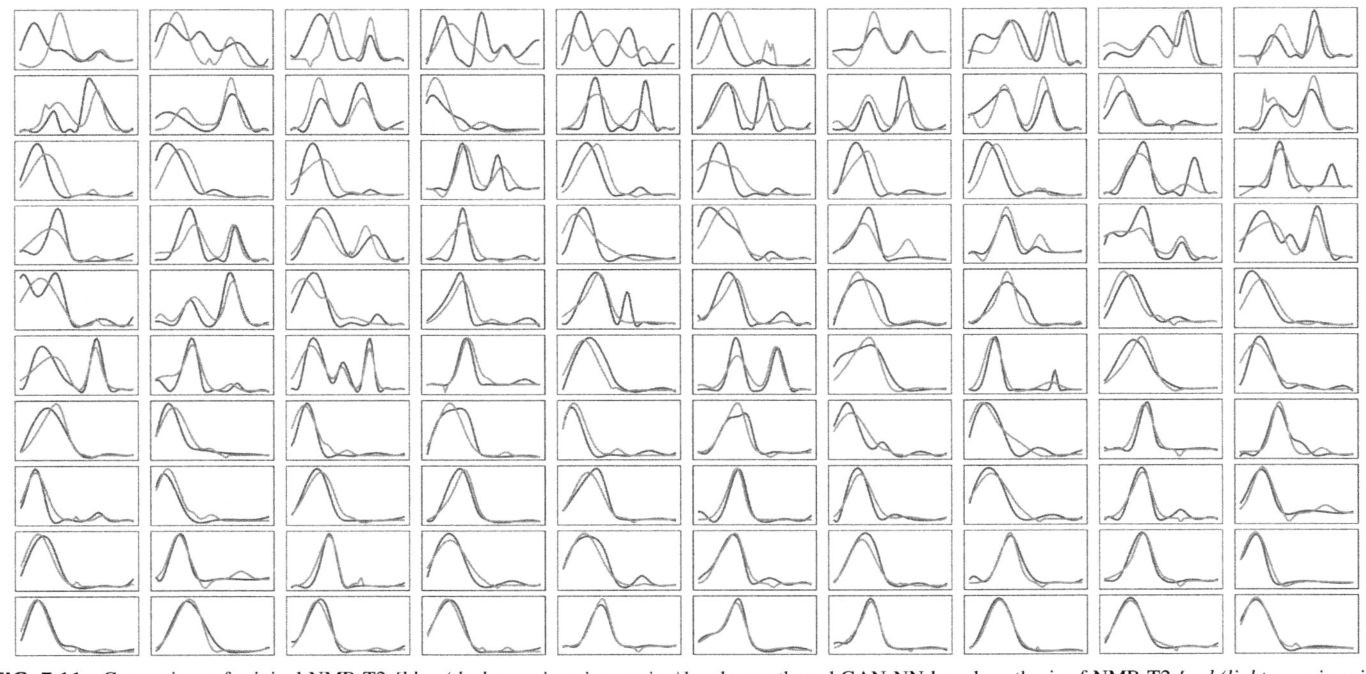

FIG. 7.11 Comparison of original NMR T2 *[blue (dark gray in print version)]* and smoothened GAN-NN-based synthesis of NMR T2 *[red (light gray in print version)]* for 100 depths from the testing dataset. x-axis represents the T2 relaxation time from 0.3 to 3000 ms; y-axis is normalized NMR signal amplitude ranging from 0 to 0.25 (unitless).

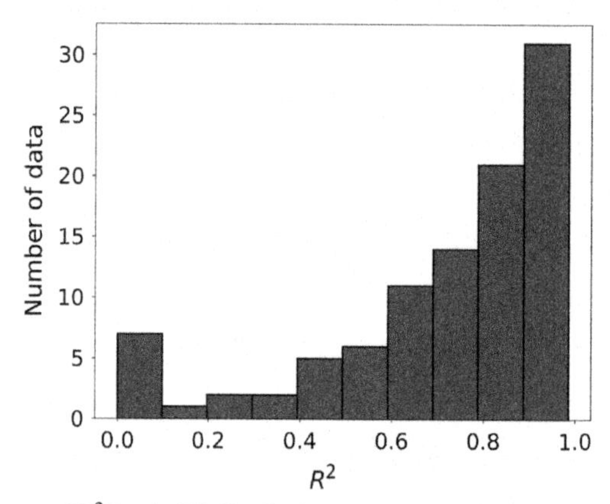

FIG. 7.12 Histogram of R^2 for the T2 distributions synthesized for the 100 discrete depths of the testing dataset when using the GAN-NN model.

synthesized by the GAN in the first stage of training the GAN-NN. As a consequence of limited training data, synthetic T2 is not as smooth as the measured T2. Gaussian fitting is performed on synthetic T2 to make them more realistic (Fig. 7.10). After the first stage of training, a third neural network is built with 4 layers having 2 hidden layers containing 30 neurons each. The third NN has 10- and 6-dimensional input and output layers, respectively. The third NN is trained together with the frozen generator from the trained GAN (Fig. 7.4), similar to the second training stage of VAE-NN (Fig. 7.3).

The training and testing data used in this section are the same as those for VAE-NN. The smoothened T2 distributions predicted by the GAN-NN are plotted against the measured distributions (Fig. 7.11). Like VAE-NN, GAN-NN has good prediction accuracy for T2 distributions with single peaks. R^2 and NRMSD of GAN-NN-based synthesis for the 100 testing data are 0.8 and 14%, respectively. GAN-NN prediction performance is slightly better than that of VAE-NN. Fig. 7.12 shows that the histogram of R^2 for GAN-NN-based synthesis is centered around 0.9 with a few below 0.6. Moreover, 65 out of the 100 testing depths have R^2 higher than 0.7.

7 Application of the VAEc-NN model

In the first stage of training, VAEc generalizes the NMR T2 in the training data set, and the decoder generates a typical NMR T2 by randomly sampling from the three-dimensional latent vector. Fig. 7.13 is a 2D sample from the three-dimensional latent space generated by the encoder. Fig. 7.13 is a slice plane from the 3D latent space. It demonstrates how the latent vectors are related to the NMR T2 distributions. The VAEc learns from the training dataset and generalizes the typical NMR T2 shape in the latent space. The decoder can

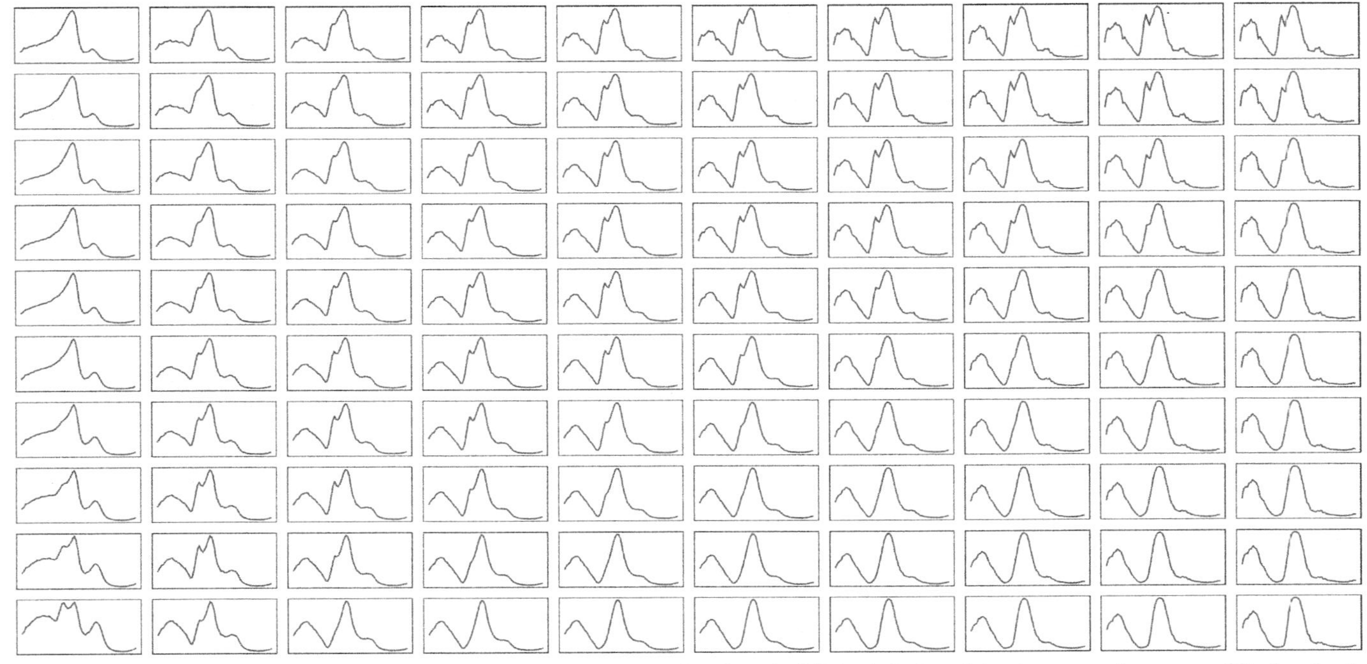

FIG. 7.13 Manifold of T2 in the three-dimensional latent space as learned by VAEc from the 460 depths in the training dataset. x-axis represents T2 relaxation time from 0.3 to 3000 ms; y-axis is normalized NMR signal amplitude ranging from 0 to 0.25 (unitless).

decode a latent vector into an NMR T2 distribution. By sampling from the latent space, we can visualize VAEc's understanding (also referred as abstraction) of a typical NMR T2 distribution. In Fig. 7.13, the learned T2 distribution changes gradually in the 3D latent space with slight successive variations in the number of peaks, smoothness of the distribution, the positions of the peaks, and heights and widths of the peaks. The VAEc learning ensures that the gradual changing characteristic of the learned NMR T2 (as plotted in Fig. 7.13) is in accordance with the characteristic of NMR T2 in the formation. NMR T2 in adjacent formations should be similar to each other with gradual changing characteristics. After the first stage of training the VAEc-NN that focuses on the training VAEc, the second stage trains the four-layered neural network that learns to relate the learned features in NMR data to the 10 inversion-derived logs.

The trained VAEc-NN is tested on 100 depths of the testing dataset. The testing performance of the VAEc-NN is shown in Fig. 7.14. The average R^2 of NMR T2 synthesis for the 100 testing depths is 0.75. In Fig. 7.14, the randomly selected NMR T2 testing data show a variety of shapes. About 70% of 100 samples are single-peak NMR T2, and 30% are double-peak. For single-peak NMR T2, the predicted NMR T2 and true NMR T2 almost overlay with each other. For double-peak NMR T2, the accuracy is lower than single-peak data. Although the VAEc-NN model can predict the peak position of double-peak NMR T2 with high accuracy, the model has low prediction accuracy for the height and shape of double peaks.

Histogram of R^2 of NMR synthesis for the 100 testing depths is shown in Fig. 7.15. R^2 achieved for 60 out of the 100 depths is above 0.7. In the 300-ft thick shale formation studied in this chapter, close to one-third of the depths have two-peaked T2 distribution. Variation in the shape of T2 distribution indicates changes in formation characteristics and changes in the relationships between T2 and formation mineral content and fluid saturation. In the absence of large volume of suitable training data, the deep neural networks cannot accurately determine the statistical relationships between T2 distributions and the inversion-derived logs.

8 Application of the LSTM network model

LSTM network is trained and tested with the same dataset as the one used for the VAEc-NN model, described in the previous section. The average accuracy of the LSTM-based synthesis of NMR T2 in terms of R^2 is 0.78. LSTM-based synthesis of T2 distributions are shown in Fig. 7.16. Similar to other models, LSTM better synthesizes the single-peaked T2 distribution as compared with the double-peaked T2. Histogram of R^2 in Fig. 7.17 indicates that LSTM synthesizes at $R^2 > 0.9$ for 33 out of the 100 depths, which higher than the performance of VAEc-NN model (Fig. 7.15). LSTM model performance for single-peaked T2 is better than that of the VAEc-NN model, whereas an

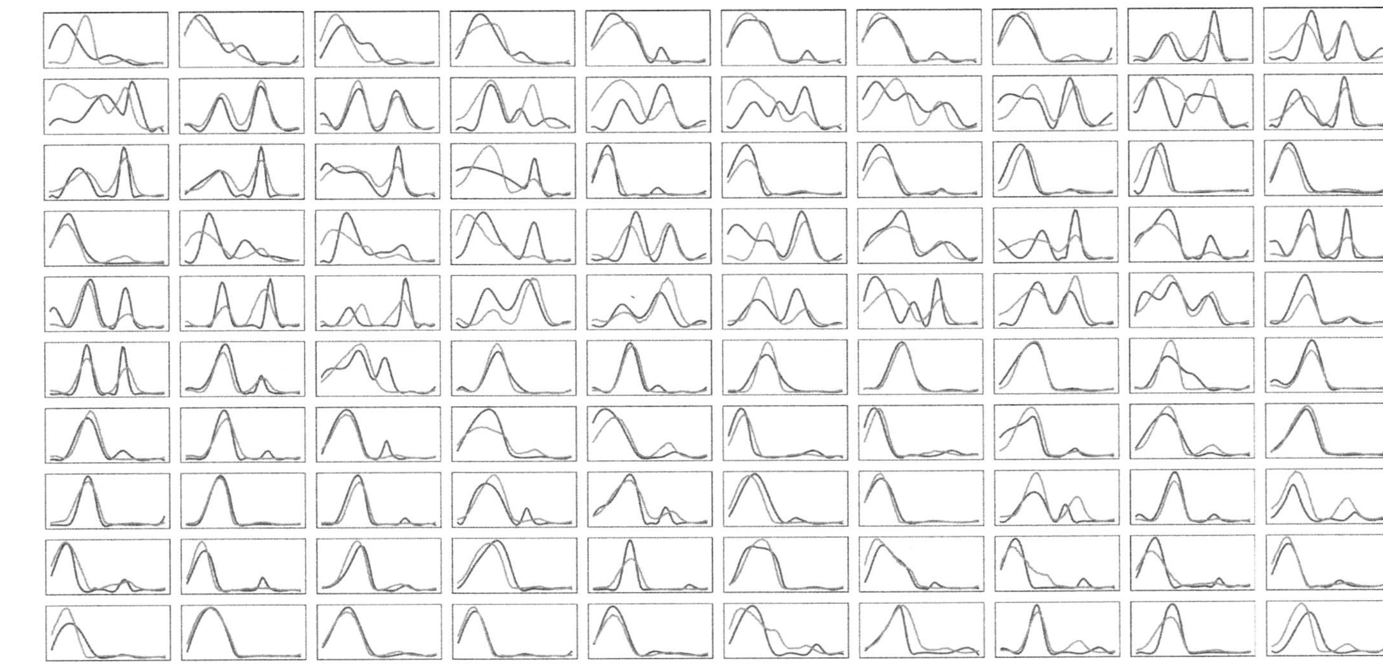

FIG. 7.14 Comparison of original NMR T2 *[blue (dark gray in print version)]* and VAEc-NN-based synthesis of NMR T2 *[red (light gray in print version)]* for 100 depths from the testing dataset. *x*-axis represents the T2 relaxation time from 0.3 to 3000 ms; *y*-axis is normalized NMR signal amplitude ranging from 0 to 0.25 (unitless).

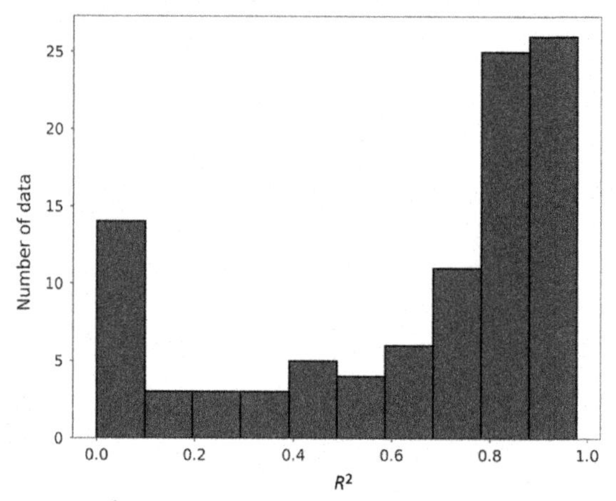

FIG. 7.15 Histogram of R^2 for the T2 distributions synthesized for the 100 discrete depths of the testing dataset when using the VAEc-NN model having a three-dimensional latent layer.

opposite trend is observed for double-peaked T2. Notably, several LSTM syntheses of single-peaked NMR T2 do not demonstrate the typical T2 shape; for example, the fifth T2 distribution in the last row and the sixth T2 distribution in the fourth row.

Compared with the VAEc-NN model, the LSTM model has more freedom resulting in higher accuracy with the possibility to overfit. Unlike LSTM the decoder part of the VAEc is pretrained, which constrains the synthesis capability of the VAEc-NN model. The decoder trained in the first stage of VAEc-NN training when used in the second stage of the training constrains the capacity of the VAEc-NN model by forcing the synthesis through the decoder.

9 Conclusions

Four deep neural network architectures were successfully trained to synthesize the NMR T2 distributions, comprising 64 discrete amplitudes corresponding to the 64 T2 bins ranging from 0.3 to 3000 ms. NMR T2 approximate the fluid-filled pore size distributions of hydrocarbon-bearing or water-bearing geological formations. The deep neural network models were trained to synthesize the entire 64-dimensional NMR T2 distribution by processing seven mineral content logs and third fluid saturation logs derived from the inversion of conventional logs acquired in the shale formation. GAN-NN was slightly better in the NMR T2 synthesis than the VAE-NN. R^2 and NRMSD for the VAE-NN model were 0.77 and 15%, respectively, whereas those for GAN-NN were 0.8 and 14%, respectively. VAEc-NN, VAE-NN, and GAN-NN model are trained in two stages. In the first stage a decoder/generator is trained to accurately reconstruct the NMR T2 distribution by

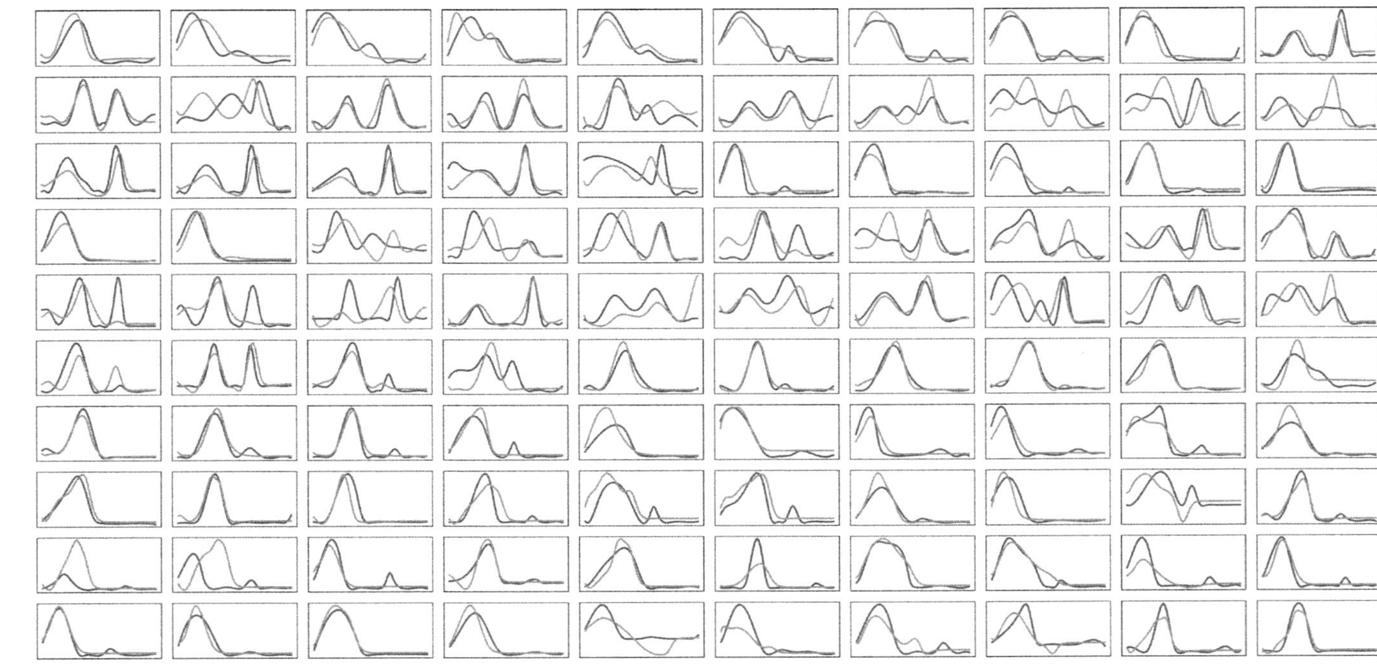

FIG. 7.16 Comparison of original NMR T2 [*blue (dark gray in print version)*] and LSTM-based synthesis of NMR T2 [*red (light gray in print version)*] for 100 depths from the testing dataset. *x*-axis represents the T2 relaxation time from 0.3 to 3000 ms; *y*-axis is normalized NMR signal amplitude ranging from 0 to 0.25 (unitless).

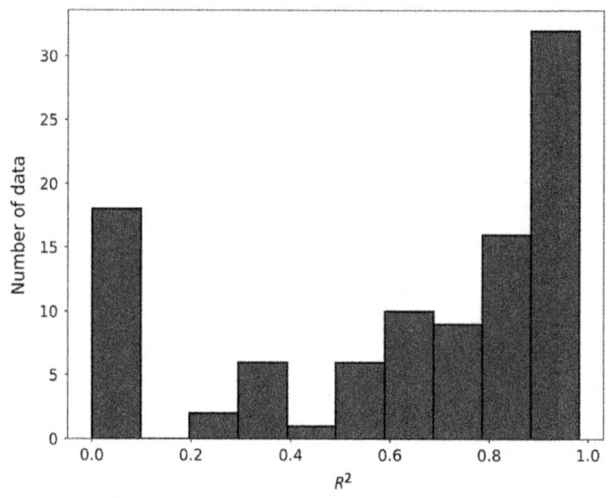

FIG. 7.17 Histogram of R^2 for the T2 distributions synthesized for the 100 discrete depths in the testing dataset when using LSTM network.

processing the measured NMR T2 in the training dataset. In the second stage a simple neural network is connected with the decoder/generator trained in the first stage to synthesize the NMR T2 by processing the 10 inversion-derived logs. On the other hand, the LSTM model treats the 10 inversion-derived logs as an input sequence and the 64-dimensional NMR T2 as an output sequence. LSTM performs sequence-to-sequence modeling by representing the input sequence as an intermediate/context vector using the LSTM encoder network, which is then decoded by the LSTM decoder network to sequentially generate the entire NMR T2 distribution, one bin at a time. LSTM and VAEc-NN models have a similar number of parameters, and LSTM requires only one training stage; however, LSTM requires one order of magnitude higher training time due to the complexity in sequentially encoding and decoding the internal state. The accuracy values of NMR T2 synthesis for the VAEc-NN and LSTM models in terms of R^2 are 0.75 and 0.78, respectively, and the training times are 43 and 580 s, respectively. This study opens up the possibility of applying deep neural networks to enhance reservoir characterization and improve project economics by characterizing the fluid-filled pore size distribution of the subsurface geological formations.

References

[1] Han Y, Misra S, Wang H, Toumelin E. Hydrocarbon saturation in a Lower-Paleozoic organic-rich shale gas formation based on Markov-chain Monte Carlo stochastic inversion of broadband electromagnetic dispersion logs. Fuel 2019;243:645–58.

[2] Tathed P, Han Y, Misra S. Hydrocarbon saturation in Bakken Petroleum System based on joint inversion of resistivity and dielectric dispersion logs. Fuel 2018;233:45–55.

[3] Guo C, Liu RC. A borehole imaging method using electromagnetic short pulse in oil-based mud. IEEE Geosci Remote Sens Lett 2010;7(4):856–60.

[4] Baron L, Holliger K. Constraints on the permeability structure of alluvial aquifers from the poro-elastic inversion of multifrequency P-wave sonic velocity logs. IEEE Trans Geosci Remote Sens 2011;49(6):1937–48.
[5] Hong D, Yang S, Yang S. A separately determining anisotropic formation parameter method for triaxial induction data. IEEE Geosci Remote Sens Lett 2014;11(5):1015–8.
[6] Wong PM, Gedeon TD, Taggart IJ. An improved technique in porosity prediction: a neural network approach. IEEE Trans Geosci Remote Sens 1995;33(4):971–80.
[7] Chang H-C, Chen H-C, Fang J-H. Lithology determination from well logs with fuzzy associative memory neural network. IEEE Trans Geosci Remote Sens 1997;35(3):773–80.
[8] Xu C, Misra S, Srinivasan P, Ma S, editors. When petrophysics meets big data: what can machine do? SPE middle east oil and gas show and conference. Society of Petroleum Engineers; 2019.
[9] He J, Misra S. Generation of synthetic dielectric dispersion logs in organic-rich shale formations using neural-network models. Geophysics 2019;84(3):1–46.
[10] Li H, He J, Misra S, editors. Data-driven in-situ geomechanical characterization in shale reservoirs. SPE annual technical conference and exhibition. Society of Petroleum Engineers; 2018.
[11] Freedman R. Advances in NMR logging. J Petrol Technol 2006;58(1):60–6.
[12] Goodfellow I, Bengio Y, Courville A. Deep learning. MIT Press; 2016.
[13] Kingma DP, Welling M. Auto-encoding variational bayes [arXiv preprint arXiv:13126114]; 2013.
[14] Doersch C. Tutorial on variational autoencoders [arXiv preprint arXiv:160605908]; 2016.
[15] Radford A, Metz L, Chintala S. Unsupervised representation learning with deep convolutional generative adversarial networks [arXiv preprint arXiv:151106434]; 2015.
[16] Reed S, Akata Z, Yan X, Logeswaran L, Schiele B, Lee H. Generative adversarial text to image synthesis [arXiv preprint arXiv:160505396]; 2016.
[17] Wang S, Liu W, Wu J, Cao L, Meng Q, Kennedy PJ, editors. Training deep neural networks on imbalanced data sets. 2016 international joint conference on neural networks (IJCNN). IEEE; 2016.

Chapter 8

Comparative study of shallow and deep machine learning models for synthesizing in situ NMR T2 distributions

Siddharth Misra* and Hao Li†

*Harold Vance Department of Petroleum Engineering, Texas A&M University, College Station, TX, United States, †The University of Oklahoma, Norman, OK, United States

Chapter outline

Nomenclature

ϕ	SVR kernel function
θ	coefficients
ANN	artificial neural network
GAN	generative adversarial network
***k*NNR**	*k*-nearest neighbor regressor

Machine Learning for Subsurface Characterization. https://doi.org/10.1016/B978-0-12-817736-5.00008-9

LASSO	least absolute shrinkage and selection operator
LSTM	long short-term memory
OLS	ordinary least squares
SVC	support vector classifier
SVR	support vector regression
VAE	variational autoencoder
w	coefficient vector
X	input log matrix
Y	output log matrix
y_i	original log response at depth i

Subscripts

i	formation (i)

1 Introduction

Economical reservoir development relies on accurate geological and petrophysical characterizations. One characterization technique is based on core samples acquired from the subsurface. Core samples provide a direct way to analyze petrophysical and geomechanical properties of subsurface formations. Acquiring core samples is expensive and is restricted by operational constraints. Another characterization technique is based on well logs that measure formation responses corresponding to various geophysical phenomena. Different well logging tools utilizing different physical principles are deployed in the wellbore environment for measuring the subsurface formation responses. Certain well logs, such as density, natural radiation, and resistivity logs, are "easy-to-acquire" measurements because of the relatively simple tool design, tool physics requirements, and operational protocols. On the other hand, measurements of NMR logs and imaging logs tend to be expensive and prohibitive due to the tool size, complex tool physics, intricate operational procedures, and slow logging speed. Such logs can be categorized as "hard-to-acquire" logs. Well logs can be used in the raw form, such as resistivity or density, or can be inverted/processed to obtain estimates of certain desired physical properties, such as fluid saturations and mineral composition. Formation properties like permeability and pore size distribution can be inverted from the raw NMR logs.

NMR logging tool excites the hydrogen nuclei in the in situ subsurface fluids by applying an external magnetic field, and the relaxation of the hydrogen nuclei, upon the removal of the external magnetic field, generates a relaxation signal that is inverted to obtain the NMR T2 distribution comprising 64 T2 amplitudes corresponding to 64 T2 bins. NMR T2 distribution is the relaxation time distribution of relaxing hydrogen nuclei in the formation fluid at a specific formation depth. NMR T2 distribution can be processed to estimate the pore size distribution and fluid mobility of the near-wellbore formation volume. The pore size distribution information contained in the NMR logs is critical to evaluate the pore-scale phase behavior of formation fluid, and no other well logs can provide this valuable information. However, due to financial or operational considerations, NMR logging tool cannot be deployed in all the wells drilled in a reservoir. To overcome the limited access to the "hard-to-acquire" NMR logs, we processed easy-to-acquire conventional logs using the shallow-learning and deep learning methods to synthesize the in situ NMR T2 distributions.

At any depth, various well logs exhibit strong to intermediate correlations because they are the responses from the same fluid-filled porous material. Well logs measure various responses of a formation by utilizing various physical excitations/fields/phenomena, such as electrical, acoustic, electromagnetic, chemical, and thermal processes. Although different physical phenomena are utilized, the responses measured by different well logging tools at the same formation depth are from the same material, and each well log records certain similar aspects of the formation properties. Consequently, there are several complex relationships, dependencies, and distinct patterns between the "easy-to-acquire" logs and the "hard-to-acquire" logs. Machine learning can find hidden relationships and patterns in large data. Machine learning methods can synthesize "hard-to-acquire" well logs by learning the relationships between the "hard-to-acquire" well logs and the "easy-to-acquire" well logs.

Shallow-learning models have been widely applied in oil and gas industry for prediction, classification, and regression tasks. For example, regression-type models trained using supervised learning have been applied in pseudo-log generation and petrophysical property prediction to improve reservoir characterization. Among all shallow models, artificial neural network (ANN) is the most widely used model due to its capability to account for nonlinear trends; nonetheless, ANN predictions are hard to generalize, interpret and explain. ANN has been successfully applied in pseudo-log generation [1–3] and petrophysical property prediction [4–6]. Support vector regression has been applied in the prediction of unconfined compressive strength from other petrophysical properties [7]. LASSO and *k*-nearest neighbor regression models have been applied in drilling and reservoir engineering tasks [8, 9].

Compared to shallow learning models, deep learning models are suited for learning hierarchical features from large-sized high-dimensional dataset without explicit feature extraction prior to the model training [10]. Deep learning models have been applied to log synthesis [11, 12] and petrophysical characterization [13]. Various architectures for deep learning can be designed specific to the dataset and learning task. For example, convolutional neural networks are suitable for spatial data and have been applied to image analysis; while recurrent neural networks are suitable for sequential data and have been applied to production data analysis [14]. This chapter compares the performances of shallow-learning and deep learning models trained to synthesize NMR T2 distribution by processing 12 "easy-to-acquire" conventional logs.

2 Dataset

The goal of this study is to generate NMR T2 distribution from other easy-to-acquire subsurface information and compare the performance of various models in this log-synthesis task. The dataset used for to the comparison contains raw and inverted logs acquired from 575 depth points in a vertical well drilled in a shale formation. The 300-ft of shale formation of interest for our investigation comprises seven distinct geological formations, as shown in Fig. 8.1. Formations F1 to F3 are source rock shale, and formations F4 to F7 are clay-rich dolomudstone (Fig. 8.1).

Two distinct types of features/inputs are used in this study, namely, raw logs and inverted logs. For purposes of comparison, we separately build the shallow and deep models on each of the two types of features. Twelve raw logs used in this investigation are five array-induction resistivity log at various depths of investigation (AF10, AF20, AF30, AF60, and AF90), caliper (DCAL), compressional sonic (DTCO), shear sonic (DTSM), gamma ray (GR), neutron porosity (NPOR), PEFZ, and formation density (RHOZ) logs. Ten inverted logs used in this investigation are anhydrite, calcite, chlorite, dolomite, illite, K-feldspar, quartz, free water, oil, and bound-water logs. The inverted logs are computed by data inversion of the raw resistivity, neutron, density, and gamma ray logs. Inverted logs can be considered as specially engineered features to facilitate the model training. In comparison to raw logs, when using raw logs, deeper networks are required for the same learning tasks, which demands more computation time and memory. However, the raw logs benefit from the fact that no preprocessing infrastructure is required. Inverted composition logs were used in NMR T2 synthesis [15, 16].

3 Shallow-learning models

Six shallow-learning models were trained using supervised learning to synthesize the 64 discrete T2 amplitudes that constitute the entire NMR T2 distribution. These regression-type models include ordinary least squares (OLS), least absolute shrinkage and selection operator (LASSO), and ElasticNet that are simple linear regression models. Support vector regression (SVR), an extension of the support vector classifier (SVC), was used as the fourth regression model. SVR model can only predict one target at a time. Fifth model is the *k*-nearest neighbor regressor (*k*NNR), which is an extension of the *k*NN classifier, can simultaneously predict multiple targets. The sixth model is a simple neural network based on multilayer perceptron for simultaneous prediction of multiple targets.

3.1 Ordinary least squares

OLS is the simplest linear regression model. OLS model is built by minimizing the cost/loss function shown in Eq. (8.1) formulated as square of L2 norm

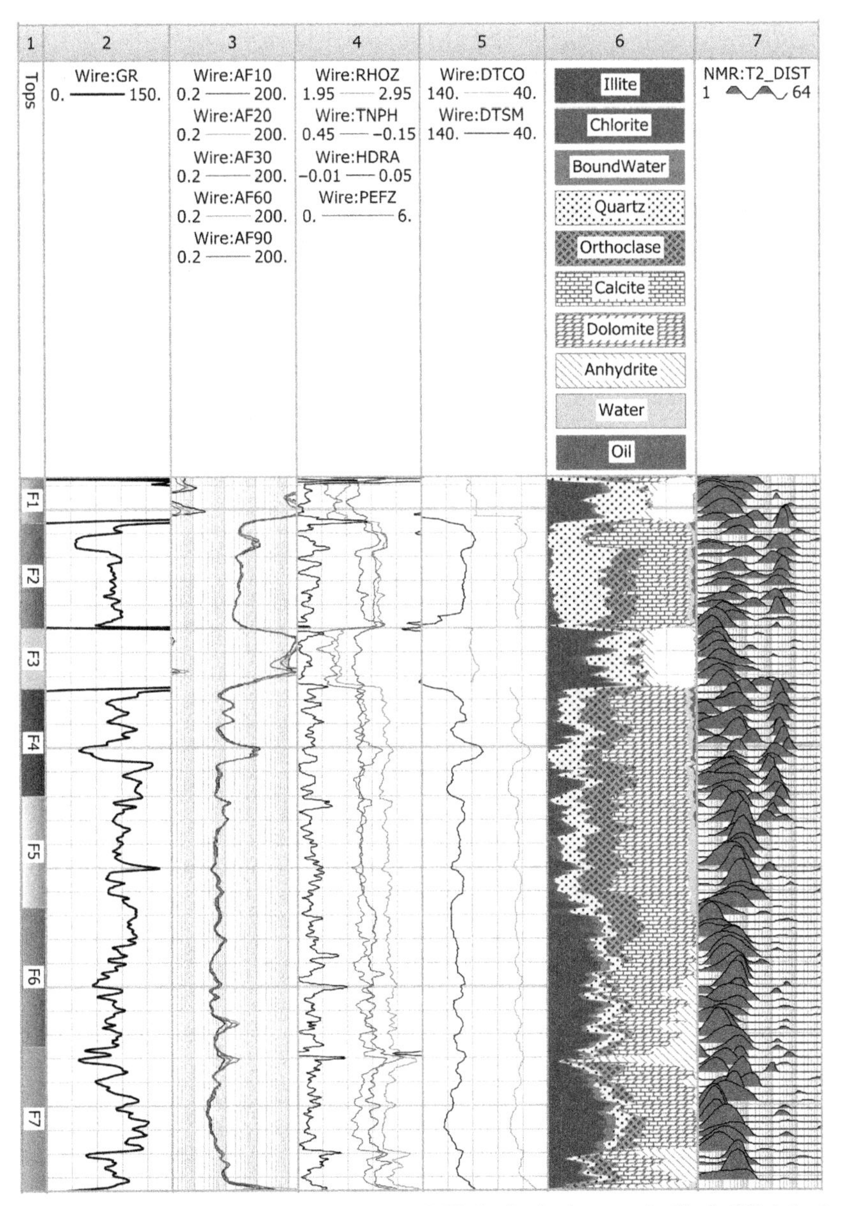

FIG. 8.1 Raw logs, inverted logs, and target NMR T2 distribution logs acquired in the 300-ft depth interval of a shale formation. Track 1, formation tops; track 2, gamma ray; track 3, resistivity; track 4, density and neutron porosity; track 5, sonic travel times; track 6, mineral contents and fluid saturations; and track 7, NMR T2 distribution.

of errors between predicted values $X\theta$ and measured values Y, which is expressed as

$$\min_{\theta} \|X\theta - Y\|_2^2 \tag{8.1}$$

where the feature X is a 2D array of raw/inverted logs for 475 depth points, where each depth is considered as a sample having either 10 inverted logs or 12 raw logs as features. The target Y is a 2D array of T2 amplitudes measured across 64 T2 bins, which constitutes the NMR T2 distribution, for the 475 depth points. Coefficient vector θ is a 1D array of coefficients/parameters of the model that are computed by minimizing Eq. (8.1) during the model training. OLS model learns to predict by minimizing the cost function shown in Eq. (8.1), which requires the minimization of the square of L_2 norm of errors in the model prediction ($|X\theta - Y|$). θ is the consequence of supervised learning of the OLS model. OLS model does not have any hyperparameter.

3.2 Least absolute shrinkage and selection operator

LASSO model is an extension of the OLS model, where a regularization/penalty term $\alpha\|\theta\|_1$ is added to the cost/loss function of the OLS model:

$$\min_{\theta} \|X\theta - Y\|_2^2 + \alpha\|\theta\|_1 \tag{8.2}$$

where α is a hyperparameter referred as the penalty parameter and $\|\theta\|_1$ is L_1 norm of the coefficient/parameter vector. The penalty term prevents overfitting and ensures that the LASSO model neglects correlated features. According to Eq. (8.2), when minimizing the cost function for building the LASSO model, L_1 norm of coefficient vector θ needs to be minimized along with the minimization of square of L_2 norm of errors in the model prediction ($|X\theta - Y|$). Minimization of L_1 norm requires reduction of majority of coefficients in the coefficient vector θ to zero.

3.3 ElasticNet

ElasticNet is an extension of OLS and LASSO formulations. The cost function of the ElasticNet includes L_1 and L_2 norms of the coefficient vector θ comprising coefficients/parameters learnt by the ElasticNet model. The cost function of the ElasticNet model is formulated as

$$\min_{\theta} \|X\theta - Y\|_2^2 + \alpha\rho\|\theta\|_1 + \alpha(1-\rho)\|\theta\|_2^2 \tag{8.3}$$

where ρ balances the overall penalty due to L_1 and L_2 norms of model parameters and $\|\theta\|_2$ is L_2 norm of the coefficient/parameter vector. Compared with LASSO, ElasticNet model is more unique and does not severely penalize correlated features. Unlike ElasticNet, LASSO model drastically reduces the parameters corresponding to most of the correlated features to zero resulting in nonunique, less generalizable model dependent on very few features.

3.4 Support vector regressor

Support vector regression is based on support vector classifier (SVC). SVC classifies the dataset by finding a hyperplane and decision boundaries that maximize the margin of separation between data points belonging to different classes/groups. Unlike SVC, SVR is used for regression tasks. SVR processes data to learn the coefficients that define the hyperplane such that cost associated with data points certain distance away from the hyperplane is the minimum. Regression model produced by SVR depends only on a subset of the training data, because the cost function for building the model ignores any training data within a certain margin around the hyperplane. Only the point outside the margin contributes to the final cost associated with the model. OLS minimizes the error in model prediction, whereas SVR fits the error in model prediction within a certain threshold. SVR model is built by minimizing the mismatch of the predicted and true target values while keeping the mapping function as smooth as possible, which is formulated as

$$\text{minimize}\ \frac{1}{2}\|w\|^2 + C\sum_{i=1}^{n}(\xi_i + \xi_i^*) \tag{8.4}$$

$$\text{subject to} \begin{cases} y_i - w^T\phi(x_i) - b \leq \varepsilon + \zeta_i \\ w^T\phi(x_i) + b - y_i \leq \varepsilon + \zeta_i^* \\ \zeta_i, \zeta_i^* \geq 0 \end{cases} \tag{8.5}$$

where ε is the error we can tolerate in high-dimensional space that defines the margin around the hyperplane, ξ_i and ξ_i^* are slack variables introduced for cases when the optimization in high-dimensional space with error limit of ε is not feasible, and ϕ is a kernel function that maps input dataset from current space to higher-dimensional space, like the kernel function in SVC. We use the radial basis function (RBF) as the kernel function. The SVR algorithm can only predict one target. We trained 64 different SVR models to generate the entire NMR T2 distribution comprising T2 amplitudes for 64 bins, such that each SVR model predicts T2 amplitude for one of the 64 NMR T2 bins.

3.5 *k*-Nearest neighbor regressor

k-Nearest neighbor regressor (*k*NNR) is based on the nearest neighbor classifier. *k*NN classifier first calculates distances of all the training data points from an unclassified data point, then selects the *k*-nearest points to the unclassified data point, and finally assigns a class to the unclassified data point depending on the class that is the most common among the *k*-nearest neighbors. *k*NNR predicts the target for a new data point as a weighted average of the target values for the *k*-nearest neighbors to the new data point. The weights are the inverse of the distance between the training data points and the new data point for which the target value needs to be predicted. *k*NNR does not build a model and does not need a training phase. *k*NNR requires all the training data points to be available during the deployment, which is a drawback of this method. *k*NNR synthesizes the log by computing the weighted average of targets for the k-nearest training points around the testing points.

3.6 Artificial neural network

Artificial neural network (ANN) is a collection of connected computational units or nodes called neurons arranged in multiple computational layers. Each neuron linearly combines its inputs and then passes it through an activation function, which can be a linear or nonlinear filter. Linear combination of inputs is performed by summing up the products of weights and inputs. ANN generates the target through feed-forward data flow and then updates the weights of each neuron by backpropagation of errors during the training iterations. For purposes of NMR synthesis, we implement four-layered ANN with two hidden layers. Each hidden layer has 200 neurons.

3.7 Comparisons of the test accuracy and computational time of the shallow-learning models

The six models implemented are shallow-learning regression models. A grid search was performed to find hyperparameters that optimize the models. Table 8.1 shows the overall coefficient of determination, R2, that measures the fit between original and synthesized NMR T2 distributions in the test dataset from the entire 300 ft of the shale formation. Table 8.2 shows the computational time for training the shallow-learning models. R2 and computational time are evaluated for both inverted logs and raw logs.

TABLE 8.1 Median R2 of the NMR T2 synthesis by the shallow-learning models on the testing dataset

	OLS	LASSO	ElasticNet	SVR	*k*NNR	ANN
Inverted log	0.60	0.60	0.60	0.74	0.63	0.67
Raw log	0.63	0.63	0.61	0.46	0.68	0.59

TABLE 8.2 Computational times (in seconds) for training the shallow-learning models

	OLS	LASSO	ElasticNet	SVR	*k*NNR	ANN
Inverted log	0.002	0.33	0.009	0.33	0.0010	0.32
Raw log	0.005	0.34	0.009	0.21	0.005	0.20

Median R2 for NMR T2 synthesis using OLS, LASSO, and ElasticNet models are relatively similar. For NMR T2 synthesis, grid search was performed to find the best penalty parameters (hyperparameter) for LASSO and ElasticNet models. Best performance was achieved when the penalty parameters are very small values, which indicates that none of the features are redundant, trivial, or highly correlated. SVR model performs well on inverted well logs but not on raw logs. This indicates that the feature space created by the inverted logs lead to a better containment of the data points within the predefined margin. *k*NNR synthesizes the log by computing the weighted average of targets for the *k*-nearest training points around the testing points; nonetheless the algorithm can be computationally intensive for large-sized, high-dimensional dataset due to the numerous distance calculations that are performed to find the neighbors. With raw logs, *k*NNR exhibits the best log-synthesis performance with a R2 of 0.68, and SVR performs the best on inverted logs with R2 of 0.74. The computational times for model training range from 0.001 to 0.3 s. The models are run on a Dell workstation with Intel 6-core Xeon CPU and 32-GB RAM. *k*NNR and OLS are the fastest on the raw log with the best accuracy. ANN, SVR, and LASSO models require large computational time, especially when processing the inverted log.

Figs. 8.2 and 8.3 compare synthesized and measured NMR T2 distributions. The OLS, LASSO, and ElasticNet have intermediate performance. When processing inverted log, SVR and ANN exhibit good log-synthesis performance (Fig. 8.2). However, SVR's log-synthesis performance drops to 0.46 when using raw logs as input (Fig. 8.3). ANN seems to overfit with raw logs. ANN-generated NMR T2 is not smooth. OLS, LASSO, and ElasticNet models cannot handle nonlinear relationships required to synthesize the NMR T2 by processing the easy-to-acquire logs. On the other hand, ANN model with 200 neurons each in two hidden layers has relatively large estimation capacity but ignores the basic spatial dependency among the 64 bins of the NMR T2 spectra. In the following section, we use four different deep learning models to synthesis NMR T2 distribution with an emphasis on learning the spatial and sequential features of typical NMR T2 distribution and then relating these multiscale features with the easy-to-acquire logs.

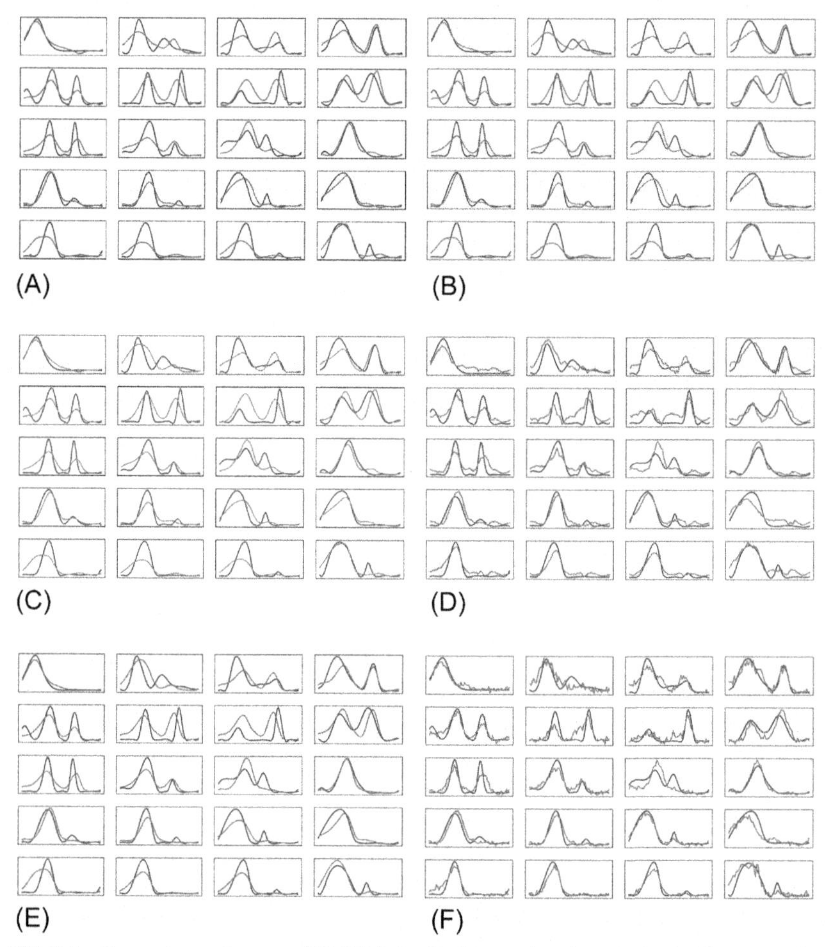

FIG. 8.2 Comparisons of measured NMR T2 distributions against those synthesized by processing inverted logs in the testing dataset using (A) OLS, (B) LASSO, (C) ElasticNet, (D) SVR, (E) *k*NNR, and (F) ANN models.

4 Deep learning models

Most deep learning models are based on deep neural network architecture comprising several hidden layers. There is no clear boundary between shallow- and deep learning models. The definition of deep learning evolves with the development computation speed and data size. Unlike shallow-learning models, deep learning models can be designed to have specific architecture that facilitates better approximation, abstraction, and generalization of information in training dataset. In this chapter, we apply four distinct deep learning models for the desired NMR T2 synthesis. Three of the four deep models first learn features and abstractions essential for

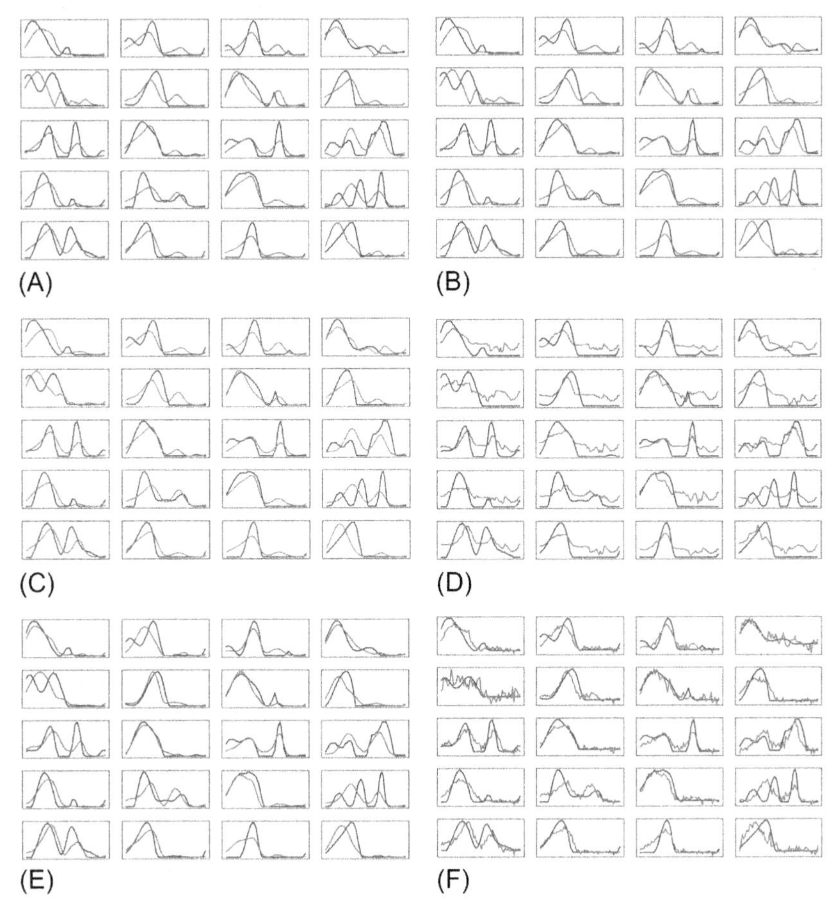

FIG. 8.3 Comparisons of measured NMR T2 distributions against those synthesized by processing raw logs in the testing dataset using (A) OLS, (B) LASSO, (C) ElasticNet, (D) SVR, (E) *k*NNR, and (F) ANN models.

reconstructing the NMR T2 distribution; following that, the learned abstractions and features of NMR T2 are related to the easy-to-acquire logs for the synthesis of NMR T2 distribution. The three models are variational autoencoder assisted neural network (VAE-NN), variational autoencoder with convolutional layer assisted neural network (VAEc-NN), and generative adversarial network assisted neural network (GAN-NN). These three models adopt a two-step training process. In the first step, VAE-NN and VAEc-NN models build a decoder network and the GAN-NN model builds a generator network for purposes of learning to reconstruct NMR T2 distribution. In the second step, a simple ANN with 3–5 hidden layers is connected to the trained decoder/generator for learning the complex hidden relationship between the "easy-to-acquire" input logs and the NMR T2 distribution. This two-step training process increases the robustness of the log synthesis. The fourth model is long short-term memory network that processes the easy-to-acquire logs as a sequence and tries to find a corresponding sequence of NMR T2 distribution.

4.1 Variational autoencoder assisted neural network

VAE-NN is trained in two steps. First step uses the variational autoencoder (VAE) network comprising encoder and decoder networks. VAE learns to abstract and reconstruct the NMR T2 distribution. The encoder projects the NMR T2 from training dataset to a 2D or 3D latent space, and the decoder reconstructs the NMR T2. The decoder takes the encoded latent vector as input and decodes it into NMR T2 distribution. The goal of the first step is to reproduce the NMR T2 distribution. The generation of the latent variable involves a sampling process from Gaussian distribution, the VAE is trained to project NMR T2 distribution with similar features to similar space, which reduces the cost of reconstruction. After the first step of training, the decoder network has learnt to generate a typical NMR response by processing the latent vectors.

Decoder trained in the first step of training is frozen to preserve the VAE's learning related to the NMR T2 distributions in the training dataset. In the second step of training, a simple fully connected ANN with 3–5 hidden layers is connected to the trained decoder (Fig. 8.4). The "easy-to-acquire" input logs are fed to the ANN, and the ANN is trained to generate the latent vector for the decoder network. The latent vector will be decoded into NMR T2 distribution by the decoder. In doing so, the second step of training relates the easy-to-acquire logs to the NMR T2 distribution. Fig. 8.5 illustrates the manifold learned by the VAE in the first step of training. It basically is an abstraction learnt by the decoder network from the NMR T2 distributions in training set. The gradual and smooth changing NMR T2 in each of the subplots is what the VAE learned from the training set in the first step of training. This learnt manifold represents the essential features of the NMR T2 distribution.

4.2 Generative adversarial network assisted neural network

GAN-NN model follows a two-step training process, with some similarity to VAE-NN. In the first step, generative adversarial network (GAN) learns from the NMR T2 distribution in the training dataset; this involves training a generator network to reconstruct NMR T2 distributions with large similarity to those in the training dataset. This requires a discriminator network that evaluates the NMR synthesis achieved by the generator network. In the first step, the generator and discriminator networks are trained alternatively using only the NMR T2 distributions in the training dataset. First a random vector

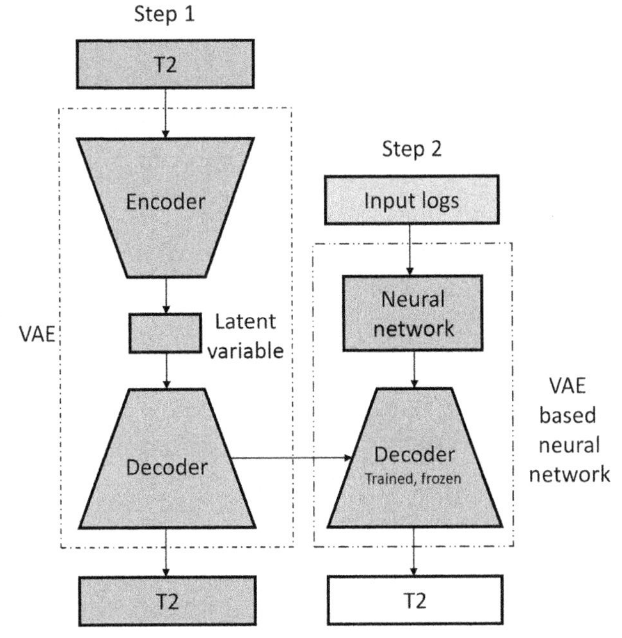

FIG. 8.4 Two-step training process schematic of the VAE assisted neural network.

z is fed as input to the generator network to generate synthetic NMR T2 distribution. After the synthetic NMR T2 are generated, the discriminator takes 50% synthetic NMR T2 and 50% real NMR T2 as inputs and determines whether a input is synthetic or real NMR T2. As the iterative training of the generator and discriminator networks proceeds, generator learns to reconstruct NMR T2 by minimizing the error in the NMR T2 reconstruction, and the discriminator learns to identify the synthetic NMR T2 generated by the generator by maximizing its accuracy in distinguishing real and synthetic NMR T2. In other words, the generator is trained to fool the discriminator, and the discriminator is trained to not be fooled by generator. The two neural networks compete with each other during the first step of training the GAN-NN. Gradually the generator network learns to generate the NMR T2 with large similarity to the real NMR T2 in the training dataset (Fig. 8.6).

After the first step of training, the generator network gains the ability to synthesize physically consistent NMR T2 distributions. For the second step of training, similar to the VAE-NN training process, the generator is frozen and connected to a simple ANN with 3–5 hidden layers to associate the "easy-to-acquire" logs with the NMR T2 distribution. The training of GAN is more complex as compared with VAE because the generator and discriminator networks need to be balanced to achieve the desired learning. The result generated by the generator is further smoothened using a Gaussian fitting method.

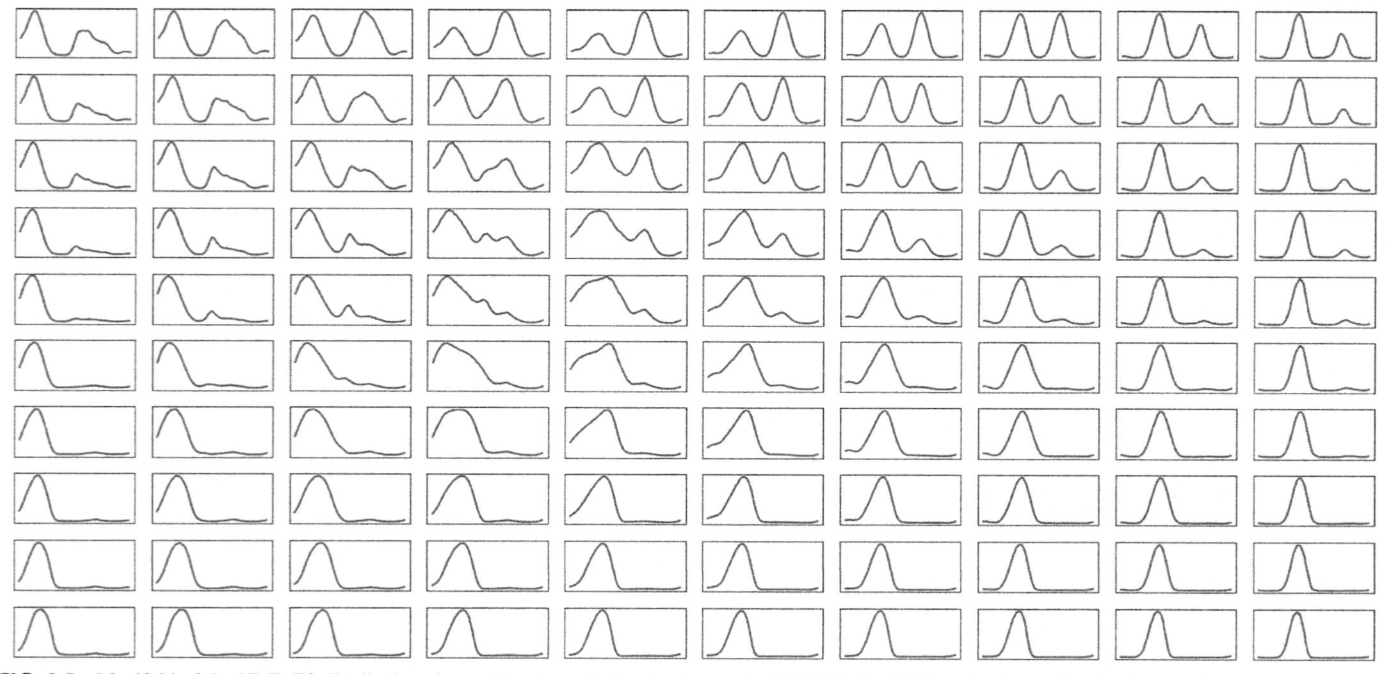

FIG. 8.5 Manifold of the NMR T2 distributions learned by the variational autoencoder (VAE) in the VAE-assisted neural network.

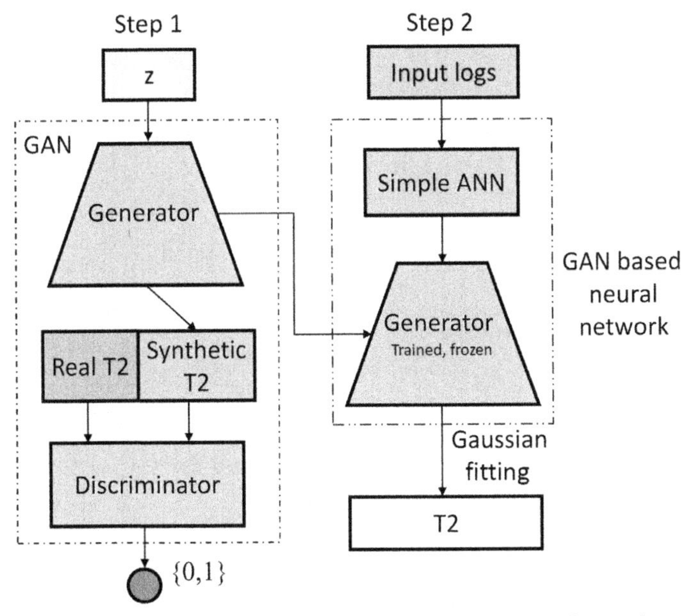

FIG. 8.6 Two-step training process schematic of the GAN assisted neural network.

4.3 Variational autoencoder with convolutional layer assisted neural network

VAEc-NN model incorporates convolutional layers within VAE architecture for the first step of training. Convolutional layers extract spatial/topological features in the NMR T2. The details of the model are shown in Fig. 8.7.

4.4 Encoder-decoder long short-term memory network

The encoder-decoder long short-term memory (LSTM) network performs sequence-to-sequence (many-to-many) mapping similar to language translation. In this study, the sequence of "easy-to-acquire" logs at a specific depth is processed by the LSTM network to generate the sequence of NMR T2 distribution at that depth. The encoder LSTM learns to process the sequence of "easy-to-acquire" logs to compute an internal state representative of the variations in the input sequence. The decoder network learns to process the encoded internal state to obtain the sequence of 64 T2 amplitudes, constituting the NMR T2 distribution. The encoder and decoder networks are designed by implementing one or more LSTM layers. As the LSTM recurrently processes the sequential information, it generates an internal state that summarizes previous information in the sequence. LSTM model is good at solving the long-term dependency in the input and output

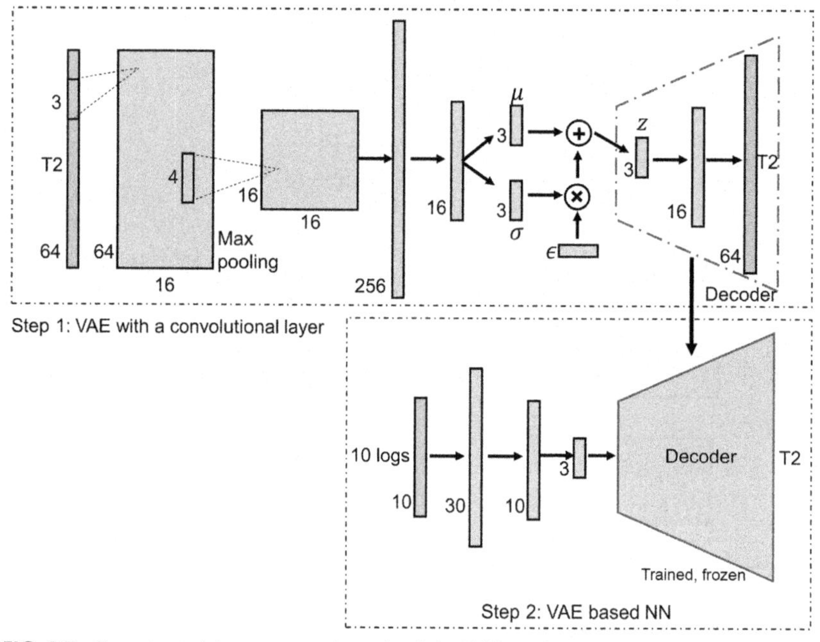

FIG. 8.7 Two-step training process schematic of the VAEc assisted neural network.

sequences. LSTM model treats the NMR T2 distribution as a sequential data as a function of T2 time. The encoder and decoder network in this model are different from those in VAE. The encoder here encodes the input logs into a context vector. The encoded vector is repeated 64 times and recurrently decoded to 64-dimensional NMR T2 distribution. The training of the implemented encoder-decoder LSTM model is a one-step process (Fig. 8.8). They are capable of learning the complex dynamics within the ordering of input/output sequences.

4.5 Comparisons of the accuracy and computational time of the deep learning models

Table 8.3 shows the overall coefficient of determination, R2, that quantifies the fit between original and synthesized NMR T2 distributions for the entire 300 ft of the shale formation when using four deep learning models. Table 8.4 lists the computation times required to train the models. R2 and computational time are evaluated for both inverted logs and raw logs. Median R2 for NMR T2 synthesis using the deep learning models is relatively similar. Deep models perform slightly better when processing the inversion-derived logs. LSTM model has the lowest performance and one-order higher computational time. The inversion-derived logs act as features extracted from well logs.

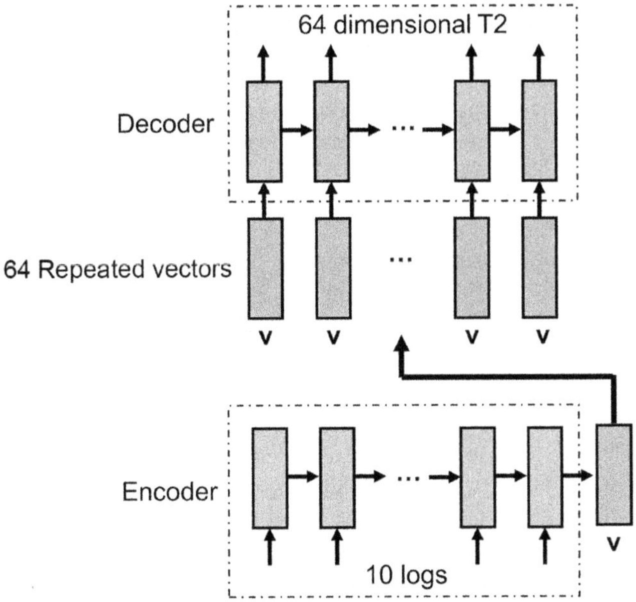

FIG. 8.8 One-step training process schematic of the encoder-decoder LSTM network model.

TABLE 8.3 Median R2 of the NMR T2 synthesis by the deep learning models on the testing dataset

	VAE-NN	GAN-NN	VAEc-NN	LSTM
Inverted log	0.75	0.80	0.78	0.75
Raw log	0.76	0.75	0.77	0.69

TABLE 8.4 Computation times (in seconds) for training the deep learning models

	VAE-NN	GAN-NN	VAEc-NN	LSTM
Inverted log	47.55	22.91	47.61	569.14
Raw log	47.34	22.62	37.67	589.12

Figs. 8.9 and 8.10 compare the original and synthesized NMR T2 distributions obtained by processing the inversion-derived and raw logs, respectively, using the deep learning models. Deep learning models exhibit better NMR T2 synthesis as compared to the shallow-learning models. The synthetic NMR T2 is smooth and appears physically consistent.

4.6 Cross validation

Because of the limited size of the training dataset, the deep learning models were trained with fixed training steps. To ensure the deep learning models do not overfit, it is recommended to use the cross-validation method. In our study, we test the efficacy of the cross validation on the VAE-NN model. A validation set is built by randomly selecting 100 samples from the 300-ft interval of the shale formation. The training set contains 375 samples. The validation loss is monitored during training, and when the validation loss increases, the training is automatically stopped to prevent overfitting. The VAE-NN model so trained exhibits a R2 score of 0.60 on the testing dataset containing 100 samples. Notably the cross validation led to decrease in the testing/generalization performance due to the reduction in data size under the limited data available to us for this study. For our case, having a separate validation set reduces the training set size resulting in the poorly trained, highly biased deep learning model. Cross validation is suitable when the model has access to large training dataset; in those cases the model can use the validation set to monitor overfitting.

5 Comparison of the performances of the deep and shallow models

Both shallow- and deep learning models can be implemented for NMR T2 log synthesis. Shallow models are easier to apply due to their simplicity in architecture, lower training-resource requirements, and relatively easier explainability. Log-synthesis performance of shallow models in terms of the R2 score is in the range of 0.6–0.75. A simple ANN with two hidden layers exhibits the best performance among the shallow-learning models in terms of R2; nonetheless, ANN performs poorly in terms of the generalization of the NMR T2 distribution to ensure that the synthesis is physically consistent (Figs. 8.2 and 8.3).

The four deep learning models implemented in this study process the same dataset as the shallow-learning models. The two-step training process used for VAE-, VAEc-, and GAN-based deep neural networks leads to stable and physically consistent NMR T2 distributions, which resemble those in the training set (Figs. 8.5 and 8.6). The performance of the four deep learning models is better than that of shallow-learning models. The average R2 score for the log synthesis accomplished by the deep models ranges from

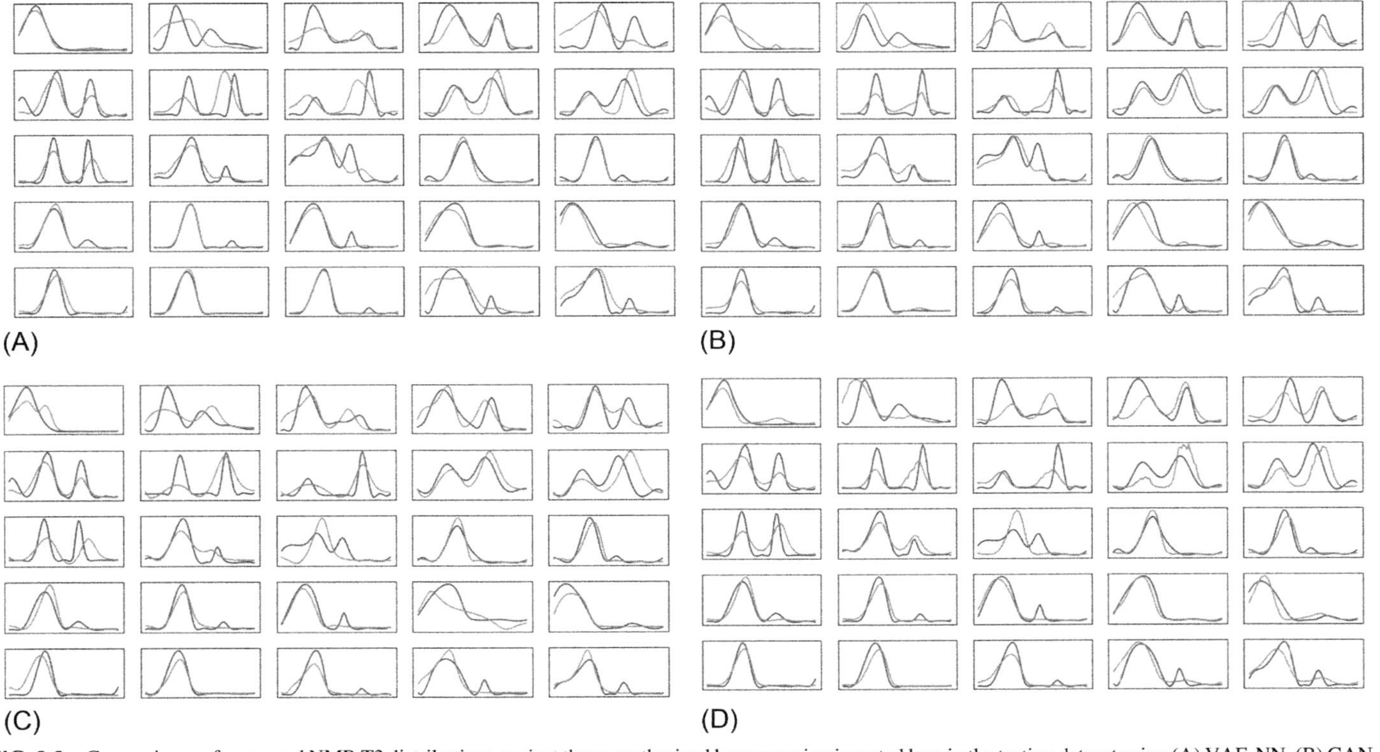

FIG. 8.9 Comparisons of measured NMR T2 distributions against those synthesized by processing inverted logs in the testing dataset using (A) VAE-NN, (B) GAN-NN, (C) VAEc-NN, and (D) LSTM.

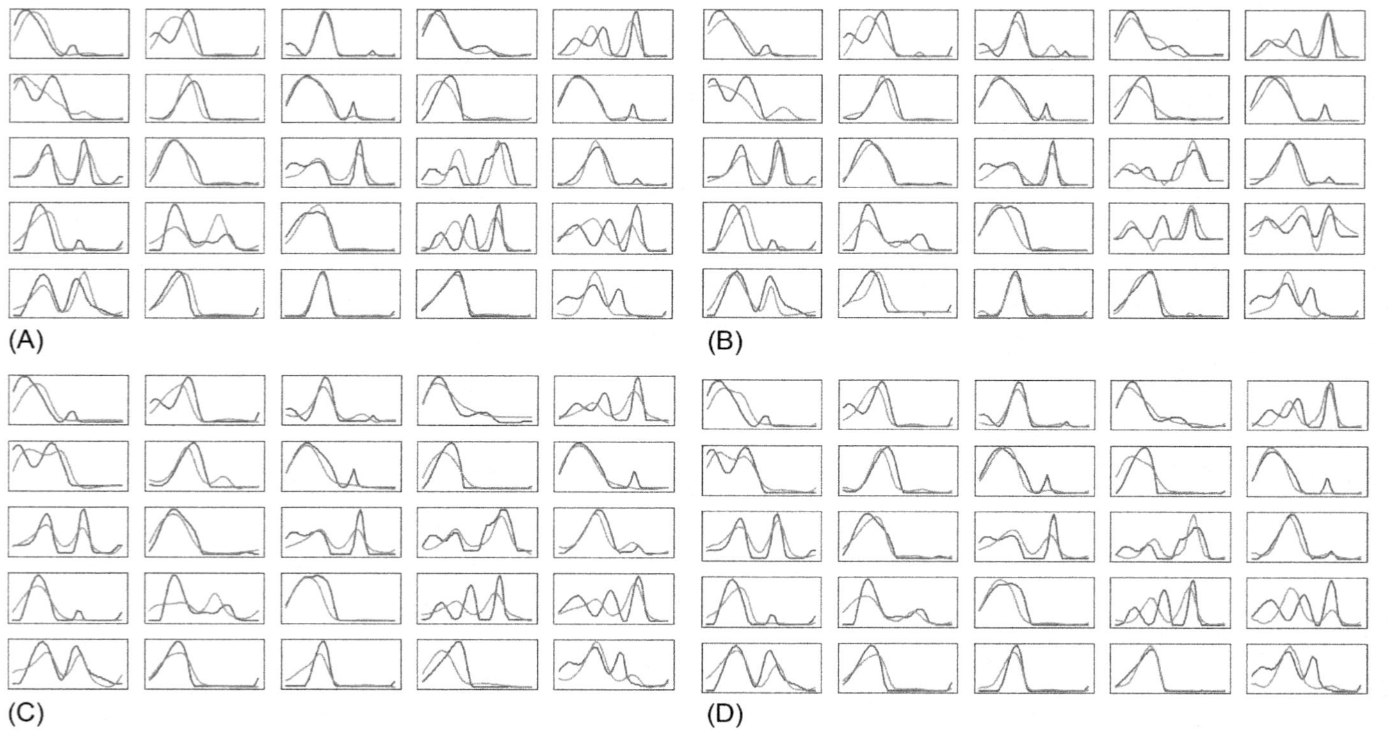

FIG. 8.10 Comparisons of measured NMR T2 distributions against those synthesized by processing raw logs in the testing dataset using (A) VAE-NN, (B) GAN-NN, (C) VAEc-NN, and (D) LSTM.

0.75 to 0.8. However, the training time of the deep learning models is 2–3 orders of magnitude more as compared with the shallow models. Among the four model, LSTM model does not perform as good as the other three models, probably, because the input "easy-to-acquire" logs are inherently not sequences.

Two types of the input logs (feature set) are used for training the shallow- and deep learning models. The first type of input logs is the raw form of the "easy-to-acquire" logs, while the second type of input logs is the inversion-derived fluid saturation and mineral composition logs obtained by inverting the raw form of the "easy-to-acquire" logs. Best performing shallow-learning models, SVR and ANN, perform much better on the inverted logs; however, deep learning models exhibit relatively performance when processing inversion-derived logs and raw logs. LSTM performance drops significantly when processing the raw logs. In the raw form, the logs have strong correlations/co-linearity because the raw logs are influenced by similar physical properties, fluid saturations, and mineral compositions of the formation. When data inversion is performed on the raw logs to obtain the mineral composition and fluid saturation logs, the inversion-derived logs are less correlated and better suited for the robust training of the deep learning models and nonlinear shallow-learning models, like SVR and ANN.

6 Discussions

In the oil and gas industry, several well logs are acquired in the borehole thousands of feet below the ground for purposes of accurate reservoir characterization. We applied shallow- and deep learning models to synthesize the NMR T2 distributions from inversion-derived and raw well logs. Permeability, residual saturations, and pore size distribution can be estimated from the synthesized NMR T2 distribution. These petrophysical estimations are critical for reservoir characterization. This study demonstrates the potential of shallow- and deep learning models to synthesize NMR T2 distribution. Limitations of this study are as follows:

(1) Dataset available for this study was limited in size;
(2) Due to limited data size, we could not perform a robust cross validation of deep learning models to prevent overfitting;
(3) NMR T2 distributions in the available dataset were primarily single-peak spectra, and there were limited depths with two-peak spectra; consequently, our models tend to have poor performance in depths with two-peak NMR spectra; and
(4) The log-synthesis performance in different wells could not be assessed for purposes of blind testing due to data unavailability.

NMR logs, specifically NMR T2 distributions, are hard to acquire due to financial and operational constrains. Under such data constraints, deep learning models can be used to reliably synthesize the NMR T2 distributions to facilitate improved reservoir characterization.

7 Conclusions

Six shallow- and four deep learning models were used to process the easy-to-acquire well logs for synthesizing the NMR T2 distributions for a 300-ft interval of a shale formation. Both the raw form of "easy-to-acquire" well logs and the inversion-derived formation mineral and fluid composition logs (obtained by processing the "easy-to-acquire" logs) were used for the synthesis of the NMR T2 distribution logs. Log-synthesis performances of the deep learning models quantified in terms of R2 score range from 0.75 to 0.8, whereas the performances of shallow-learning models range from 0.6 to 0.75 in terms of R2. Two-step training of deep neural networks based on variational autoencoder, generative adversarial network, and variational autoencoder with convolutional layers resulted in robust deep learning models that exhibit physically consistent reconstruction. Deep learning models and nonlinear shallow-learning models, like support vector regressor and artificial neural network, perform better NMR T2 synthesis by processing the inversion-derived formation mineral and fluid composition logs. Inversion-derived logs can be considered as specially engineered features extracted from the raw logs; consequently, the inversion-derived logs are less correlated and have more independent, relevant information that boosts the performance of the deep learning and nonlinear shallow-learning models.

References

[1] Alzate GA, Arbelaez-Londono A, Naranjo Agudelo AJ, Zabala Romero RD, Rosero Bolanos MA, Rodriguez Escalante DL, et al. Generating synthetic well logs by artificial neural networks (ANN) using MISO-ARMAX model in cupiagua field. In: SPE Latin America and caribbean petroleum engineering conference, Maracaibo, Venezuela, May 21, 2014. SPE: Society of Petroleum Engineers; 2014.

[2] Alloush RM, Elkatatny SM, Mahmoud MA, Moussa TM, Ali AZ, Abdulraheem A. Estimation of geomechanical failure parameters from well logs using artificial intelligence techniques. In: SPE Kuwait oil & gas show and conference, Kuwait City, Kuwait, October 15, 2017. SPE: Society of Petroleum Engineers; 2017.

[3] Li H, Misra S, He J. Neural network modeling of in situ fluid-filled pore size distributions in subsurface shale reservoirs under data constraints. Neural Comput Appl 2019;1–13.

[4] Akande KO, Olatunji SO, Owolabi TO, Abdulraheem A. Feature selection-based ANN for improved characterization of carbonate reservoir. In: SPE Saudi Arabia section annual technical symposium and exhibition, Al-Khobar, Saudi Arabia, June 4, 2015. SPE: Society of Petroleum Engineers; 2015.

[5] Moghadasi L, Ranaee E, Inzoli F, Guadagnini A. Petrophysical well log analysis through intelligent methods. In: SPE Bergen one day seminar, Bergen, Norway, April 5, 2017. SPE: Society of Petroleum Engineers; 2017.

[6] Olayiwola T. Application of artificial neural network to estimate permeability from nuclear magnetic resonance log. In: SPE annual technical conference and exhibition, San Antonio, TX, October 9, 2017. SPE: Society of Petroleum Engineers; 2017.

[7] Negara A, Ali S, AlDhamen A, Kesserwan H, Jin G. Unconfined compressive strength prediction from petrophysical properties and elemental spectroscopy using support-vector

regression. In: SPE Kingdom of Saudi Arabia annual technical symposium and exhibition, Dammam, Saudi Arabia, June 1, 2017. SPE: Society of Petroleum Engineers; 2017.

[8] Barone A, Sen MK. An improved classification method that combines feature selection with nonlinear Bayesian classification and regression: a case study on pore-fluid prediction. In: 2017 SEG international exposition and annual meeting, Houston, TX, October 23, 2017. SEG: Society of Exploration Geophysicists; 2017.

[9] Onwuchekwa C. Application of machine learning ideas to reservoir fluid properties estimation. In: SPE Nigeria annual international conference and exhibition, Lagos, Nigeria, August 6, 2018. SPE: Society of Petroleum Engineers; 2018.

[10] LeCun Y, Bengio Y, Hinton G. Deep learning. Nature 2015;521(7553):436–44.

[11] Korjani M, Popa A, Grijalva E, Cassidy S, Ershaghi IA. New approach to reservoir characterization using deep learning neural networks. In: SPE Western regional meeting, Anchorage, AK, May 23, 2016. SPE: Society of Petroleum Engineers; 2016.

[12] Korjani MM, Popa AS, Grijalva E, Cassidy S, Ershaghi I. Reservoir characterization using fuzzy kriging and deep learning neural networks. In: SPE annual technical conference and exhibition, Dubai, UAE, September 26, 2016. SPE: Society of Petroleum Engineers; 2016.

[13] Odi U, Nguyen T. Geological facies prediction using computed tomography in a machine learning and deep learning environment. In: SPE/AAPG/SEG unconventional resources technology conference, Houston, TX, August 9, 2018. URTEC: Unconventional Resources Technology Conference; 2018.

[14] Tian C, Horne RN. Recurrent neural networks for permanent downhole gauge data analysis. In: SPE annual technical conference and exhibition, San Antonio, TX, October 9, 2017. SPE: Society of Petroleum Engineers; 2017.

[15] Li H, Misra S. Prediction of subsurface NMR T2 distributions in a shale petroleum system using variational autoencoder-based neural networks. IEEE Geosci Remote Sens Lett 2017;PP(99):1–3.

[16] Li H, Misra S. Prediction of subsurface NMR T2 distribution from formation-mineral composition using variational autoencoder. In: SEG technical program expanded abstracts 2017; 2017. p. 3350–4.

Chapter 9

Noninvasive fracture characterization based on the classification of sonic wave travel times

Siddharth Misra* and Hao Li[†]

*Harold Vance Department of Petroleum Engineering, Texas A&M University, College Station, TX, United States, [†]The University of Oklahoma, Norman, OK, United States

Chapter outline

Machine Learning for Subsurface Characterization. https://doi.org/10.1016/B978-0-12-817736-5.00009-0

1 Introduction

1.1 Mechanical discontinuities

Mechanical discontinuity in the material is generally referred as crack or fracture. Predicting and monitoring the geometry, distribution, and condition of mechanical discontinuities are critical for structural health monitoring, rock mechanics, geotechnical projects, geothermal reservoir development, sequestration, and hydraulic fracturing. For example, in the oil and gas industry, geometry and direction of induced fracture systems are critical to the hydrocarbon production rate. Monitoring, description, and prediction of the state and behavior of discontinuities are an important research topic.

Mechanical discontinuity is created when the stress inside the material exceeds the strength of the material, causing the material to lose cohesion along its weakest plane. Discontinuity can be formed due to compression, tension, or shear stress. Mode I (opening) discontinuity develops when tensile stress acts normal to the plane of the discontinuity; the discontinuity planes move away from each other. Mode II discontinuity develops due to sliding, where in-plane shear stresses act parallel to the plane of the discontinuity and perpendicular to the discontinuity front. Mode III discontinuity develops due to tearing, where out-of-plane shear stress acts parallel to the plane of the discontinuity and parallel to the discontinuity front. Brittleness is an important characteristic that affects the rate of development of discontinuity. Brittle material breaks without significant plastic deformation. At low temperature, materials are brittle due to the constrained molecular motion. High confining pressure hinders the generation of discontinuities and thus makes materials ductile.

1.2 Characterization of discontinuities

Different characterization techniques have been used to evaluate the location, orientation, and density of discontinuities in various materials. Ultrasonic measurement and CT scanning are the most commonly used nondestructive testing (NDT) methods applied to the static characterization of discontinuities in materials. A popular NDT method for dynamic characterization of discontinuities is the acoustic emission (AE) testing to determine locations and modes when discontinuities initiate or propagate in material [1]. NDT methods are widely used in the oil and gas industry, geological characterization, and civil engineering to characterize the discontinuities. Common NDT methods can be divided into two categories: mechanical methods and electromagnetic methods. Both methods utilize wave propagation to investigate the interior structure of material. Mechanical methods include seismic, ultrasonic, hammer, and acoustic emission methods. Electromagnetic methods include radar, galvanic resistivity, induction, and dielectric methods. These characterization techniques differ in the spatial scale of investigation, such that finer structures can be assessed by decreasing the wavelength of the propagating wave. In geology and geophysics, the investigation scale is from hundred meters to kilometers; in civil engineering, the scale is from centimeters to tens of meters; in the oil and gas industry, the investigation scale is from millimeters to meters.

One widely used NDT method is ultrasonic testing, which is a centimeter-scale measurement to characterize the mechanical discontinuity in material. Ultrasonic testing relies on the propagation of compressional (P-wave) and transverse (S-wave) sonic wave through the material. Ultrasonic testing can be based on reflection, transmission, or refraction of waves as they propagate from the source to the sensor. Ultrasonic pulse velocity (UPV) measurement utilizes wave transmission, whereas ultrasonic pulse echo (UPE) measurement uses wave reflection. Ultrasonic testing method was used to investigate the development of discontinuities in rock samples under uniaxial compressive test by Martínez-Martínez et al. [2] and Ramos et al. [3]. They found that the proportion of the high-frequency to low-frequency component decreases with the increase in the intensity of discontinuities. Another NDT method is acoustic emission (AE) testing that can be used to characterize the general mode and 3D location of cracks in material, but it may not able to reconstruct the 3D geometry. CT scanning can be used to reconstruct the 3D geometry of the discontinuity, but it is more expensive than AE testing. Unlike AE testing, it is harder to deploy CT scanning to acquire real-time fracture signatures. Watanabe et al. [4] applied CT scanning to build a 3D numerical model of rock fractures. Cai et al. [5] used CT scanning to reconstruct the 3D fracture network in coal samples during cyclic loading. CT scanning and AE testing are often used in tandem to quantify the fracture geometry in materials during geomechanical experiments, such as laboratory fracturing or triaxial experiments. Unlike CT, AE, and ultrasonic methods, resistivity and electromagnetic methods have not shown much potential in the characterization of discontinuities. In our study, we intend to develop data-driven classification methods to facilitate static characterization of discontinuities at centimeter scale by processing compressional wavefront travel times recorded at multiple sensors/receivers when emitted by a single source/transmitter.

1.3 Machine learning for characterization of discontinuities

Most popular supervised machine learning algorithms used for the characterization of discontinuities are artificial neural network (ANN), random forest, support vector machine (SVM), and convolutional neural network (CNN). Zhou et al. [6] applied ANN and SVM to classify rock fracture and blast events based on AE signals from the rock fracturing experiment. Liu et al. [7] used ANN to predict rock types from AE measurements because different rocks have different failure modes and different failure modes generate different types of AE signals. Farhidzadeh et al. [8] used SVM to classify the AE signals as being emitted from tensile fracture or shear fracture. Wang et al. used machine learning algorithms on laboratory data describing the stress filed, like stress ratio and intensity factor, to predict the crack growth rate. Miller et al. [9] applied CNN to process 2D images acquired from rock fracturing simulation. The 2D fracture image is processed as a weighted graph so that the CNN learns the graph features that are most predictive of final fracture length distribution. CNN has been applied on a large dataset of labeled raw seismic waveforms to learn a compact representation that can discriminate seismic noise from earthquake signals [10]. Also, CNN can process labeled seismic data to output probabilistic locations of earthquake sources [10]. Loutas et al. [11] applied nonhomogeneous hidden semi-Markov model (NHHSMM) to process AE data for predicting the fatigue life of composite material under laboratory fatigue in real time. In terms of unsupervised methods, studies have successfully applied clustering methods to group the acoustic emission data with different signatures into different groups (clusters) to facilitate the characterization of fracture modes [1, 12]. Apart from analyzing the measured data, several researches have analyzed simulated data to monitor and predict the fracture evolution process. Moore et al. [13] gathered simulated data from the finite-discrete element model to train ANN to predict whether two fractures will coalesce or not based on the parameters of fracture orientations, distances between two fractures, and the minimum distance from one of the fractures to its nearest boundary, to name a few. Rovinelli et al. [14] applied Bayesian network on simulated data to identify relevant micromechanical and microstructural variables that influence the direction and rate of crack propagation [14].

2 Objective

Characterization of mechanical discontinuity (crack or fracture) is a challenging problem in material evaluation, geophysics, civil engineering, rock mechanics, and oil and gas industry, to name a few. The geometry of the embedded discontinuities is governed by the geomechanical properties and the stress distributions. Popular methods for the characterization of discontinuities include acoustic emission, ultrasonic imaging, and CT scanning. AE method reconstructs the fracture system by detecting the acoustic events during fracture propagation. CT scanning method is based on the X-ray absorption phenomenon, such that open fractures absorb relatively less compared with surrounding materials. The CT scanning process is a relatively time-consuming and high-cost characterization technique. Ultrasonic imaging is another way for fracture characterization under laboratory condition.

Sonic wave propagation phenomenon is influenced by the mechanical discontinuities (fractures or cracks) in the material. Discontinuity can be investigated by measuring the compressional wave and shear wave travel times. This study explores the feasibility of applying classification methods to process the compressional wavefront travel times for the noninvasive characterization of mechanical discontinuity in material. In this study, we perform three major tasks in chronological order: (1) create thousands of numerical models of material containing mechanical discontinuities of various spatial properties, such as orientation, distribution, and dispersivity; (2) perform fast-marching (FM) simulations of compressional wavefront propagation through each numerical model of material containing discontinuities to create a dataset of multipoint compressional wavefront travel times recorded at multiple sensors when emitted by a single source along with a label that categorically represents the overall spatial properties of the discontinuities embedded in the material; and (3) train several data-driven classification methods on the dataset of compressional wavefront travel times with associated labels representing the spatial characteristics of the embedded static discontinuities to learn to characterize (categorize) the materials containing discontinuities. The configurations/arrangement of 1 sonic source and 28 sensors/receivers for the measurement of travel times are inspired by real laboratory experiments [15].

2.1 Assumptions and limitations of the proposed data-driven fracture characterization method

a. The method is developed for two-dimensional rectangular materials, assuming homogeneity in the vertical direction.
b. A discontinuity is a linear element, and the embedded discontinuities can be represented as discrete fracture network (DFN).
c. The method is developed for compressional wavefront travel times ignoring the reflection, scattering, refraction, phase change, later arrivals, and dispersion.
d. The mechanical properties and the velocity of sonic wave propagation are assumed to be homogeneous and isotropic for the background material.
e. The mechanical properties and the velocity of sonic wave propagation are assumed to be homogeneous and isotropic for each discontinuity.
f. Discontinuities in material are assumed to follow certain statistical distributions of spatial properties.
g. Wavefront travel times are simulated using the fast-marching method (FMM), which can have errors in the presence of large contrasts due mechanical discontinuities.
h. The method is developed for scenarios when the discontinuities and pores are filled with air.

2.2 Significance and relevance of the proposed data-driven fracture characterization method

a. Numerical models of material containing discontinuities and simulations of wavefront travel times are inspired by real-world laboratory experiments.
b. As a proof of concept, a limited number of source-receiver pairs were used to generate the labeled travel-time dataset for developing the classification-assisted fracture characterization method. Characterization performance

will increase with the increase in number of source-receiver pairs. In the real world, the number of source-receiver pairs can be five times or higher as compared with that implemented in this study.

c. As a proof of concept, each receiver detects only the first arrival (i.e., wavefront travel time) of compressional wave, and the characterization is accomplished only with compressional travel-time data, which are very limited information related to the wave propagation through a fractured material. Characterization performance will increase with the use of shear wavefront travel times or full compressional/shear waveforms.

d. Feature importance is performed using feature permutation importance to identify the optimal placement of receivers/sensors for accomplishing the desired static characterization of discontinuities by processing the multipoint measurements of compressional wavefront travel times.

e. The proposed method can approximate certain statistical parameters of the system of discontinuities, like dominant orientation and intensity of distribution, with simple source-receiver configurations.

3 Fast-marching method (FMM)

3.1 Introduction

The fast-marching method (FMM) is developed to solve the eikonal equation, expressed as

$$|\nabla u(\boldsymbol{x})| = \frac{1}{f(\boldsymbol{x})} \quad \text{for}\, \boldsymbol{x} \in \Omega \tag{9.1}$$

$$u(\boldsymbol{x}) = 0 \quad \text{for}\, \boldsymbol{x} \in \partial\Omega \tag{9.2}$$

where $u(\boldsymbol{x})$ represents the wavefront travel time (i.e., time of first arrival) at location $\boldsymbol{x}$, $f(\boldsymbol{x})$ represents the velocity function for the heterogeneous material, Ω is a region with a well-behaved boundary, $\partial\Omega$ is the boundary, and $\boldsymbol{x}$ is a specific location in the material. Eikonal equation characterizes the evolution of a closed surface Ω through a material with a specific velocity function f. FMM is similar to the Dijkstra's algorithm. Both algorithms monitor a collection of nodes defining the region Ω and expand the collection of nodes by iteratively including a new node just outside the boundary $\partial\Omega$ of the region Ω that can be reached with the least travel time from the nodes on the boundary $\partial\Omega$. FMM has been successfully applied in modeling the evolution of wavefront in heterogeneous geomaterial [16]. In this study, FMM is used to simulate the compressional wavefront propagation in fractured material.

3.2 Validation

FMM predictions are validated against analytical solutions and against the predictions of k-Wave (MATLAB toolbox) in materials with and without fractures.

An extensive validation was performed to ensure the reliability of the simulated dataset generated using FMM for purposes of developing/evaluating data-driven models for the noninvasive static characterization of mechanical discontinuities.

3.2.1 Numerical model of homogeneous material at various spatial discretization

Discretization governs the accuracy of the FMM and k-Wave simulations [17]. Spatial discretization of a numerical model is defined by the number of grids (or grid size) implemented in the model. With the increase in the number of girds, the accuracy and computational time of the simulation of wavefront travel time will increase till a certain optimal number of grids. Compressional velocity of the background material is 4000 m/s. Fig. 9.1 represents two distinct configurations of one source and multiple sensors. At initial time, the compressional wavefront originates from the source marked on Fig. 9.1. In Case #1 (Fig. 9.1A), 1 source is placed in the center of the material, and 32 sensors are placed symmetrically around the source. In Case #2 (Fig. 9.1B), 1 sensor is placed at the lower left corner of the material, and 40 sensors are placed at different angles with respect to x-axis. Four different discretizations are selected for the simulation of the compressional wave propagation in terms of number of grids in x- and y-axes: 50 by 50, 75 by 75, 100 by 100, and 500 by 500. The simulation results from FMM, k-Wave, and analytical solution method are shown in Figs. 9.2 and 9.3.

As demonstrated in Figs. 9.2 and 9.3, FMM predictions are in good agreement with analytical solution at discretization of 75 grids by 75 grids or more. The k-Wave simulation deviates from the analytical solution when the discretization is smaller than 500 by 500 grids for a material of dimension 150 mm

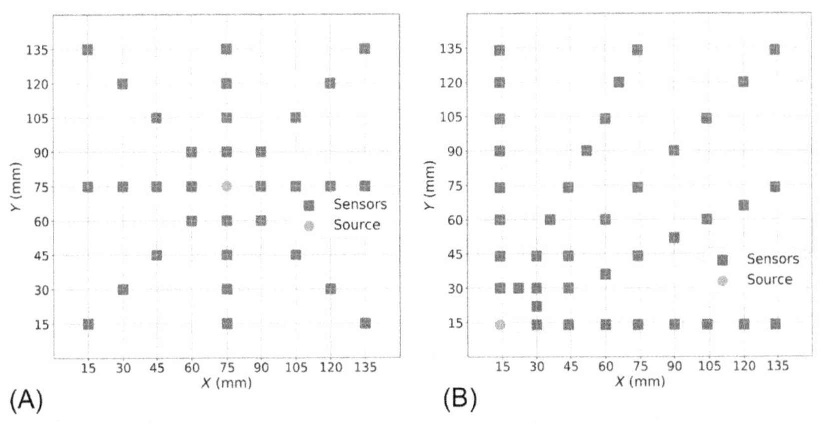

FIG. 9.1 Two distinct source-sensor configurations for FMM validation on material of dimension 150 mm by 150 mm when discretized using different numbers of grids in x- and y-axes. (A) Case #1 and (B) Case #2.

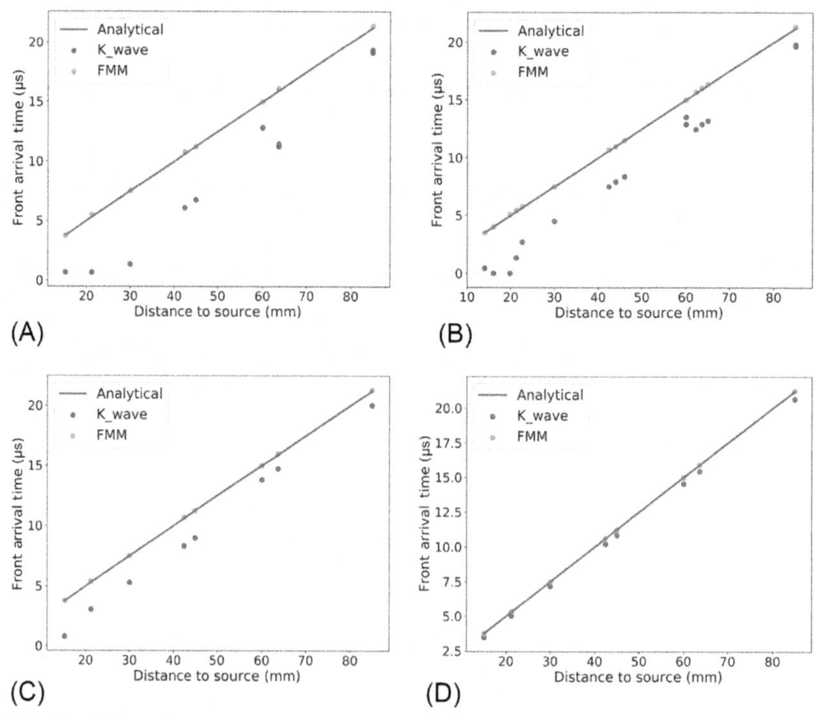

FIG. 9.2 Effect of number of grids (or grid size) on FMM and k-Wave simulations. Comparison of compressional wavefront travel time at each sensor calculated using the FMM, k-Wave, and analytical method for Case #1, as shown in Fig. 9.1A, for four different numbers of grids. x-Axis represents the distance between the source and the sensor/receiver; y-axis represents the wavefront travel time at receiver/sensor. (A) Number of grids: 50 by 50, (B) number of grids: 75 by 75, (C) number of grids: 100 by 100, and (D) number of grids: 500 by 500.

by 150 mm (i.e., grid size of 0.3 mm by 0.3 mm). FMM only simulates the wavefront, and the FMM calculation process is fast and efficient for simple cases. k-Wave simulation quantifies the whole process of wave propagation resulting in the calculation of full waveform at the sensors. Based on our numerical experiments, when the number of grids is 500 by 500 for the material of dimension 150 mm by 150 mm, both k-Wave and FMM predictions of travel times coincide with the analytical solutions. In all subsequent numerical experiments and dataset generation, a grid number of 500 by 500 will be used to discretize the materials of any dimension (i.e., grid size smaller than 0.3 mm by 0.3 mm).

3.2.2 *Material with large contrasts in compressional wave velocity*

FMM simulation and other numerical simulations tend to be unstable and inaccurate in the presence of large contrasts in material properties. In this section, a numerical model of material dimension 150 mm by 300 mm is created with 60 alternating layers that result in large contrasts (Fig. 9.4) and discretized using

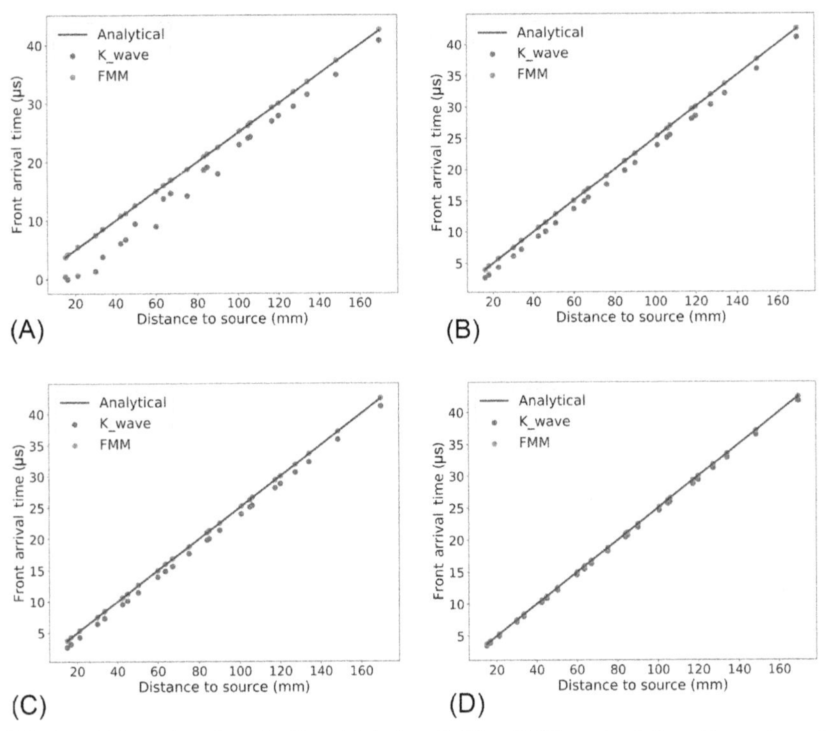

FIG. 9.3 Effect of number of grids (grid size) on FMM and k-Wave simulations. Comparison of compressional wavefront travel time at each sensor calculated using the FMM, k-Wave, and analytical method for Case #2, as shown in Fig. 9.1B, for four different numbers of grids. *x*-Axis represents the distance between the source and the sensor; *y*-axis represents the wavefront travel time at receiver/sensor. (A) Number of grids: 50 by 50, (B) number of grids: 75 by 75, (C) number of grids: 100 by 100, and (D) number of grids: 500 by 500.

500 by 1000 grids. Each alternating layer is 5 mm in thickness, and each grid had a length and breadth of 0.3 mm. The source is located at the center of the left boundary. Ten equally spaced sensors are placed in the middle of the material on a horizontal line through the center of the material. Two cases with different wave propagation velocities and contrasts are investigated in this section (Fig. 9.4). For Case #1, the compressional wave velocity of the material alternates between 4000 and 2000 m/s with a contrast of 2. For Case #2, the compressional wave velocity of the material alternates between 4000 and 1000 m/s with a contrast of 4.

FMM and k-Wave predictions for Case #1 exhibit good agreement with analytical solutions (Fig. 9.5). For Case #2, FMM predictions are accurate, but the k-Wave predictions deviate from the analytical solution as the wave travels farther from the source. In the k-Wave simulation, the sonic wave undergoes multiple reflections between the two interfaces of layers and gradually loses energy during the propagation; consequently, it is harder to detect the first arrivals at the receivers. Unlike k-Wave simulation, FMM simulation only accounts for the

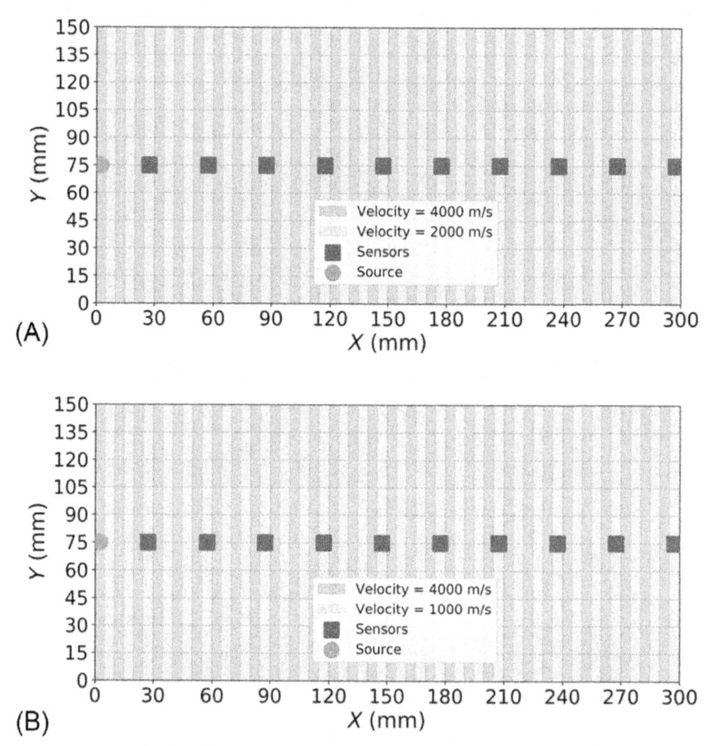

FIG. 9.4 Source-sensor configuration for FMM validation on a material of dimension 150 mm by 300 mm with 60 alternating layers. The changes in compressional wave velocity result in large contrasts. (A) Case #1: compressional wave velocity of the material alternates between 4000 and 2000 m/s. (B) Case #2: compressional wave velocity of the material alternates between 4000 and 1000 m/s.

wavefront propagation while ignoring the multiple wave reflections and the loss of energy due to reflection/refraction. The deviation between analytical and k-Wave predictions increases with increase in the distance of sensor from the source (Fig. 9.5).

3.2.3 *Material with parallel discontinuities*

In this section, FMM model is validated on two materials containing parallel discontinuities oriented along the *y*-axis and the wave propagation along the *x*-axis. Compressional velocity of discontinuities in Case #1 is 45 m/s (unrealistic), and that in Case #2 is 450 m/s. Compressional velocity of the background material is 4000 m/s. The material containing parallel discontinuities is shown in Fig. 9.6. The fractured material has a dimension of 150 mm by 300 mm and discretized using 500 by 1000 grids. Three hundred parallel and vertical discontinuities are embedded into the material. Each discontinuity is 0.3 mm in thickness. The source-sensor configuration is similar to that used in the previous

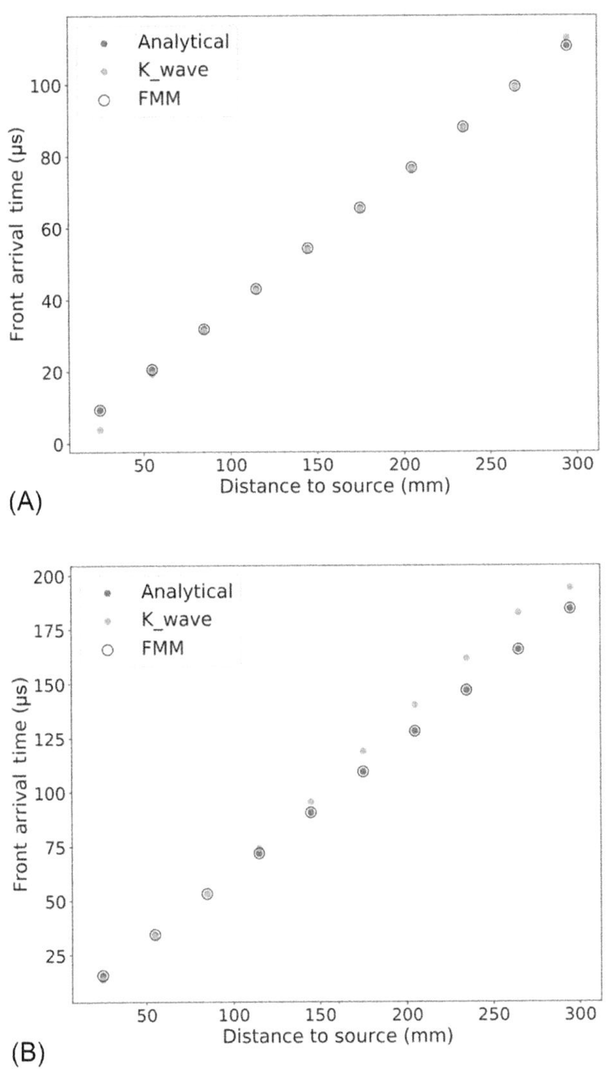

FIG. 9.5 Effect of large contrasts on FMM and k-Wave simulations. Comparison of compressional wavefront travel time at each sensor calculated using the FMM, k-Wave, and analytical method for two cases with alternating compressional velocity shown in Fig. 9.4. x-Axis represents the distance between the source and the sensor; y-axis represents the time of arrival of the wavefront at the receiver/sensor. (A) Case #1 and (B) Case #2.

section (Fig. 9.4). The source is located at the center of the left boundary. Ten equally spaced sensors are placed in the middle of the material on a horizontal line through the center of the material. This numerical experiment investigates the validity of the FMM predictions in the presence of large contrasts due to discontinuities.

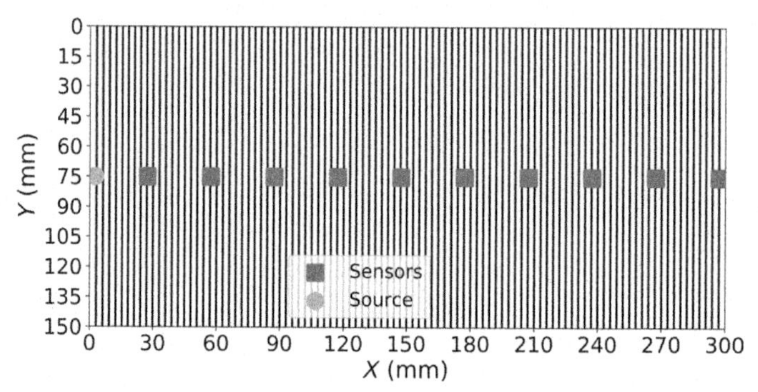

FIG. 9.6 Source-sensor configuration for FMM validation on a material of dimension 150 mm by 300 mm with 300 embedded parallel discontinuities of 0.3-mm thickness. Compressional velocity of discontinuities in Case #1 is 45 m/s, and that in Case #2 is 450 m/s. Compressional velocity of the background material is 4000 m/s.

Fig. 9.7 shows the wavefront travel time calculated using FMM and analytical solution. The x-axis is the distance of the 10 sensors from the source; the y-axis is the wavefront arrival time at each sensor. FMM predictions of travel times are not adversely affected by the presence of large contrasts due to discontinuities and by the presence of high density of discontinuity. However, the k-Wave predictions are severely affected due to the discontinuities of Case #1 and Case #2. k-Wave simulation is extremely slow for these cases, and k-Wave predictions are not added to Fig. 9.7.

3.2.4 *Material with smoothly varying velocity distribution*

In this section, we compare the FMM predictions of travel time with the analytical solutions for compressional wave propagation through materials exhibiting smooth spatial variation of compressional wave velocity across the entire material (Fig. 9.8). For certain functional forms of compressional velocity in terms of the coordinates x and y, FMM predictions of travel time can be represented in an analytical form in terms of the coordinates x and y. For such cases, the arrival of the wavefront at any location (x, y) can be expressed in terms of x and y (Fig. 9.9). For purposes of validation, in this section, the smoothly varying velocity of the material is expressed as

$$f = \frac{1}{\left((-2x)^2 + (-2y)^2\right)^{0.5}} \tag{9.3}$$

The corresponding analytical solution for arrival time $t(x,y)$ is

$$t = x^2 + y^2 \tag{9.4}$$

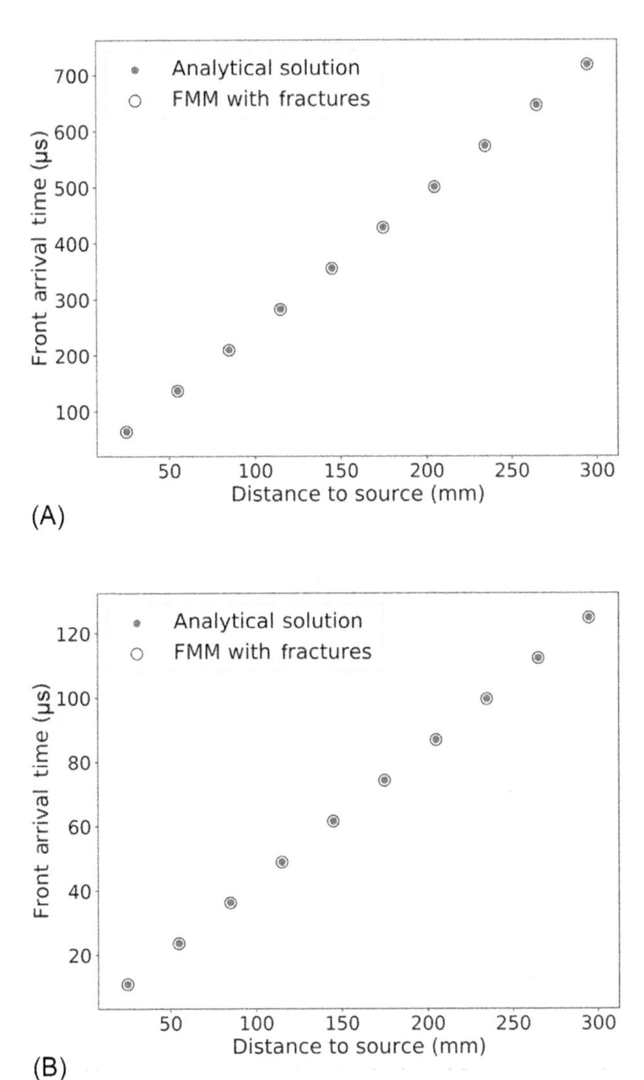

FIG. 9.7 Effect of high density of large contrasts on FMM simulations. Comparison of compressional wavefront travel time at each sensor calculated using the FMM and analytical method for two cases with different compressional velocity of parallel discontinuities/fractures, as shown in Fig. 9.6. *x*-Axis represents the distance between the source and the sensor; *y*-axis represents the time of arrival of the wavefront at the receiver/sensor. (A) Case #1. Compressional velocity of fracture: 45 m/s. (B) Case #2. Compressional velocity of fracture: 450 m/s.

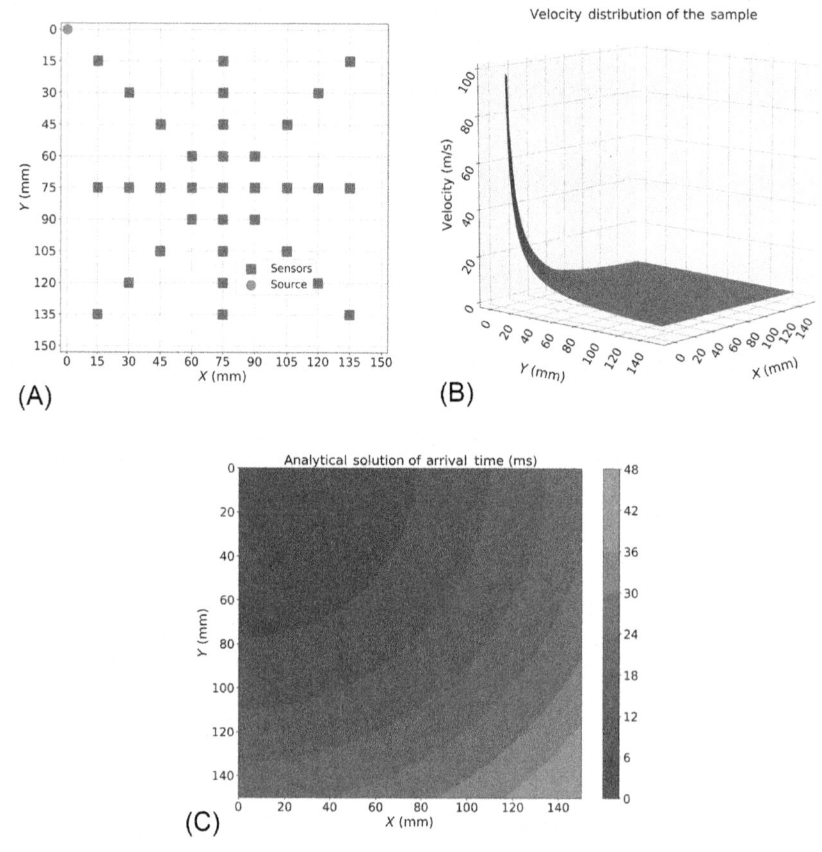

FIG. 9.8 (A) Source-sensor configuration and (B) compressional velocity distribution for FMM validation on a material of dimension 150 mm by 150 mm with smoothly varying compositional velocity, as shown in (B). (C) Analytical solution of the arrival time of a compressional wave starting from (0,0), located at the top left corner in (A), for the material with compressional velocity as shown in (B).

The source-sensor configuration, velocity distribution in the material, and the analytical solution of the travel time are shown in Fig. 9.8A–C. Fig. 9.9 compares the analytical solution with the FMM predictions.

3.3 Fast-marching simulation of compressional wavefront travel time for materials containing discontinuities

In this section, FMM is used to simulate compressional wavefront propagation through a material containing discontinuities, which are randomly distributed in the material. FMM simulations are generated for two cases with distinct network of discontinuities. In Case #1, 100 nonintersecting horizontal discontinuities are randomly distributed in the material (Fig. 9.10A). In the Case #2, 100 nonaligned discontinuities are randomly distributed in the material, such that

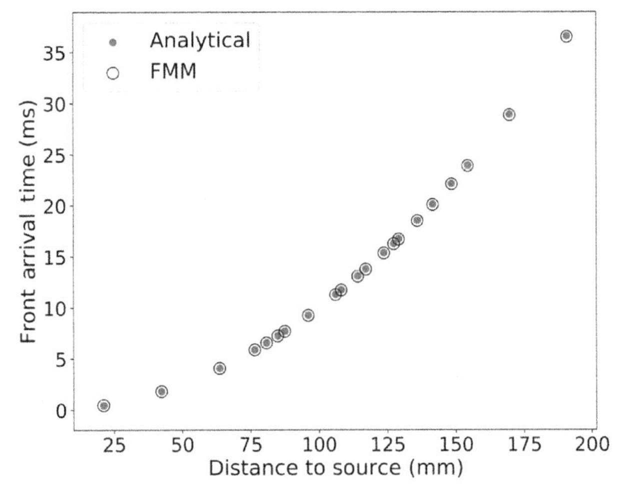

FIG. 9.9 Effect of smoothly varying compositional velocity on FMM simulations. Comparison of compressional wavefront travel time at each sensor, as shown in Fig. 9.8A, calculated using the FMM and analytical method for material with smoothly varying compositional velocity, as shown in Fig. 9.8B. x-Axis represents the distance between the source and the sensor; y-axis represents the time of arrival of the wavefront at the receiver/sensor.

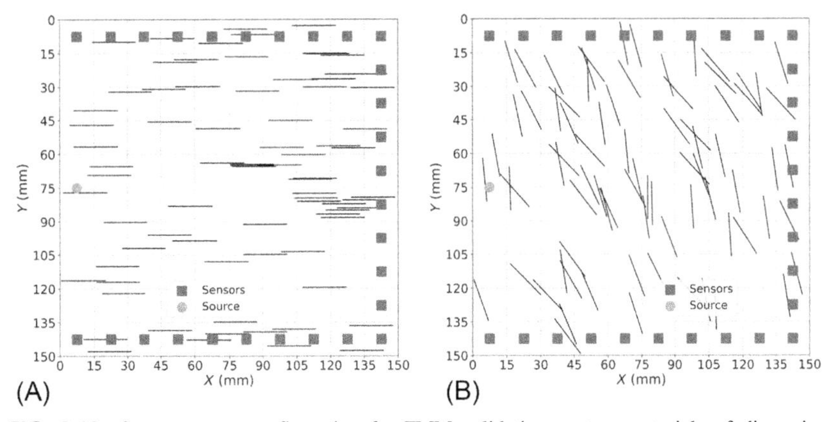

FIG. 9.10 Source-sensor configuration for FMM validation on two materials of dimension 150 mm by 150 mm containing 100 randomly distributed, embedded discontinuities. In Case #1, all discontinuities are parallel and horizontal. In Case #2, discontinuities are dispersed around the main orientation of −45 degrees to the x-axis. Compressional wave velocity in the background material and embedded discontinuities are 4000 and 400 m/s, respectively. (A) Case #1 and (B) Case #2.

the discontinuities exhibit certain dispersion around the main orientation of −45 degrees with respect to the x-axis (Fig. 9.10B). The two materials containing discontinuities are shown in Fig. 9.10. The compressional wavefront travel times are measured at 28 sensors/receivers located on the boundary of the material. The boundary on which transmitter/source is placed is known as the source-bearing boundary. There are no receivers on the source-bearing boundary.

The receivers are referred using indices ranging from 0 to 27. The receivers located on the upper boundary adjacent to the source-bearing boundary have index ranging from 0 to 9. The receivers located on the lower boundary adjacent to the source-bearing boundary have index ranging from 18 to 27. The receivers located on the boundary opposite to the source-bearing boundary have index ranging from 9 to 18. On each of the three boundaries, the receivers are placed incrementally in the order of indices. The receiver with index of 0 is located at the top left corner, the receiver with index of 9 is located at the top right corner, the receiver with index of 27 is located at the bottom right corner, and the receiver with index of 18 is located at the bottom left corner.

Fig. 9.11 demonstrates the effect of randomly distributed discontinuities and their primary orientations on the wavefront propagation computed using FM simulation. The arrival times computed using FMM for the two materials containing discontinuities (Case #1 and Case #2) can be compared against the FMM predictions of arrival times for an unfractured material. For Case #1, travel times with and without discontinuities are similar for the receivers 10–17 (opposite to the source-bearing boundary) and slightly different for the rest of the receivers on the adjacent boundaries. This indicates, for Case #1, thin horizontal open discontinuities do not affect the wavefront propagation along x-axis and slightly hinder propagation along y-axis. For Case #2, travel times with and without discontinuities are similar for receivers 18–27 (lower boundary to the transmitter-bearing boundary) and different for the rest of the receivers. This indicates the primary orientation of discontinuities of −45 degrees with respect to the x-axis does not affect wave propagation to lower boundary; however, the primary orientation of discontinuities of −45 degrees with respect to the x-axis affects the wave propagation to upper boundary and substantially affects wave propagation to the opposite boundary. The differences in travel times for

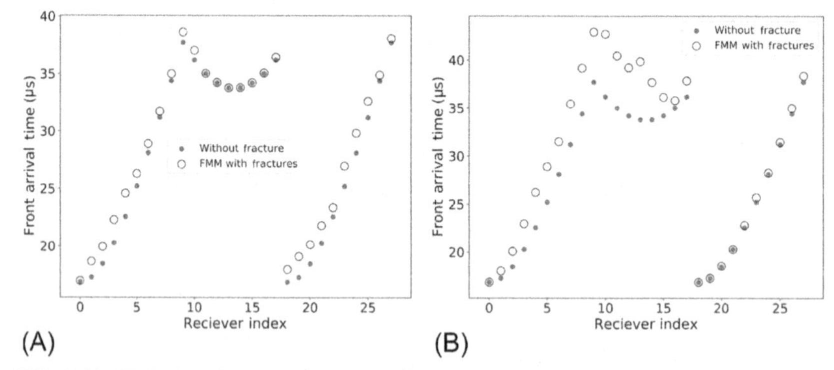

FIG. 9.11 FMM-based travel-time predictions for the 28 receivers placed around the fractured material. Comparison of compressional wavefront travel time at each sensor/receiver calculated using the FMM for (A) Case #1 and (B) Case #2, as shown in Fig. 9.10. Analytical solution for material without discontinuities is provided as a benchmark to determine the effect of discontinuities. x-Axis represents the sensor/receiver index ranging from 0 to 27; y-axis represents the time of arrival of the wavefront at the receiver/sensor. (A) Case #1 and (B) Case #2.

materials with and without discontinuities indicate the velocity anisotropy of the material containing discontinuities.

4 Methodology for developing data-driven model for the noninvasive characterization of static mechanical discontinuities in material

We perform three major tasks in chronological order: Step 1 create thousands of numerical models (realizations) of material containing various network of static discontinuities (Figs. 9.12 and 9.13A); Step 2 perform fast-marching simulations of compressional wavefront that starts from one source and propagates through the material containing discontinuities to create a dataset of compressional wavefront travel times measured at multiple receivers (Fig. 9.13B), which is combined with user-assigned label that categorically represents the overall spatial characteristics of the embedded network of static discontinuities; and Step 3 train several data-driven classification methods (Fig. 9.13C) on the dataset of compressional wavefront travel times with associated user-assigned labels to learn to noninvasively characterize the material containing discontinuities using only the compressional wavefront travel times measured at various receivers/sensors.

In this study, we focus on noninvasive characterization of materials containing embedded network of static discontinuities using limited sonic measurements.

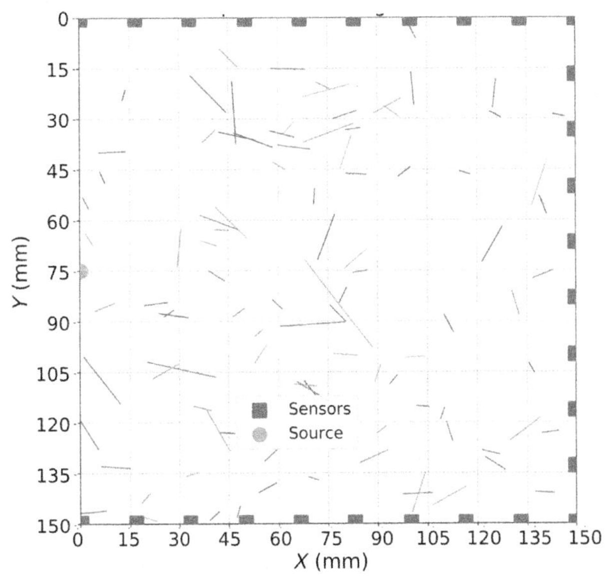

FIG. 9.12 A realization of material containing discontinuities generated using DFN model and the locations of 1 source and 28 sensors along the boundary of the material. FMM is used to simulate the compressional wavefront propagation from the source to sensors through the material.

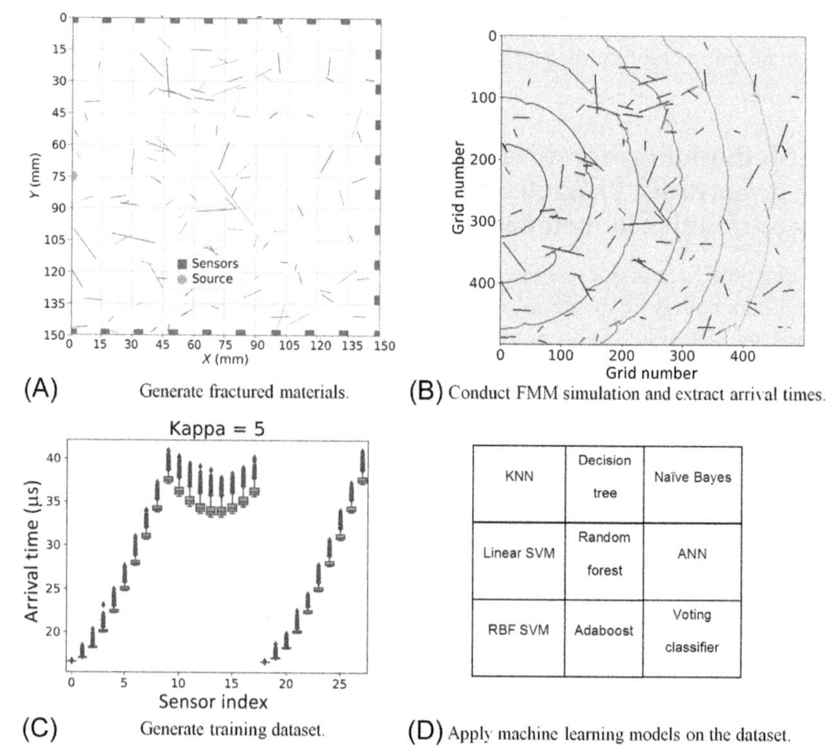

FIG. 9.13 Methodology for developing data-driven model for the noninvasive characterization of static mechanical discontinuities in material: (A) one realization of 2D numerical model of material containing discontinuities generated using DFN model and the location of one source/transmitter and the 28 receivers/sensors, (B) FMM simulation of compressional wavefront propagation through material shown in (A), (C) the arrival times computed at each sensor for 10,000 realizations, and (D) nine data-driven classifiers are trained and tested on the dataset to learn to relate the 28-dimensional feature vector to the user-assigned label of a realization.

Our hypothesis is that the data-driven models can be developed and deployed for noninvasive static fracture characterization under constrained sonic-measurement scenario. To that end, we only focus on the arrival times of compressional wavefront (due to single source) at 28 receiver/sensor locations. To develop models under data-constrained scenario, we do not use shear wave, full waveforms, wave reflections and phase changes, multiple sources/transmitters at various locations, and hundreds of sensors/receivers. When data-driven models perform well in a desired task under constrained data scenario, there is high likelihood that the data-driven approach will perform significantly better when exposed to varied measurements and larger datasets.

Fig. 9.12 presents one realization of 2D numerical model of material containing discontinuities and the location of 1 source/transmitter and the 28 receivers/sensors. Numerical models (realizations) of material containing static discontinuities of various primary orientations, distributions, and dispersions around the primary orientation are generated using a discrete fracture network (DFN) model. DFN model is used to generate multiple realizations of material containing discontinuities with specific spatial properties. Fast-marching method (FMM) is then used to calculate the compressional wavefront travel times from a single transmitter to 28 receivers/sensors placed along the boundaries of the fractured material. A compressional wavefront travel-time dataset is built by collecting the FMM simulations of travel times for each realization of fractured material. Each fractured material has a dimension of 150 mm by 150 mm and is discretized using 500 by 500 grids for accurate FMM simulation of wavefront propagation. Each grid has a length and breadth of 0.3 mm. Each realization of material containing discontinuities also has a user-assigned label corresponding to the overall spatial characteristics of the embedded network of discontinuities. For each user-assigned label, DFN model generates 10,000 realizations. FMM simulations of wavefront propagation for 10,000 realizations of fractured material containing discontinuities of specific orientation, distribution, and dispersion (similar to that shown in Fig. 9.12) take around 2 h on Dell workstation with 3.5 GHz Intel Xeon CUP and 32 GB RAM.

The set of travel times computed for the 28 receivers for each realization of fractured material along with the corresponding user-assigned label of the realization constitutes the labeled compressional wavefront travel-time (LCT) dataset (Fig. 9.13B). Data-driven models are developed on the LCT dataset to learn to relate a set of FMM-derived travel times with the user-assigned label of the realization, which describe the spatial characteristics of the embedded network of static discontinuities. When generating the multiple realizations of material containing discontinuities with a specific label (e.g., material containing randomly distributed discontinuities of random lengths exhibiting ∓ 10 degrees dispersion around the primary orientation of 60 degrees), the statistical properties of the embedded discontinuities are the same for all the realizations having the same label but differ from the statistical properties of the embedded discontinuities for realizations corresponding to other labels.

The wavefront propagation is affected by a variety of mechanical and physical properties. For purposes of generating the LCT dataset, we assume the background material is clean sandstone. The compressional wave velocity in clean sandstone is around 4500 m/s. Each discontinuity is assumed to be open and filled with air. The compressional wave velocity in discontinuity is assumed to be 340 m/s. The width of each discontinuity is 0.3 mm. The embedded network of discontinuities is generated using the discrete fracture network (DFN) method. When creating multiple realizations using DFN method, the network of discontinuities is defined by the probability distribution of location, primary orientation, dispersion, and length of the discontinuities. The location, orientation, and length of the discontinuities are characterized using various probability distributions, similar to the work done by Fadakar Alghalandis [18]. The detailed

TABLE 9.1 Set of parameters defining the numerical models (realizations) of material containing discontinuities.

Parameters	Values
Material dimension	150 mm by 150 mm
Number of sources	1
Number of receivers	28
Number of discontinuities	100
Length of discontinuities	0.3–3 mm (follows the exponential distribution)
Orientation of discontinuities	−20 to 20 or −50 to 50 degrees (follows the von Mises distribution)
Location/distribution of discontinuities	Modeled using intensity functions, such as random, Gaussian distribution
Compressional wave velocity of the background material	4500 m/s (assuming clean sandstone)
Compressional wave velocity of each discontinuity	340 m/s (assuming filled with air)

methods and parameters used to generate the numerical models (realizations) of material containing discontinuities are listed in Table 9.1.

FMM simulation is conducted on each realization (Fig. 9.13A) to simulate the compressional wave propagation (Fig. 9.13B) originating from the single source. The wavefront arrival time is computed at each sensor location. For purposes of developing data-driven model, each realization is considered as a sample, the arrival times computed at each sensor for each realization are considered as the features, and the user-assigned label corresponding to each realization is considered as target. Feature is a 28-dimensional vector. Travel times for 10,000 realizations (i.e., samples) are computed for each label (Fig. 9.13C), which represent a network of discontinuities with specific spatial characteristics. Nine data-driven classifiers (Fig. 9.13D) are trained and tested on the LCT dataset to learn to relate the 28-dimensional feature vector to the 1-dimensional target.

4.1 Classification methods implemented for the proposed fracture characterization workflow

Nine classification methods are trained on the labeled compressional wavefront travel-time dataset to learn to characterize materials containing discontinuities by using multipoint compressional wavefront travel-time measurements. Hyperparameter optimization is performed using grid search along with five-fold cross validation to ensure the generalization of the trained classifiers. The classification models will facilitate the categorization of materials containing discontinuities of various spatial characteristics solely based on multipoint measurements of compressional wavefront arrival times.

4.1.1 K-nearest neighbors (KNN) classifier

For a new, unseen, unlabeled sample, KNN selects K neighboring samples from the training/testing dataset that are most similar to the new sample. Similarity between two points is measured in terms of Minkowski distance between them in the feature space. Following that, the new unlabeled sample is assigned a class based on the majority class among the K neighboring samples selected from the training/testing dataset. KNN does not require an explicit training stage. KNN is a nonparametric model because the model does not require to learn any parameter (e.g., weight or bias). Nonetheless, to ensure the generalizability of KNN model, certain hyperparameters, such as K and distance metric, need to be adjusted so as to obtain good performance on the testing dataset. Increase in K leads to underfitting of the KNN model (high bias). When K is low, the KNN model is easily affected by outliers, resulting in model overfitting (high variance) (Fig. 9.14).

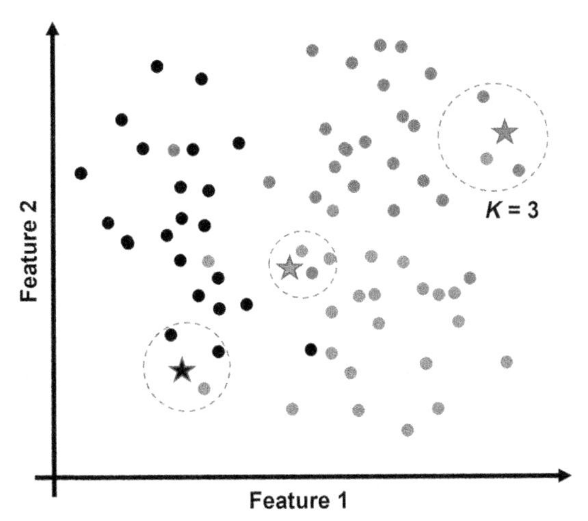

FIG. 9.14 Implementation of KNN classifier with $K = 3$ on a dataset that has two features and three classes. All training/testing samples are represented as circles. Three new, unseen, unlabeled samples are represented as stars. KNN algorithm finds three training/testing samples that are closest to the new sample and then assigns the majority class of the neighbors as the class for the new sample.

4.1.2 Support vector machine (SVM) classifier

SVM was originally designed to create highly generalizable classifier for binary classification. SVM transforms the original feature space into a higher-dimensional space based on a user-defined kernel function and then finds support vectors to maximize the separation (margin) between two classes. SVM first approximates a hyperplane that separates both the classes. Accordingly, SVM selects samples from both the classes, referred as support vectors, that are closest to the hyperplane. The total separation between the hyperplane and the support vectors is referred as margin. SVM then iteratively optimizes the hyperplane and supports vectors to maximize the margin, thereby finding the most generalizable decision boundaries. When the dataset is separable by nonlinear boundary, certain kernels are implemented in the SVM to appropriately transform the feature space. For a dataset that is not easily separable, soft margin is used to avoid overfitting by giving less weightage to classification errors around the decision boundaries. In this study, we use two SVM classifiers, one with linear kernel and the other with a radial basis function kernel (Fig. 9.15).

4.1.3 Decision tree (DT) classifier

Decision tree is a nonparametric classification method suited for multiclass classification. DT classifier processes the training dataset to build a tree-like decision structure, which starts from a root node and ends at several leaves (Fig. 9.16). All nodes between the root node and leaves are called internal

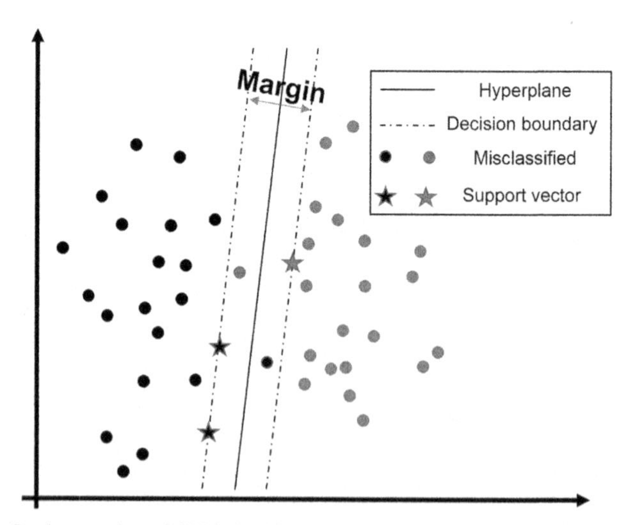

FIG. 9.15 Implementation of SVM classifier without kernel function on a dataset that has two features and two classes. All training samples are represented as circles or stars. Support vectors (denoted as *stars*) are from the training samples such that they are closest to the hyperplane among the other training samples for each of the two classes. Two training samples have been misclassified because they lie on the wrong side of the hyperplane.

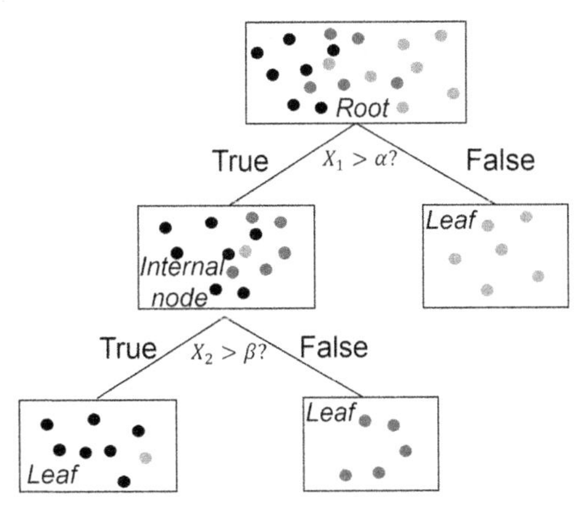

FIG. 9.16 Implementation of DT classifier on a dataset that has two features (X_1 and X_2) and three classes. Two of the three leaves in the tree are pure leaves. At each node, DT classifier finds the feature and corresponding feature threshold to perform a split, such that there is an increase in purity of the dataset after the split.

nodes. Each node is split into internal nodes and/or leaves such that there is an increase in purity of the dataset after the split, that is, each split should cause the dataset to be separated into groups that contain samples predominantly belonging to one class. At each node, the algorithm selects a feature and a threshold value of the corresponding feature, such that there is a maximum drop in entropy or impurity when the node is split using the chosen feature and the corresponding threshold. The best-case scenario during splitting is to obtain a pure leaf, which contains samples belonging to only one class. The DT algorithm does not require feature scaling. DT classifier is sensitive to noise in data and selection of training dataset due to the high variance of the method. Hyperparameter optimization is required to lower the variance at the cost of high bias. Bias of the DT classifier can be reduced at the cost of increasing variance by allowing the tree to grow till greater depth (i.e., more splits) or by allowing the leaves to contain fewer samples, such that more splits are made to obtain pure leave nodes. Nonetheless, the decision tree model is easy to interpret because the decision-making process during training and deployment can be easily understood by following the decision flow in the tree-like decision structure.

4.1.4 Random forest (RF) classifier

RF classifier is an ensemble method that trains several decision trees in parallel with bootstrapping followed by aggregation, jointly referred as bagging (Fig. 9.17). Bootstrapping indicates that several individual decision trees are trained in parallel on various subsets of the training dataset using different subsets of available features. Bootstrapping ensures that each individual decision

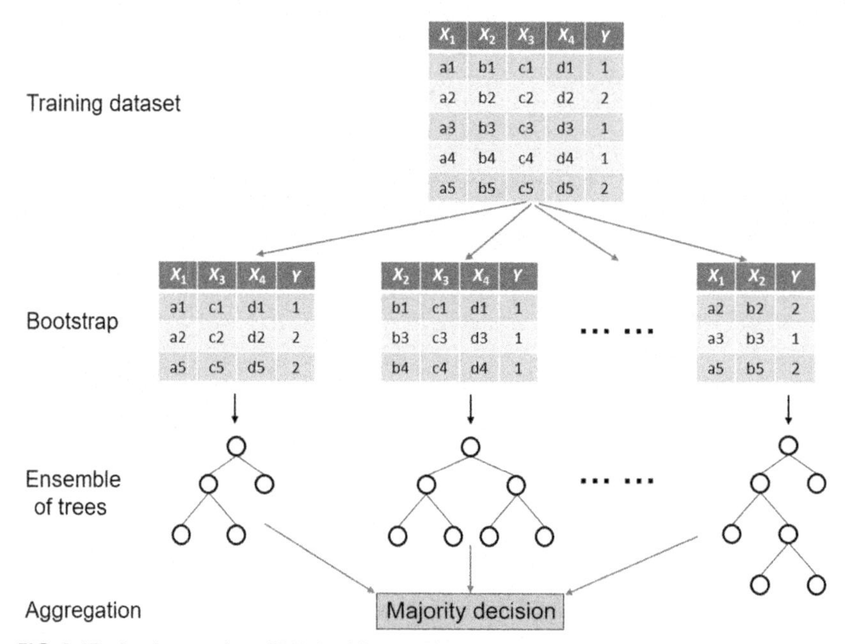

FIG. 9.17 Implementation of RF classifier on a dataset that has four features (X_1, X_2, X_3, and X_4) and two classes ($Y = 1$ and 2). RF classifier is an ensemble method that trains several decision trees in parallel with bootstrapping followed by aggregation. Each tree is trained on different subsets of training sample and features.

tree in the random forest is unique, which reduces the overall variance of the RF classifier. For the final decision, RF classifier aggregates the decisions of individual trees; consequently, RF classifier exhibits good generalization. RF classifier tends to outperform most other classification methods in terms of accuracy without issues of overfitting. Like DT classifier, RF classifier does not need feature scaling. Unlike DT classifier, RF classifier is more robust to the selection of training samples and noise in training dataset. RF classifier is harder to interpret but easier to tune the hyperparameter as compared with DT classifier.

4.1.5 AdaBoost classifier

AdaBoost is an ensemble method that trains and deploys trees in series. AdaBoost implements boosting, wherein a set of weak classifiers is connected in series such that each weak classifier tries to improve the classification of samples that were misclassified by the previous weak classifier. In doing so, boosting combines weak classifiers in series to create a strong classifier. The decision trees used in boosting methods are called "stump" because each decision tree tends to be shallow models that do not overfit but can be biased. An individual tree is trained to pay specific attention to the weakness of only the previous tree.

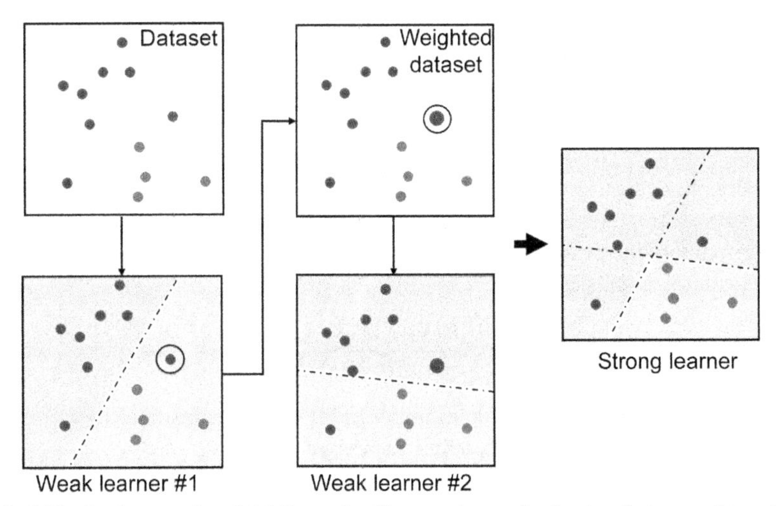

FIG. 9.18 Implementation of AdaBoost classifier on a dataset that has two features and two classes. Weak learner #2 improves on the mistake made by weak learner #1, such that the decision boundaries learnt by the two weak learners can be combined to form a strong learner. In this case, each weak learner is a decision tree, and AdaBoost classifier (i.e., strong learner) combines the weak learner in series.

The weight of a sample misclassified by the previous tree will be boosted so that the subsequent tree focuses on correctly classifying the previously misclassified sample. The classification accuracy increases when more weak classifiers are added in series to the model; however, this may lead to severe overfitting and drop in generalization capability. AdaBoost is suited for imbalanced datasets but underperforms in the presence of noise. AdaBoost is slower to train. Hyperparameter optimization of AdaBoost is much more difficult than RF classifier (Fig. 9.18).

4.1.6 Naïve Bayes (NB) classifier

Naïve Bayes classifier is a probabilistic classifier based on Bayes' theorem, which assumes that each feature makes an independent and equal contribution to the target class. NB classifier assumes that each feature is independent and does not interact with each other, such that each feature independently and equally contributes to the probability of a sample to belong to a specific class. NB classifier is simple to implement and computationally fast and performs well on large datasets having high dimensionality. NB classifier is conducive for real-time applications and is not sensitive to noise. NB classifier processes the training dataset to calculate the class probabilities $P(y_i)$ and the conditional probabilities, which define the frequency of each feature value for a given class value divided by the frequency of instances with that class value. NB classifier best performs when correlated features are removed because correlated features will be voted twice in the model leading to the overemphasis of the importance of the correlated features.

For purposes of explanation, NB classifier will be applied on training samples, such that each sample has three features (X_1, X_2, X_3) and a single label (y_i), where $i = 1$ or 2. Therefore, NB classifier needs to accomplish the binary classification task of assigning a single label y, either y_1 or y_2, to a sample based on its feature values. As the first goal, the algorithm processes the training dataset to approximate the probability of a class y_i for a given set of feature values (X_1, X_2, X_3) that is expressed as

$$P(y_i|X_1, X_2, X_3) = \frac{P(X_1|y_i)P(X_2|y_i)P(X_3|y_i)P(y_i)}{P(X_1)P(X_2)P(X_3)} \tag{9.5}$$

For a specific dataset, the denominator in Eq. (9.5) is constant. So, Eq. (9.5) can be simplified to a proportionality expressed as

$$P(y_i|X_1, X_2, X_3) \propto P(X_1|y_i)P(X_2|y_i)P(X_3|y_i)P(y_i) \tag{9.6}$$

In Eq. (9.6), individual $P(X_j|y_i)$, where $j = 1, 2$ or 3, can be calculated based on the assumption of the distributions of the features. For discrete features, the feature distribution is assumed to follow multinomial distribution, whereas, for continuous-valued features, the feature distribution is assumed to follow Gaussian distribution. To calculate the statistical parameters (such as mean and variance) of the feature distributions, the dataset is first segmented by the class, and then, the parameters are calculated for each class to enable the calculation of $P(X_j|y_i)$. Finally, the algorithm estimates the probability of a given sample with known feature values to belong to a certain class by picking the y_i that leads to the largest value of $P(X_1|y_i)P(X_2|y_i)P(X_3|y_i)P(y_i)$. This statement is mathematically represented as

$$y = \text{argmax}_{y_i} P(X_1|y_i)P(X_2|y_i)P(X_3|y_i)P(y_i) \tag{9.7}$$

This is referred to as the maximum a posteriori decision rule; in other words, pick the hypothesis that is most probable.

4.1.7 Artificial neural network (ANN) classifier

ANN is composed of consecutive layers, where each layer contains several computational units in parallel (Fig. 9.19). Each computational unit is referred as the neuron. The layer of ANN that reads the features is called the input layer, while the layer of ANN that generates the final targets is called the output layer. In our case, the input layer has 28 neurons to read the 28 travel-time measurements for each sample. The output layer has either four or eight dimensions based on the number of classes to be assigned. Any layer between the input layer and output layer is called the hidden layer. The output of the previous layer is taken as input for the next layer. In a densely connected network, each neuron in a layer is connected to all the neurons in

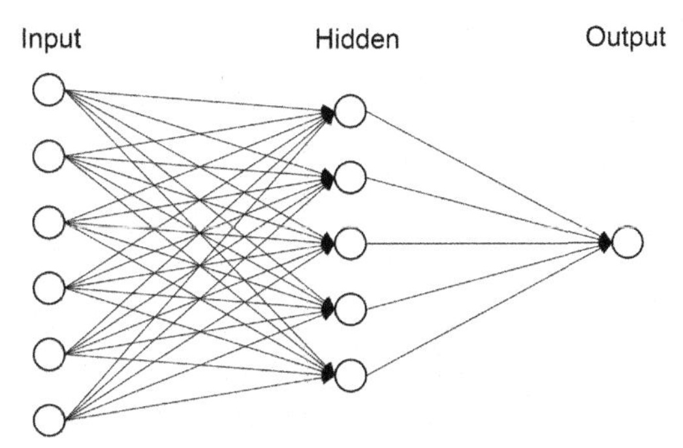

FIG. 9.19 Artificial neural network with an input layer, one hidden layer, and an output layer. ANN is fed six-dimensional feature vector. Hidden and output layers have five and one neurons, respectively.

the immediately previous layer (input layer does not have a preceding layer). A neuron is a mathematical transformation that sums all the inputs to the neuron from the neurons of the previous layer multiplied by corresponding weights of the corresponding connections between the neurons and then applies a nonlinear filter or activation function (generally having a sigmoid shape) on the summation to generate an output (Fig. 9.20). For example, if the activation function of the output neuron is a sigmoid function ranging from 0 to 1, the output will be in the range [0, 1]. The neurons in hidden layers generally use the rectified linear unit (ReLU) or tanh as the activation

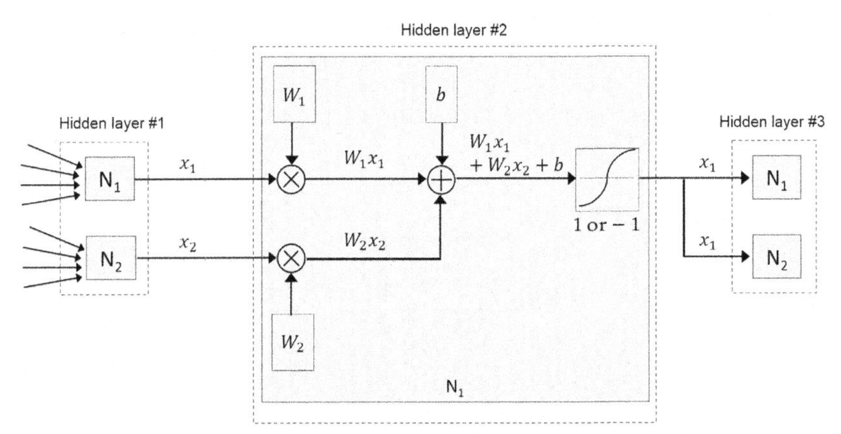

FIG. 9.20 An illustration of three hidden layers of an artificial neural network. Hidden layers #1, #2, and #3 contain 2, 1, and 2 neurons, respectively. Two neurons (N_1 and N_2) in hidden layer #1 are connected to four neurons in the previous layer. Outputs of the two neurons (N_1 and N_2) of hidden layer #1 are fed to one neuron (N_1) in hidden layer #2. The single output of hidden layer #2 is fed to the two neurons (N_1 and N_2) of the hidden layer #3. The inner working of the neuron (N_1) of hidden layer #2 is presented for better explanation of the neuron as a computation unit.

function, whereas the activation function of neurons in the output layer for a regression task is usually a linear filter and for a classification task is usually a softmax filter.

When a feature vector is fed to the input layer of ANN model, the input values are propagated forward through the network, neuron by neuron and layer by layer, until the sequence of mathematical transformations (weight, bias, and activation function) applied on the feature vector reaches the output layer. Activation functions are added to each neuron to allow the ANN to account for nonlinear behavior in the training dataset. Stacking of the layers facilitates the nonlinear capabilities of ANN for complex function approximation. The flexibility of ANN leads to overfitting issues that can be handled through hyperparameter tuning. ANN training involves feedforward of data signals to generate the output and then the backpropagation of errors for gradient descent optimization. During the process of training an ANN model, first, the weights/biases for the neurons in the output layer are updated; then, the weights/biases of the neurons of each preceding layer are iteratively updated, till the weights/biases of neurons of the first hidden layer are updated.

4.1.8 Voting classifier

Voting classifier is a metaclassifier that makes prediction by combining the predictions of several individual classifiers based on certain predefined strategy of voting (Fig. 9.21). This is also referred as stacking or metaensembling, which is a technique involving the combination of several first-level predictive models to generate a second-level predictive model, which tends to outperform all of them. Second-level model utilizes the strengths of each first-level model where they perform the best while averaging the impact of their weaknesses in other parts of the dataset. Each classifier in the voting classifier is trained and tuned on the same dataset in parallel. Voting classifier makes the final prediction by combining the predictions of individual classifiers using soft voting or hard voting.

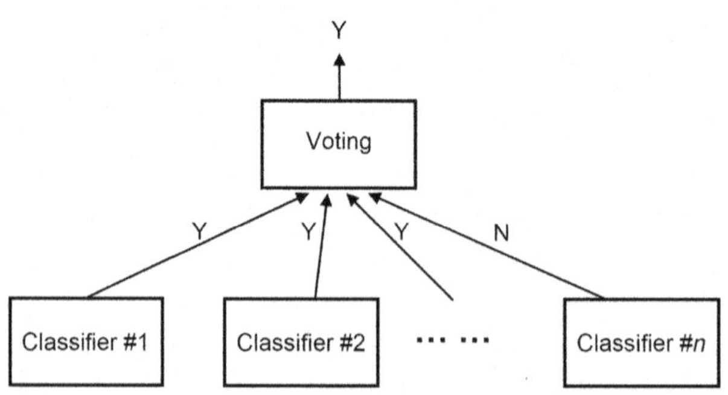

FIG. 9.21 Implementation of voting classifier on a dataset that has two classes (Y and N). Voting classifier combines the predictions of n classifiers using hard voting. Consequently, the final decision of the voting classifier is Class Y because the total number of classifiers that predicted Class Y exceeds $n/2$.

Soft voting can lead to different decision as compared with hard voting. Soft voting entails computing a weighed sum of the predicted probabilities of all models for each class. Hard voting uses predicted class labels for majority rule voting, which is a simple majority vote for accuracy. For example, a voting classifier ensembles three classifiers. When the individual classifiers 1, 2, and 3 predict Class Y, Class N, and Class N, respectively, hard voting will lead to Class N as the ensemble decision because 2/3 classifiers predict the Class N. When the individual classifiers 1, 2, and 3 predict Class Y with probability of 90%, 45%, and 36%, soft voting will lead to final prediction of Class Y because the average probability is $(90 + 45 + 36)/3 = 57\%$. Soft voting can improve on hard voting because it takes into account more information; it uses each classifier's uncertainty in the final decision. We use hard voting in this study.

5 Results for the classification-based noninvasive characterization of static mechanical discontinuities in materials

5.1 Characterization of material containing static discontinuities of various dispersions around the primary orientation

5.1.1 Background

In this section, nine classifiers (discussed in Section 4.1) process compressional wavefront travel times to categorize materials containing discontinuities in terms of the dispersion around the primary orientation. The von Mises distribution is used to generate 100 discontinuities with various dispersions around a primary orientation. The von Mises probability density function for the orientation of discontinuity can be expressed as [19]

$$f(x|\,\mu,k) = \frac{e^{k\cos(x-\mu)}}{2\pi I_0(k)} \tag{9.8}$$

where x is the orientation of discontinuity, μ is the mode of the orientations of the discontinuities, k is the concentration (inverse of dispersion) of the orientations around the mode, and I_0 is the modified Bessel function of order 0. The distribution is clustered around μ and the dispersion around the mode is expressed as $1/k$. The dispersion $1/k$ controls the deviation of the orientations around the mode μ. Mode and dispersion $(1/k)$ are statistical parameters of the von Mises distribution analogous to mean and standard deviation of Gaussian distribution. von Mises distribution is an approximation to the wrapped normal distribution resulting from the "wrapping" of a normal distribution around a unit circle. Three types of material containing discontinuities are generated with concentration parameter (k) of 0, 5, and 1000 (Fig. 9.22).

Fig. 9.22 illustrates the materials containing discontinuities (upper row) and the corresponding distribution of dispersion around the primary orientation of

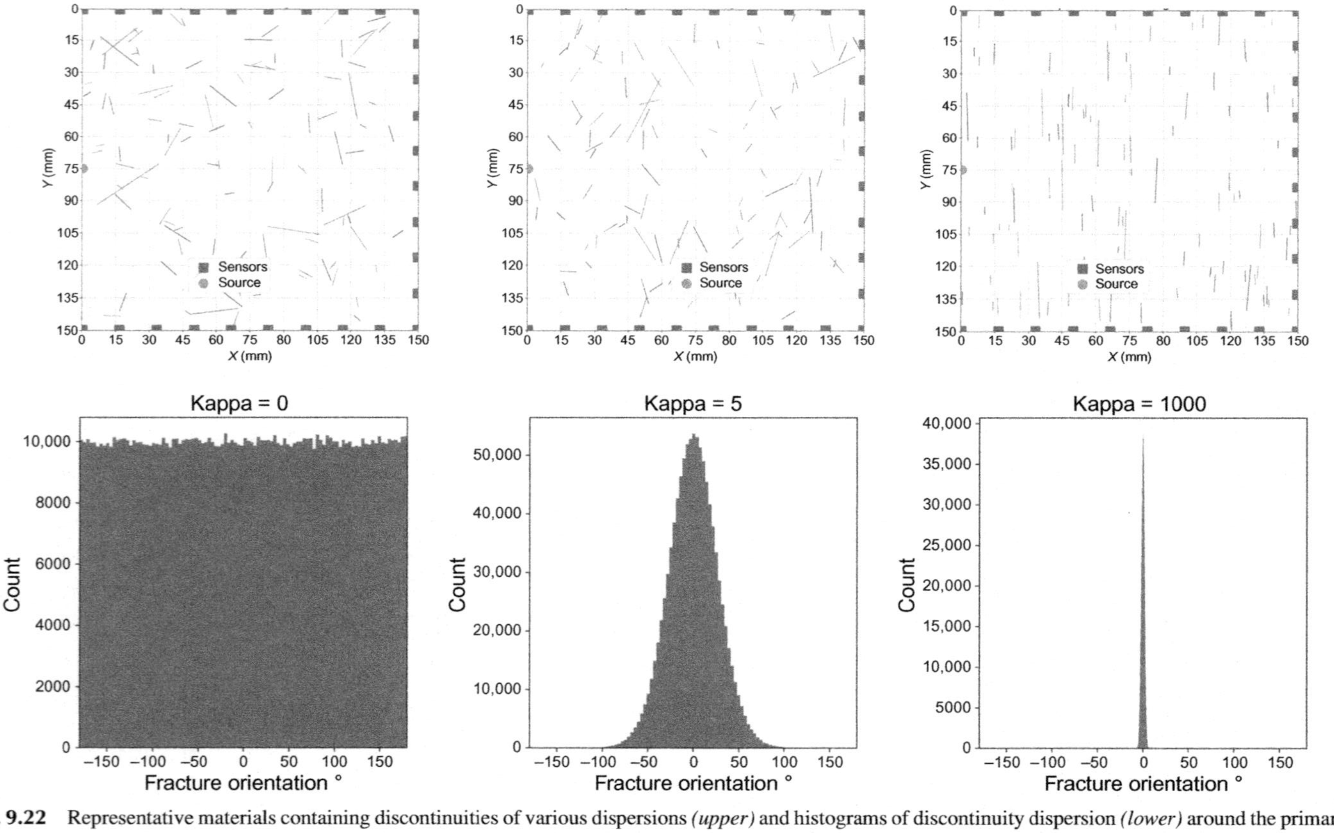

FIG. 9.22 Representative materials containing discontinuities of various dispersions *(upper)* and histograms of discontinuity dispersion *(lower)* around the primary orientation parallel to *y*-axis. For kappa = 0 *(left)*, the dispersion is large, and the dispersion is a uniform distribution. For kappa = 5 *(middle)*, the dispersion is mostly between −50 and 50 degrees around the primary orientation of discontinuity. For kappa = 1000 *(right)*, the dispersion is negligible, and the orientations of discontinuity are almost parallel to each other and to the primary orientation parallel to *y*-axis.

discontinuities (lower row). Three concentration parameters (kappa, k, and inverse of dispersion) equal to 0, 5, and 1000 are selected for this study. When the kappa equals 0, the orientations of discontinuity follow a uniform distribution. When the kappa equals 5, the orientations are centered at 0 degree and primarily dispersed between -50 and 50 degrees. When the dispersion parameter equals 1000, all the fractures are almost parallel to the y-axis, which is 0 degree. These three types of materials represent high, intermediate, and low dispersivity of discontinuities around the primary orientation.

When compressional wave propagates through such materials containing discontinuities, the multipoint measurements of compressional wavefront travel times are too complicated for humans to recognize and differentiate in terms of the dispersion of the embedded network of discontinuities that give rise to the measured travel-time signature. There do not exist simple simulation techniques and physical law to characterize such embedded network of discontinuities based on the measurements of the travel times. Machine learning provides a possible way to extract complex relationships/patterns from the multipoint wavefront travel-time measurements and relate them to the dispersion of the embedded network of discontinuities. To that end, DFN model was used to generate 10,000 samples (realizations) for each of the three types of dispersion around the primary orientation. In total, the training/testing dataset contains 30,000 samples. The compressional wavefront travel times originate from one source and are measured at 28 sensors/receivers located on the three boundaries of the material, as shown in Fig. 9.22. The boundary on which one transmitter/source is placed is known as the source-bearing boundary. There are no receivers on the source-bearing boundary. The receivers are referred using indices ranging from 0 to 27. For purposes of developing data-driven models, each sample in the dataset contains 28-dimensional feature vector, representing the travel times measured at the 28 receivers. Each sample was assigned a label, denoting a material type, for purpose of model training and testing. The user-assigned label categorically represents the dispersion around primary orientation of the embedded network of static discontinuities in a sample. The dataset of compressional wavefront travel times with associated user-assigned labels is processed by the nine classifiers to learn to noninvasively characterize the material containing discontinuities in terms of dispersion around primary orientation. The classification task is to classify the 28-dimensional travel-time measurements into the four classes representing the dispersion of the embedded network of discontinuities.

5.1.2 Model accuracy

Nine classifiers are applied to the dataset: KNN, SVM with linear kernel, SVM with radial basis function (RBF) kernel, decision tree, random forest, AdaBoost, Naïve Bayes, ANN, and voting classifier. These nine classifiers cover the most popular classifiers used in machine learning and data-driven approaches.

TABLE 9.2 Classification accuracy of the nine classifiers on the test dataset, also referred as the generalization performance, for the classification-based noninvasive characterization of material containing static discontinuities of various dispersions around the primary orientation.

Classifiers	Accuracy
KNN	0.60
Linear SVM	0.65
RBF SVM	0.65
Decision tree	0.59
Random forest	0.64
AdaBoost	0.64
Naïve Bayes	0.59
ANN	0.60
Voting classifier	0.65

Greener shade is good performance, whereas redder shade is bad performance.

Seventy percent of the samples in the available dataset (30,000 samples with 28 features and one target label) are randomly selected as training samples, and the remaining dataset forms the testing samples. Grid search and cross validation method are used to tune the hyperparameters of the classifiers to avoid overfitting. The accuracy of the trained classifiers on the testing dataset, also referred as the generalization performance, is shown in Table 9.2. The nine classifiers have low generalization performance. The overall accuracy is around 0.6 for the nine classifiers. RBF, SVM, and voting classifier have the highest accuracy of 0.65. It can be concluded that classification-based noninvasive fracture characterization is not suitable for the three types of dispersions considered in this study. This poor performance is due to the similarity between the orientations of several discontinuities for the three types of dispersion studied in this section.

To better assess the performance of the classifiers, we reviewed the precision, recall, and F_1 score of the best-performing voting classifier (Table 9.3). A good classification performance exhibits values close to 1 for these three parameters. Material Type #1, material containing discontinuities of intermediate dispersion (kappa $=5$) around the primary orientation, has the lowest F_1 score due to low precision and recall. Interestingly, Material Type #0, material containing discontinuities of high dispersion (kappa $=0$) around the primary orientation, has the best precision and recall.

TABLE 9.3 Classification report of the voting classifier on the testing dataset for the classification-based noninvasive characterization of material containing static discontinuities of various dispersions around the primary orientation.

Material type	Precision	Recall	F_1 score	Support
0	0.73	0.78	0.75	3003
1	0.53	0.4	0.46	2999
2	0.66	0.78	0.72	2998
Avg/total	0.64	0.65	0.64	9000

5.2 Characterization of material containing static discontinuities of various primary orientations

5.2.1 Background

In this section, nine classifiers (discussed in Section 4.1) process compressional wavefront travel times to categorize materials containing discontinuities in terms of the primary orientations of the discontinuities. The von Mises distribution is used to generate 100 discontinuities of specific dispersion around various primary orientations. The presence of dispersion leads to randomness in the orientations of the discontinuities. Various networks of discontinuities with distinct primary orientation and dispersion are created as listed in Table 9.4. The four datasets with associated user-assigned labels, as listed in Table 9.4, are processed by the nine classifiers to learn to noninvasively characterize the material containing discontinuities in terms of the primary orientation in the presence of specific dispersions in the orientation. User-assigned labels denote various material types.

For the Datasets #1 and #2 (Table 9.4), user assigns one of the four labels, denoting four material types, to each sample because the samples are generated for materials containing discontinuities having primary orientation of 0, 45, 90, or 135 degrees and concentrations of 10 and 50, respectively, in terms of kappa. A concentration of 10 indicates dispersion of +50 to −50 degrees around the primary orientation. A concentration of 50 indicates dispersion of +20 to −20 degrees around the primary orientation. Similarly, for the Datasets #3 and #4 (Table 9.4), user assigns one of the eight labels to each sample because the samples are generated for materials containing discontinuities having primary orientation of 0, 22.5, 45, 67.5, 90, 112.5, 135, and 157.5 degrees and concentrations of 10 and 50, respectively, in terms of kappa. Dataset #2 and Dataset #4 are generated by materials containing discontinuities of lower dispersion around primary orientation as compared with Dataset #1 and Dataset #3, respectively; in other words, the materials that generate Dataset #2 and Dataset #4 have more directionally aligned discontinuities. In Fig. 9.23, the upper and

TABLE 9.4 Mode and concentration (Kappa; inverse of dispersion) of the von Mises distribution to create the four datasets to study the feasibility of classification-based noninvasive characterization of material containing static discontinuities of various primary orientations with specific dispersions around the primary orientation.

Dataset	Number of classes	Mode (primary orientation)	Kappa (orientation range)	Total number of samples
#1	4	0, 45, 90,135 degrees	10 (+50 to −50 degrees)	40,000
#2	4	0, 45, 90,135 degrees	50 (+20 to −20 degrees)	40,000
#3	8	0, 22.5, 45, 67.5, 90, 112.5, 135, 157.5 degrees	10 (+50 to −50 degrees)	80,000
#4	8	0, 22.5, 45, 67.5, 90, 112.5, 135, 157.5 degrees	50 (+20 to −20 degrees)	80,000

lower plots show representative materials containing discontinuities from Dataset #1 and #2, respectively, for the four user-defined classes (i.e., material types) having distinct primary orientations.

Machine learning provides a possible way to extract complex relationships/patterns from the multipoint wavefront travel-time measurements and relate them to the primary orientation of the embedded network of discontinuities. To that end, DFN model was used to generate 10,000 samples (realizations) for each type of primary orientation. Each type of material with certain orientation of discontinuities is embedded with 100 discontinuities. Datasets #1 and #2 contain 40,000 samples each, whereas Datasets #3 and #4 contain 80,000 samples each (Table 9.4). The compressional wavefront travel times originate from one source and are measured at 28 sensors/receivers located on the three boundaries of the material, as shown in Fig. 9.23, similar to those in Fig. 9.22. The source-receiver configuration is like the one used in Section 5.1. The classification task for the four datasets is to classify the 28-dimensional travel-time measurements into the four or eight classes representing the primary orientation of the embedded network of discontinuities.

5.2.2 Model accuracy

Nine classifiers are applied to the dataset: KNN, SVM with linear kernel, SVM with radial basis function (RBF) kernel, decision tree, random forest,

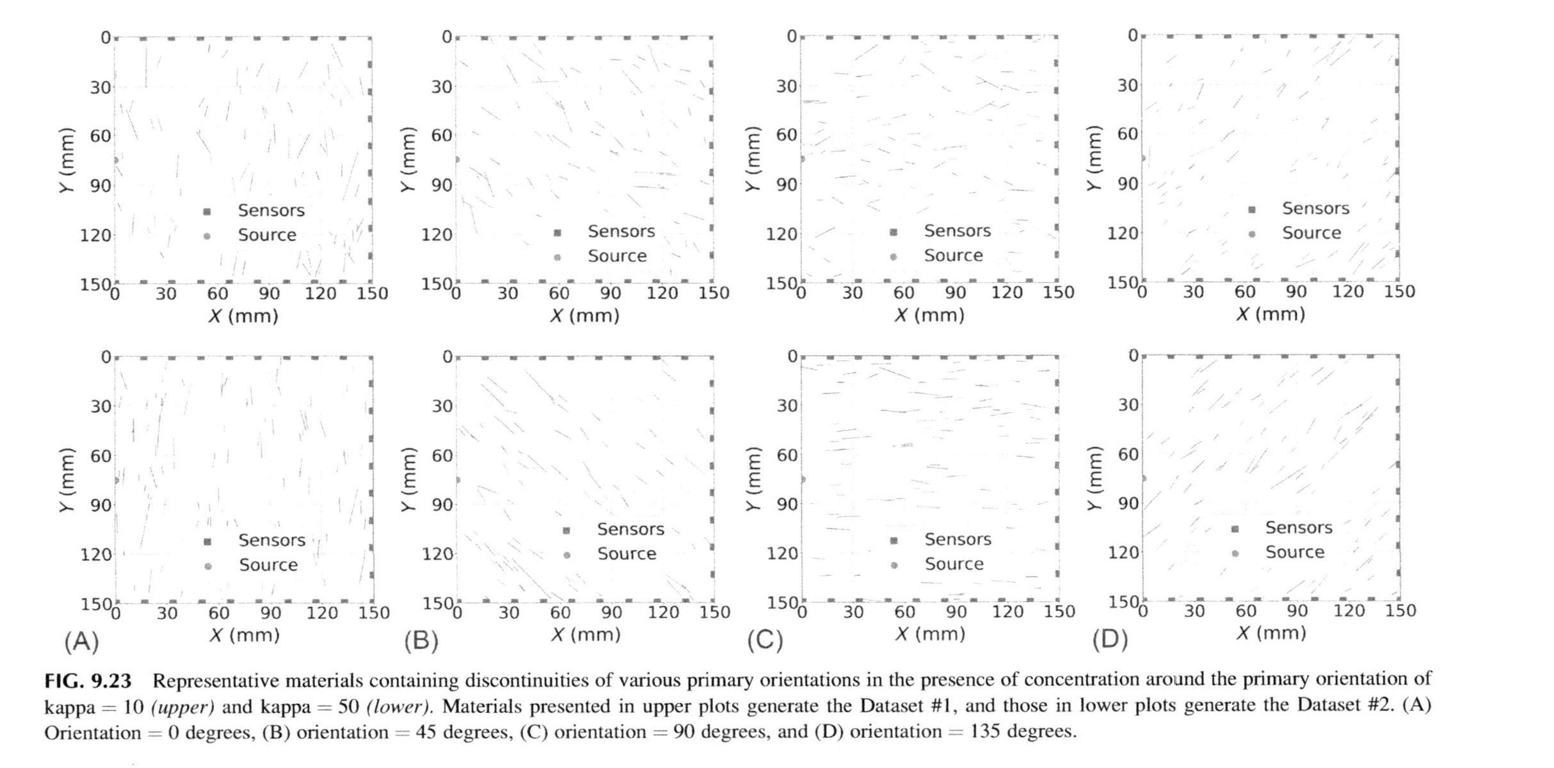

FIG. 9.23 Representative materials containing discontinuities of various primary orientations in the presence of concentration around the primary orientation of kappa = 10 *(upper)* and kappa = 50 *(lower)*. Materials presented in upper plots generate the Dataset #1, and those in lower plots generate the Dataset #2. (A) Orientation = 0 degrees, (B) orientation = 45 degrees, (C) orientation = 90 degrees, and (D) orientation = 135 degrees.

AdaBoost, Naïve Bayes, ANN, and voting classifier. Seventy percent of the samples in the each of the four dataset (40,000 or 80,000 samples with 28 features and one target label) are randomly selected as training samples, and the remaining dataset forms the testing samples. Grid search and cross validation method are used to tune the hyperparameters of the classifiers to avoid overfitting. The accuracy of the trained classifiers on the testing dataset, also referred as the generalization performance, is shown in Table 9.5. The nine classifiers have high generalization performance for the Datasets #1 and #2 and low performance for the Datasets #3 and #4. Classifiers perform near perfect for the four classes in Dataset #2 having dispersion between ∓ 20 degrees around the primary orientation, such that SVM, ANN, random forest, and voting classifiers have the best performance with a classification accuracy of 0.98. With the increase in dispersion to ∓ 50 degrees, the classification accuracy for the four classes in Dataset #1 is around 0.91. At a dispersion of ∓ 20 degrees, SVM, ANN, and voting classifiers can achieve

TABLE 9.5 Classification accuracy of the nine classifiers on the test dataset, also referred as the generalization performance, for the classification-based noninvasive characterization of material containing static discontinuities of various primary orientations.

Classifiers	Accuracy for Dataset #1: four orientation kappa = 10	Accuracy for Dataset #2: four orientation kappa = 50	Accuracy for Dataset #3: eight orientation kappa = 10	Accuracy for Dataset #4: eight orientation kappa = 50
KNN	0.87	0.95	0.57	0.69
Linear SVM	0.92	0.99	0.67	0.88
RBF SVM	0.92	0.99	0.68	0.86
Decision tree	0.82	0.95	0.55	0.77
Random forest	0.90	0.98	0.66	0.86
AdaBoost	0.91	0.98	0.64	0.85
Naïve Bayes	0.81	0.95	0.56	0.73
ANN	0.91	0.99	0.65	0.88
Voting classifier	0.92	0.99	0.69	0.89

a classification accuracy of 0.91 for the eight classes in Dataset #4. Classification of eight classes in the presence of dispersion of ∓50 degrees exhibits low generalization performance. KNN, Naïve Bayes, and decision trees are the poorest performing classifiers for all the datasets. SVM and voting classifiers are the best-performing classifiers for all the datasets.

5.2.3 Sensor/receiver importance

The compressional wavefront travel times originate from one source and are measured at 28 sensors/receivers located along the three boundaries of the material, as shown in Figs. 9.22 and 9.23. Each sample used for training or testing the classifiers comprises 28-dimensional feature vector, representing the travel times measured at the 28 receivers. Feature permutation method can be used to compute the importance of each sensor/receiver to the proposed classification-based noninvasive fracture characterization. Feature permutation method evaluates the importance of a sensor/receiver with respect to a specific classifier by sequentially replacing values of each feature (i.e., a sensor/receiver measurement) in the test dataset with random noise (noninformative data) having similar statistical distribution as the original feature values and then quantifying the drop in the performance of the trained classifier applied on the test data containing the replaced, noninformative feature values. The random noise for purposes of feature replacement is generated by shuffling the original values of a feature to be replaced; this preserves the original mean, standard deviation, and other statistical properties when replacing values corresponding to a feature. Each feature (i.e., a sensor measurement) in the dataset is replaced one by one, and the importance of the features is ranked in terms of the drop in the performance, such that important feature (i.e., a sensor measurement) causes a significant drop in the generalization performance of a pretrained classifier when applied to the test data.

The importance of sensors for the purposes of classification-based noninvasive characterization of material containing static discontinuities of various primary orientations is shown in Fig. 9.24. The importance measures are determined by computing the sensitivity of a model to random permutations of feature values (i.e., sensor measurements). The importance score quantifies the contribution of a certain feature to the predictive performance of a model in terms of how much a chosen evaluation metric deviates when the feature becomes noninformative. In Fig. 9.24, the sensors located on the boundary opposite to source are important for classification. Wavefront arrival time measured by sensors on boundaries adjacent to the transmitter-bearing boundary is not as important as those on the opposite boundary. When the network of discontinuities is horizontal (parallel to x-axis), the sonic wave minimally interacts with the discontinuities. However, when the network of discontinuities is vertical (parallel to y-axis), the sonic wave significantly interacts with discontinuities. Consequently, changes in orientation of the discontinuities will alter the

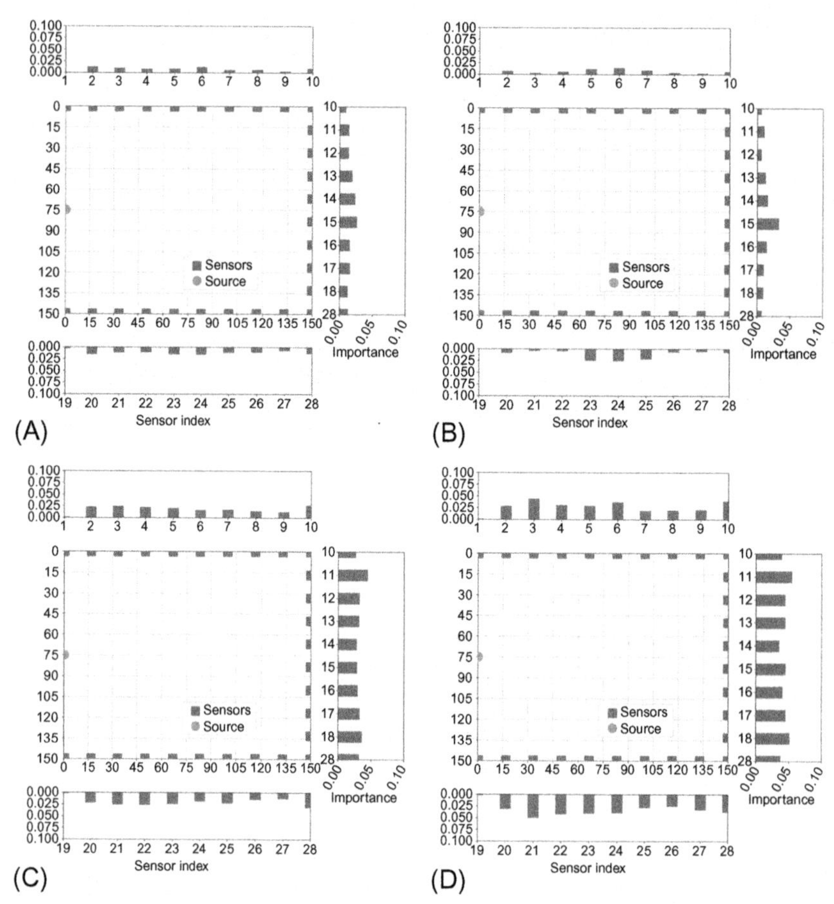

FIG. 9.24 Importance of sensor/receiver calculated using the feature permutation method for each of the four datasets/experiments. The importance values are computed as the reduction in performance of the trained voting classifier when feature values become non informative. (A) Experiment #1. Four orientations, kappa = 10, (B) Experiment #2. Four orientations, kappa = 50, (C) Experiment #3. Eight orientations, kappa = 10, and (D) Experiment #4. Eight orientations, kappa = 50.

travel time measured at various sensors/receivers. Classifiers can learn to detect the differences in travel times and relate them to the various orientations of discontinuities. Interestingly, for Dataset #2, the sensor exactly opposite to the transmitter is the most important, while those on the boundaries adjacent to transmitter-bearing boundary are the least important (Fig. 9.24B). When developing classifiers for detecting eight classes/orientations (Fig. 9.24C and D), the sensors on the adjacent boundaries are much more important than those required for four-class classification (Fig. 9.24A and B). With increase in dispersion, the sensors on the boundaries adjacent to the transmitter boundary become more important.

5.3 Characterization of material containing static discontinuities of various spatial distributions

5.3.1 Background

Due to the variations in geomechanical properties and stress, various types of spatial distribution of discontinuities develop in materials. In this section, nine classifiers (discussed in Section 4.1) process compressional wavefront travel times to categorize materials containing discontinuities in terms of the distributions of the discontinuities. A intensity function is used to define the spatial variability in the density of discontinuities for a material; following that, the locations of 100 fractures are assigned using the nonhomogeneous Poisson process [18, 20]. Intensity functions and the corresponding materials containing discontinuities are shown in Fig. 9.25.

We implement acceptance-rejection method to generate the nonhomogeneous Poisson process for assigning the location of discontinuities, as shown in the upper plots of Fig. 9.25. First, an intensity function $\lambda(x,y)$ is defined on the domain of investigation denoting the aerial extent of the material. The intensity function is normalized with respect to the maximum value of the intensity (λ^*) in the domain to obtain an acceptance probability $p(x,y) = \lambda(x,y)/\lambda^*$. Homogeneous Poisson process is used to pick a location of discontinuity in the material, which is then accepted or rejected based on the acceptance probability $p(x,y)$, thereby embedding discontinuities in the material based on a nonhomogeneous Poisson process. According to the intensity function (bottom plots of Fig. 9.25), the locations of discontinuity generated by the homogeneous Poisson process from the region with a higher intensity are more likely to be retained.

Fig. 9.25A shows a material containing discontinuities generated using constant intensity function. The distribution of discontinuities is similar to random distribution. Fig. 9.25B shows material containing discontinuities generated using linear intensity function $\lambda(x,y) = y$. The possibility of acceptance increases linearly along the y-axis toward the positive y-direction. The origin of the coordinate system is located at the upper left corner of the material. Consequently, the density of discontinuities is higher near the lower boundary adjacent to the transmitter-bearing boundary. Fig. 9.25C shows the material containing discontinuities generated using a unimodal Gaussian function as the intensity function:

$$\lambda(x,y) = c * \exp^{\left(-d*\left((x-x_0)^2 + (y-y_0)^2\right)\right)} \tag{9.9}$$

where x_0 and y_0 are the center of the Gaussian distribution, d controls the variance of the distribution, and c controls the minimum value of the intensity function. In this experiment, c is set to 1, d is set to 0.00005, and x_0 and y_0 are both set to 250, which is the center of material. The acceptance probability is high at the center of the material and drops away from the center of the material. Fig. 9.25D

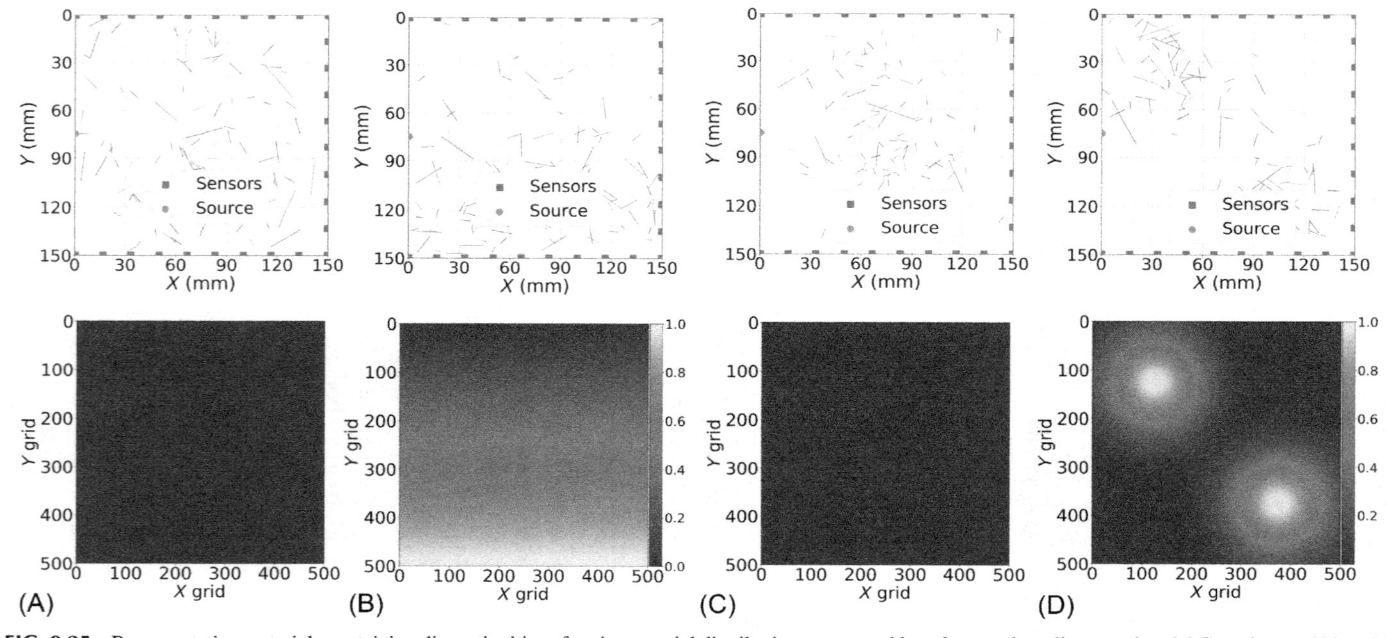

FIG. 9.25 Representative materials containing discontinuities of various spatial distributions generated based on random, linear, unimodal Gaussian, and bimodal Gaussian intensity functions. Materials similar to those presented in upper plots are used to generate the four datasets to train classifiers for the noninvasive characterization of materials containing static discontinuities of various distributions. (A) Random, (B) linear, (C) single Gaussian, and (D) double Gaussian.

shows the material containing discontinuities generated using bimodal Gaussian function as the intensity function.

Machine learning provides a possible way to extract complex relationships/patterns from the multipoint wavefront travel-time measurements and relate them to the spatial distribution of the embedded network of discontinuities. To that end, DFN model was used to generate 10,000 samples (realizations) for each type of spatial distribution. Each type of material with certain orientation of discontinuities is embedded with 100 discontinuities. The dataset used for this study contains travel times generated for the four types of materials with different types of discontinuity distributions. In total, 40,000 arrival times are gathered from four types/classes of material and serve as the training/testing dataset. The classification task is to classify the 28-dimensional travel-time measurements into the four classes representing the spatial distribution of the embedded network of discontinuities.

5.3.2 Model accuracy

Nine classifiers are applied to the dataset: KNN, SVM with linear kernel, SVM with radial basis function (RBF) kernel, decision tree, random forest, AdaBoost, Naïve Bayes, ANN, and voting classifier. Seventy percent of the samples in the available dataset (40,000 samples with 28 features and one target label) are randomly selected as training samples, and the remaining dataset forms the testing samples. Grid search and cross validation method are used to tune the hyperparameters of the classifiers to avoid overfitting. The accuracy of the trained classifiers on the testing dataset, also referred as the generalization performance, is shown in Table 9.6. The nine classifiers have good generalization performance. Voting classifier has the best performance with a classification accuracy of 0.86. SVM RBF, ANN, AdaBoost, and random forest exhibit good generalization performance close to 0.82. Naïve Bayes is the poorest performing classifier, because the features are not independent and their contributions to the outcome are also dependent.

5.3.3 Sensor importance

Changes in the spatial distribution of the discontinuities will alter the travel times measured at various sensors/receivers. Classifiers can learn to detect the differences in travel times measured at several sensors and relate them to the various spatial distributions of discontinuities. For materials with discontinuities having unimodal Gaussian distribution (Fig. 9.25C), the discontinuities are concentrated around the center, and the travel times measured by sensors #2 (top) and #20 (bottom) are lower than the travel times measured by sensors #2 and #20 for materials containing randomly distributed fractures (Fig. 9.25A). For materials with discontinuities having linear distribution (Fig. 9.25C), the travel time measured by sensor #20 is higher than that measured by sensor #2. Statistically speaking, the four types of distribution of discontinuities can be

TABLE 9.6 Classification accuracy of the nine classifiers on the test dataset, also referred as the generalization performance, for the classification-based noninvasive characterization of material containing static discontinuities of various spatial distributions.

Estimator	Accuracy
KNN	0.74
Linear SVM	0.79
RBF SVM	0.82
Decision tree	0.78
Random forest	0.84
AdaBoost	0.83
Naïve Bayes	0.69
ANN	0.83
Voting classifier	0.86

Greener shade is good performance, whereas redder shade is bad performance.

differentiated by examining the travel times measured by sensors #2 and #20. The sonic wave reaching the sensors that are closer to the transmitter, on the boundary adjacent to the transmitter-bearing boundary, interacts more with the discontinuities when the distribution of discontinuity is random as compared with the three other distributions.

The importance of sensors for the purposes of classification-based noninvasive characterization of material containing static discontinuities of various spatial distributions is shown in Fig. 9.26. The importance measures are determined by computing the sensitivity of a model to random permutations of feature values (i.e., sensor measurements). The importance score quantifies the contribution of a certain feature to the predictive performance of a model in terms of how much a chosen evaluation metric deviates when the feature becomes noninformative. In Fig. 9.26, the sensors located on the boundary opposite to source are not very important for the proposed noninvasive classification in terms of the distribution of discontinuities. Wavefront arrival time measured by sensors on boundaries adjacent to the transmitter-bearing boundary are important, especially sensors that are closer to the transmitter. Sensors closer to transmitter are important for distinguishing the four distributions, whereas the sensors on the boundary opposite to the transmitter are not capable of measuring any distinct signature facilitating the desired characterization.

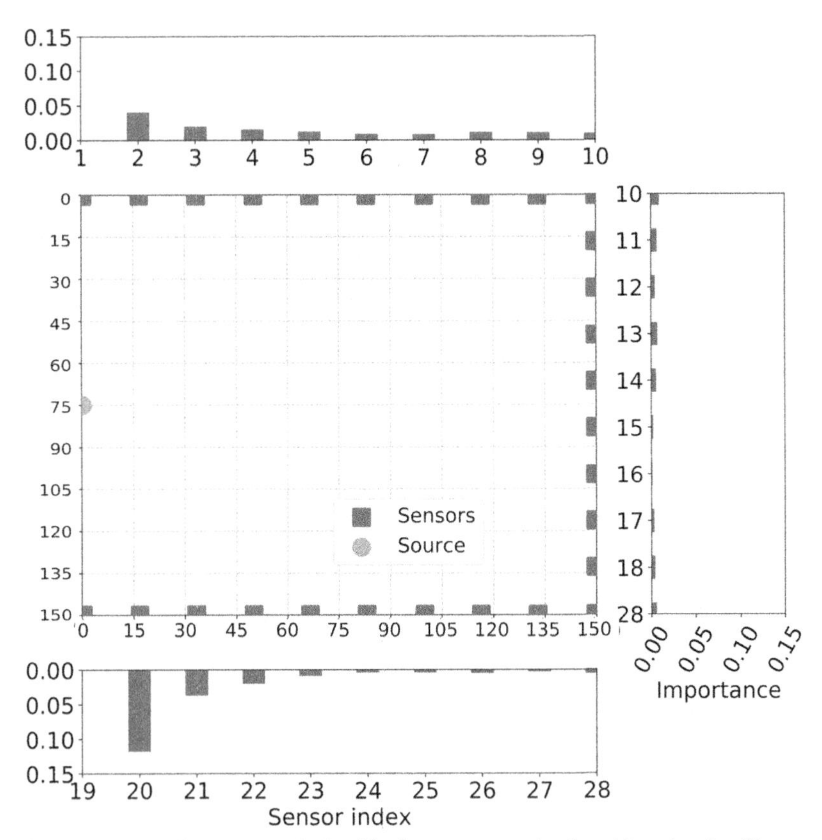

FIG. 9.26 Feature importance calculated by feature permutation for with voting classifier.

Acknowledgments

Various workflows and visualizations used in this chapter are based upon work supported by the U.S. Department of Energy, Office of Science, Office of Basic Energy Sciences, Chemical Sciences Geosciences, and Biosciences Division, under Award Number DE-SC-00019266. We thank the Integrated Core Characterization Center (IC3) and the Unconventional Shale Gas (USG) Consortium at the University of Oklahoma for providing us the shear waveforms and acoustic-emission data.

References

[1] Aggelis DG. Classification of cracking mode in concrete by acoustic emission parameters. Mech Res Commun 2011;38(3):153–7.

[2] Martínez-Martínez J, Fusi N, Galiana-Merino JJ, Benavente D, Crosta GB. Ultrasonic and X-ray computed tomography characterization of progressive fracture damage in low-porous carbonate rocks. Eng Geol 2016;200:47–57.

[3] Ramos MJ, Espinoza DN, Torres-Verdín C, Shovkun I, Grover T. Laboratory characterization and detection of fractures through combined ultrasonic and triaxial-stress testing. In: 50th US Rock Mechanics/Geomechanics Symposium. American Rock Mechanics Association; 2016.
[4] Watanabe N, Ishibashi T, Hirano N, Ohsaki Y, Tsuchiya Y, Tamagawa T, Okabe H, Tsuchiya N. Precise 3D numerical modeling of fracture flow coupled with X-ray computed tomography for reservoir core samples. SPE J 2011;16(03):683–91.
[5] Cai Y, Liu D, Mathews JP, Pan Z, Elsworth D, Yao Y, Li J, Guo X. Permeability evolution in fractured coal—combining triaxial confinement with X-ray computed tomography, acoustic emission and ultrasonic techniques. Int J Coal Geol 2014;122:91–104.
[6] Zhou Z, Cheng R, Cai X, Ma D, Jiang C. Discrimination of rock fracture and blast events based on signal complexity and machine learning. Shock Vib 2018;2018:10, 9753028.
[7] Liu X, Liang Z, Zhang Y, Wu X, Liao Z. Acoustic emission signal recognition of different rocks using wavelet transform and artificial neural network. Shock Vib 2015;2015.
[8] Farhidzadeh A, Mpalaskas AC, Matikas TE, Farhidzadeh H, Aggelis DG. Fracture mode identification in cementitious materials using supervised pattern recognition of acoustic emission features. Construct Build Mater 2014;67:129–38.
[9] Miller RL, Moore B, Viswanathan H, Srinivasan G. Image analysis using convolutional neural networks for modeling 2D fracture propagation. In: Presented at the 2017 IEEE international conference on data mining workshops (ICDMW), 18–21 November 2017; 2017. p. 979–82. https://doi.org/10.1109/ICDMW.2017.137.
[10] Perol T, Gharbi M, Denolle M. Convolutional neural network for earthquake detection and location. Sci·Adv 2018;4(2).
[11] Loutas T, Eleftheroglou N, Zarouchas D. A data-driven probabilistic framework towards the in-situ prognostics of fatigue life of composites based on acoustic emission data. Compos Struct 2017;161:522–9.
[12] Aggelis DG, Shiotani T, Terazawa M. Assessment of construction joint effect in full-scale concrete beams by acoustic emission activity. J Eng Mech 2009;136(7):906–12.
[13] Moore BA, Rougier E, O'Malley D, Srinivasan G, Hunter A, Viswanathan H. Predictive modeling of dynamic fracture growth in brittle materials with machine learning. Comput Mater Sci 2018;148:46–53.
[14] Rovinelli A, Sangid MD, Proudhon H, Ludwig W. Using machine learning and a data-driven approach to identify the small fatigue crack driving force in polycrystalline materials. npj Comput Mater 2018;4(1):35.
[15] Bhoumick P, Sondergeld C, Rai CS. Mapping hydraulic fracture in pyrophyllite using shear wave. In: 52nd US Rock Mechanics/Geomechanics Symposium. American Rock Mechanics Association; 2018.
[16] Rawlinson N, Sambridge M. Wave front evolution in strongly heterogeneous layered material using the fast marching method. Geophys J Int 2004;156(3):631–47.
[17] Treeby BE, Cox BT. k-Wave: MATLAB toolbox for the simulation and reconstruction of photoacoustic wave fields. J Biomed Opt 2010;15(2).
[18] Fadakar Alghalandis Y. Stochastic modelling of fractures in rock masses. Doctoral Dissertation; 2014.
[19] Mardia KV, Jupp PE. Directional statistics. vol. 494. John Wiley & Sons; 2009 [Reprint].
[20] Illian J, Penttinen A, Stoyan H, Stoyan D. Statistical analysis and modelling of spatial point patterns. vol. 70. John Wiley & Sons; 2008 [Reprint].

Further reading

[21] Gui G, Pan H, Lin Z, Li Y, Yuan Z. Data-driven support vector machine with optimization techniques for structural health monitoring and damage detection. KSCE J Civ Eng 2017;21(2):523–34.

[22] He M, Miao J, Feng J. Rock burst process of limestone and its acoustic emission characteristics under true-triaxial unloading conditions. Int J Rock Mech Min Sci 2010;47(2):286–98.

[23] Jingrong Z, Ke W, Yang G. Acoustic emission signals classification based on support vector machine. In: Presented at the 2010 2nd international conference on computer engineering and technology, April 16–18, 2010; 2010. p. V6-300–4. https://doi.org/10.1109/ICCET.2010.5486240.

[24] Pierson KD, Hochhalter JD, Spear AD. Data-driven correlation analysis between observed 3D fatigue-crack path and computed fields from high-fidelity, crystal-plasticity, finite-element simulations. JOM 2018;70(7):1159–67.

[25] Wang Y, Li C, Hu Y. Experimental investigation on the fracture behaviour of black shale by acoustic emission monitoring and CT image analysis during uniaxial compression. Geophys J Int 2018;213(1):660–75.

[26] Yang S-Q, Jing H-W, Wang S-Y. Experimental investigation on the strength, deformability, failure behavior and acoustic emission locations of red sandstone under triaxial compression. Rock Mech Rock Eng 2012;45(4):583–606.

[27] Zhou J, Chen M, Jin Y, Zhang G-Q. Analysis of fracture propagation behavior and fracture geometry using a tri-axial fracturing system in naturally fractured reservoirs. Int J Rock Mech Min Sci 2008;45(7):1143–52.

[28] Zhou M, Yang J. Effect of the layer orientation on fracture propagation of Longmaxi Shale under uniaxial compression using micro-CT scanning. In: Presented at the 2017 SEG international exposition and annual meeting, Houston, Texas, October 23, 2017; 2017. p. 5.

[29] Zhu Z, Burns DR, Brown S, Fehler M. Laboratory experimental studies of seismic scattering from fractures. Geophys J Int 2015;201(1):291–303.

Chapter 10

Machine learning assisted segmentation of scanning electron microscopy images of organic-rich shales with feature extraction and feature ranking

Siddharth Misra* and Yaokun Wu[†,a]

*Harold Vance Department of Petroleum Engineering, Texas A&M University, College Station, TX, United States, †Texas A&M University, College Station, TX, United States

Chapter outline

1 Introduction

Scanning electron microscope (SEM) image analysis facilitates the visualization and quantification of the microstructure, topology, morphology, and connectivity of distinct components in a porous geological material. Image segmentation is a crucial step prior to image analysis. Though manual segmentation performed by the subject-matter expert is the most reliable segmentation approach, it requires considerable time, attention, and patience to segment the high-resolution SEM

[a] Formerly at the University of Oklahoma, Norman, OK, United States.

Machine Learning for Subsurface Characterization. https://doi.org/10.1016/B978-0-12-817736-5.00010-7

images. We propose an automated image segmentation method coupled with classic feature extraction techniques.

Segmentation is the division of an image into spatially continuous, disjoint, and homogeneous regions. Traditional image segmentation methods are commonly divided into three approaches: pixel-based, edge-based, and region-based segmentation methods. Histogram thresholding–based segmentation assigns a certain class label to each pixel depending on a user-defined range of pixel intensity with an assumption that one component has a certain well-defined range of pixel intensity in the image. Images having unimodal or multimodal histograms are generally segmented using the thresholding method [1]. However, major limitations of the thresholding method include the following: (1) It requires accurate selection of threshold values and the range of pixel intensity for each component, and (2) it is unreliable when the ranges of pixel intensity for two or more components overlap. Another approach is the region-based segmentation method, which is also widely applied on SEM images. This method iteratively splits or merges various regions till continuous homogeneous regions are identified in the image. Watershed segmentation is a region-based segmentation method popular in medical image segmentation [2]. Challenges in selecting proper seed points make the region-based method prone to oversegmentation or undersegmentation. Further, this method has a long run time and is sensitive to noise.

Image segmentation aided by machine learning has shown rapid progress over the last few years. Two types of machine learning techniques, namely, supervised and unsupervised learning, are popularly employed in image segmentation. In supervised learning, a machine learns a function that relates specific features of training samples to corresponding targets (outcomes), such that the function can be later used to predict the targets for new, unseen samples when their features are known (e.g., [3, 4]). Segmentation methods using supervised learning can be divided into two broad categories: pixel-wise classification and object-based classification. Anemone et al. [5] use pixel-wise models with an artificial neural network to recognize spectral features of five different classes of land cover in remotely sensed images to find potential fossil localities. Bauer and Strauss [6] introduced an object-based method to do segmentation and classify soil cover types into stones, residues, shadow, and plants. Deep learning (e.g., [7]) also has been implemented as a powerful tool in segmentation. Wu et al. [8] developed a convolutional neural network with an encoder-decoder architecture to do semantic segmentation, where the road scene objects, like car, tree, and road, were successfully segmented with reasonable accuracy and fast runtime.

Unsupervised clustering also has been used in image segmentation. Compared with the supervised learning method where you need training data, the unsupervised learning can deal with data that have not been labeled (e.g., [9]). Unsupervised learning transforms the dataset based on the features without any predefined outcomes/targets. Shen et al. [10] introduced an extension to traditional fuzzy c-means clustering for segmenting T1 weighted magnetic resonance (MR) image of brain tissue to identify white matter, gray matter, and cerebrospinal fluid. It takes advantage of neighboring pixel intensities and locations to restrain the influence of noise. Self-organizing map (SOM) is another typical method that takes advantage of unsupervised learning. Ong et al. [11] proposed a two-stage hierarchical artificial neural network (ANN) for segmentation of color images based on SOM. Unsupervised SOM captures the dominant colors of an image to generate color clusters that are fed into second level SOM to complete the segmentation process. Jiang and Zhou [12] combined SOM with ensemble learning to improve the segmentation performance. By adopting a scheme for aligning different clusters, they can set SOM with different parameters and combine them to generate a robust segmentation result.

Image analysis is commonly used in the oil and gas industry and various geophysical applications. Tripathi et al. [13] estimated permeability from thin-section image analysis based on the Carman-Kozeny model. Budennyy et al. [14] used watershed segmentation method and statistical learning on polarized optical microscopic images to study the structure of thin section and to evaluate the properties of grain, cement, voids, and cleavage. Rahimov et al. [15] used features extracted through local binary pattern (LBP) from 3-D subsample image to classify images into prespecified six texture classes to obtain the representative permeability. Asmussen et al. [16] developed a semiautomatic region-growing segmentation workflow for image analysis of rock images for the quantification of modal composition, porosity, grain size distribution, and grain contacts. Zhao et al. [17] utilized k-means clustering and principal component analysis (PCA) to classify the remaining oil into oil film, throat retained oil, heterogeneous multipore oil, and clustered oil based on the experiments on 2-D–etched glass micromodel.

Various segmentation methods have been proposed to process SEM images to derive the information at the nanoscale. Hughes A et al. [18] identified the most relevant preprocessing, segmentation, and object classification techniques for SEM image to streamline nanostructure characterization by using the Ilastik software. They successfully used the random walk method together with the semisupervised pixel classification to classify nanoparticles into singles, dimers, flat, and piled aggregate. Tang and Spikes [19] use elemental SEM images of seven different elements from shale samples as input features into a neural network to segment these images into five phases such as calcite, feldspar, quartz, TOC, and clay/pore. However, the limitations lay in the acquisition of such elemental SEM images, and the classifiers implemented in their study were not able to distinguish between clay and pore.

The goal of our study is to use machine learning and feature extraction followed by feature ranking to automate the segmentation of SEM images of shale samples. The chapter is arranged in the following manner:

1. Description of the SEM images used in this study;
2. Overview of the workflow, including feature extraction, training dataset creation, and random forest classification;

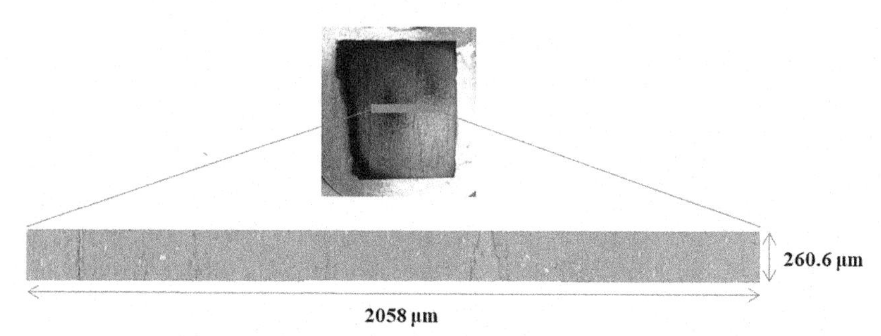

FIG. 10.1 High-resolution SEM map of a shale sample (2058 μm by 260.6 μm).

3. Comparison of the segmentation results obtained using the proposed segmentation method against other popular segmentation methods;
4. Validate and test the performance of the proposed method; and
5. Limitations of the proposed method.

2 Method

2.1 SEM images

High-resolution SEM images were acquired using the FEI Helios Nanolab 650 DualBeam FIB/SEM machine and FEI SEM MAPS software at the Integrated Core Characterization (IC^3) lab. Fig. 10.1 shows the SEM map of a thin 2058 μm by 260.6 μm section of a shale rock sample. The SEM map is divided into 1000 images, such that each image has a dimension of 20.58 μm by 26.06 μm.

2.2 Workflow

The goal of proposed machine learning-assisted SEM segmentation (Fig. 10.2) is to automate the process of identifying the four rock components, that is, whether a pixel in a SEM image represents (1) pore/crack, (2) matrix comprising clay, calcite or quartz, (3) pyrite, or (4) organic/kerogen components. The segmentation method involves two steps, feature extraction followed by classification of the extracted feature vectors. Classification algorithms, such as k-means, decision tree, random forest, and logistic regression, can be trained to relate certain features corresponding to a pixel to one of the four components. In chronological order the workflow for training and testing phase (Fig. 10.2A) involves (1) selection of training and testing pixels, (2) feature extraction, (3) compilation of features for the selected pixels to create the training and testing dataset, (4) training the classifier on the training dataset, and (5) testing the performance of the trained classifier on the testing dataset. In the deployment phase (Fig. 10.2B), the trained classifier is applied

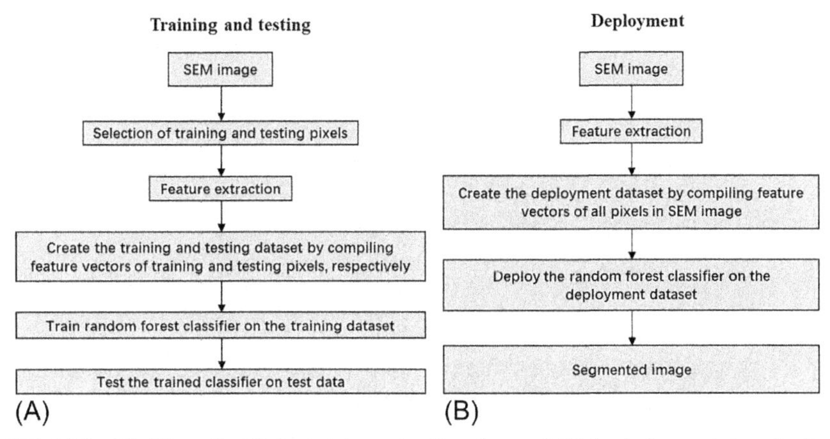

FIG. 10.2 Workflows for (A) the training and testing phase and (B) the deployment phase for the proposed machine-learning assisted SEM-image segmentation. Training and testing phase is crucial for developing robust, generalizable data-driven model. Deployment refers to the application a data-driven model on new, unseen data. No learning takes place in the deployment phase.

on SEM images on shales to perform the image segmentation. No learning takes place during the deployment.

2.3 Feature extraction

Intensity/brightness of a pixel on a gray scale can be used as a feature to distinguish various components/segments in an image. The pixel intensity of the SEM images used in this study is stored as an 8-bit integer; so, the pixel intensity in the SEM image ranges from 0 to 255, where zero is black pixel, and 255 is white pixel. Threshold-based segmentation uses only the pixel intensity to generate the segmentation. Threshold-based segmentation of SEM map, shown in Fig. 10.1, was performed by Tran et al. [20] to identify microstructures, pores, cracks, organic matter, pyrite, silica-rich clay grains, and calcite-rich clay grains. However, the pixel intensity is a weak feature when the components to be segmented have overlapping magnitudes of pixel intensity. For shale SEM images, the threshold-based segmentation has poor performance when distinguishing pore/crack component, having pixel intensity between 0 and 125, from organic/kerogen component, having pixel intensity between 80 and 130. In our study, increasing the number of relevant features facilitates the classification step.

Our extensive study indicates that seven categories of features (Fig. 10.3) are the most important for the proposed classification, namely, wavelet transform, difference of Gaussians (DoG), Hessian matrix, Sobel operator, Gaussian blur, statistical information of the neighboring pixels (local information), and pixel intensity. These features describe each pixel in comparison to its neighboring pixels based on spatial- and scale-related

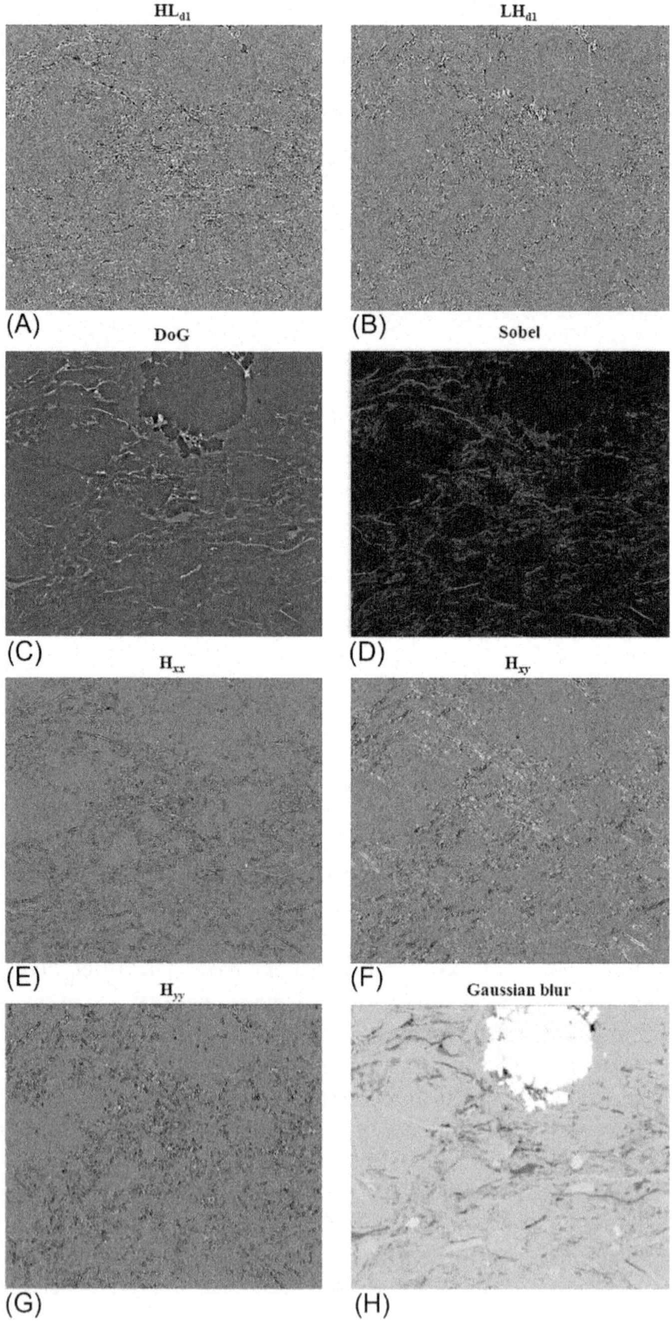

HL_{d1} LH_{d1}

(A) DoG (B) Sobel

(C) H_{xx} (D) H_{xy}

(E) H_{yy} (F) Gaussian blur

(G) (H)

FIG. 10.3 Examples of features extracted from one SEM image after the first level of processing.

information at multiple resolutions. These features are used by the classifier to assign a specific rock component to each pixel in the image. The effectiveness of the extracted features depends on the choice of parameters for the mathematical/statistical transformations used to generate the features. We select the optimum parameters based on the performance of the proposed SEM-image segmentation on the testing dataset while maintaining a low computational cost.

2.3.1 Gaussian blur (one feature)

The Gaussian blur feature is obtained by blurring (smoothing) an image using a Gaussian function to reduce the noise level, as shown in Fig. 10.3H. It can be considered as a nonuniform low-pass filter that preserves low spatial frequency and reduces image noise and negligible details in an image. It is typically achieved by convolving an image with a Gaussian kernel. This Gaussian kernel in 2-D form is expressed as

$$G_{2D}(x, y, \sigma) = \frac{1}{2\pi\sigma^2} e^{-\frac{x^2+y^2}{2\sigma^2}}$$

where σ is the standard deviation of the distribution and x and y are the location indices. The value of σ controls the variance around a mean value of the Gaussian distribution, which determines the extent of the blurring effect around a pixel. In the proposed image segmentation, we tested sigma values ranging from 0.1 to 16, such that, with an increase in sigma, the high-frequency information content reduces around the pixel. In our study, we use a standard deviation of 3 that generates the best segmentation results.

2.3.2 Difference of Gaussians (one feature)

Difference of Gaussians (DoG) is calculated as the difference between two smoothed versions of an image obtained by applying two Gaussian kernels of different standard deviations (sigma) on that image. In other words, DoG transformation of an image requires subtracting one highly blurred version of an original image from another less blurred version to act as a band-pass filter that preserves a specific spatial frequency. DoG removes high-frequency spatial components representing noise in the image through the blurring and low-frequency components representing the homogeneous region in the image through the difference operation. DoG generally serves as an edge enhancement algorithm that delineates the high-frequency content of the image free from noise. Fig. 10.3C shows a DOG-transformed image. In object detection, DoG has been used in scale-invariant feature transform (SIFT) as an approximation of Laplacian of Gaussians to find key points (local extrema) in the image. In this study, DoG is formulated as the difference between a 2-D Gaussian kernel of sigma equal to 1.414 and another 2-D Gaussian kernel of sigma equal to 1.

2.3.3 Sobel operator (one feature)

Sobel operator performs a 2-D spatial gradient operation on an image to enhance the edges. The operator consists of a pair of 3-by-3 convolution kernels (two for the two perpendicular directions) that are separately applied to an image to produce the approximate gradients for each pixel for detecting edges in vertical and horizontal directions. Edges in an image are emphasized because the gradients at the edges are usually larger than those in the homogeneous region. Fig. 10.3D shows the image obtained after application of the Sobel operator. Coefficients of the convolution kernels can be adjusted according to our requirement provided the kernel honors the properties of derivative masks. Increase in the coefficients of the middle row or column in the 3-by-3 convolution kernels increases the number of detected edges.

2.3.4 Hessian matrix (three features)

In image analysis, the Hessian matrix describes the second-order variations of local image intensity around a pixel, thereby encoding the shape information. It describes the local curvature of the spatial structures over the whole image and has been used to detect structure orientation, noise, and structure brightness and to differentiate blob-like, tubular, and sheet-like structures. Hessian matrix is suited for corner detection, which can be used for object localization and shape detection. In practice, it is computed by convolving an image with second derivatives of the Gaussian kernel in the x- and y-directions. For our purposes, this Gaussian kernel is assumed to have a standard deviation of 1, applied on each pixel in the image so that the Hessian matrix H is expressed as

$$\mathrm{H}[\mathrm{f}(x,y)] = \begin{bmatrix} \mathrm{H}_{xx} & \mathrm{H}_{xy} \\ \mathrm{H}_{yx} & \mathrm{H}_{yy} \end{bmatrix}$$

where

$$\mathrm{H}_{xx} = \frac{\partial^2 f}{\partial x^2}, \mathrm{H}_{xy} = \frac{\partial^2 f}{\partial x \partial y} = \mathrm{H}_{yx} = \frac{\partial^2 f}{\partial y \partial x}, \mathrm{H}_{yy} = \frac{\partial^2 f}{\partial y^2}$$

Three features, namely, H_{xx}, H_{xy}, and H_{yy}, are computed at each pixel location. Fig. 10.3E–G shows these three transformations obtained based on the Hessian matrix.

2.3.5 Wavelet transforms (six features)

Wavelet transforms enable multiresolution space-scale (time-frequency) analysis of signals. Both Fourier and wavelet transforms are scale/frequency-localized, but wavelets have an additional time/space-localization property. Two-dimensional wavelet transforms when applied to an image generate the coefficients with respect to certain basis function (wavelet). These coefficients can be used for image reconstruction, image denoising, image compression, feature extraction, filtering, and pattern matching. Wavelet coefficients match

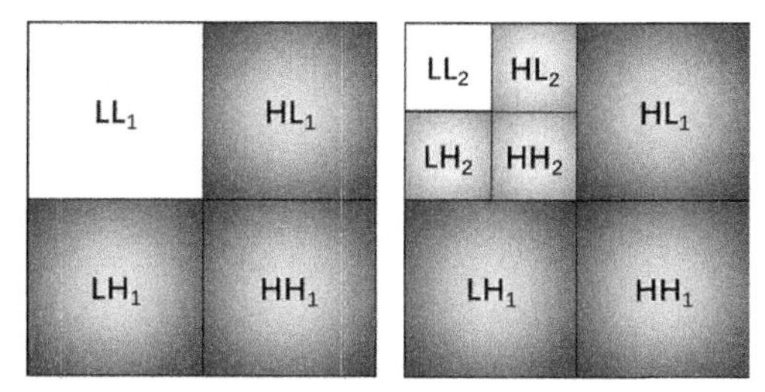

FIG. 10.4 Wavelet transforms in level 1 and level 2.

different spatial frequencies in the image at a specific scale. Low and medium spatial frequencies usually match the image content, whereas high-frequency coefficients usually represent noise or texture areas.

In practice, when a single operation of the wavelet transform is applied on the image, four subimages (set of coefficients) are generated at half the resolution of the original image. Fig. 10.4 illustrates the level 1 and level 2 decompositions obtained by the wavelet transformations. In level-1 decomposition, HL_1, LH_1, and HH_1 are three subimages obtained as high-pass filtered images, such that each subimage describes the local changes in brightness (edges) in the original image. HL_1 and LH_1 highlight the edges along vertical and horizontal directions, respectively, and HH_1 highlights the edges along the diagonal. LL_1 is an approximation of the original image obtained as a low-pass filtered, downscaled image. To obtain the next-level downscaled representations, LL_1 is further decomposed to yield LL_2, LH_2, HL_2, and HH_2.

For image segmentation the six high-pass filtered, downscaled images are modified to obtain the horizontal details (HL_{d1} and HL_{d2}), vertical details (LH_{d1} and LH_{d2}), and diagonal details (HH_{d1} and HH_{d2}), where the subscripts d1 and d2 represent the level 1 and level 2, respectively. Figs. 10.3A and B show the horizontal and vertical details. We do not use the LL_1, LL_2, or level 3 decompositions for the proposed segmentation. In doing so, for each pixel, we generate six features.

2.3.6 *Local information (three features)*

Local information includes the minimum, maximum, and mean values of pixels within a certain local neighborhood of pixels. In this study, the local information of pixel describes the 3-pixel by 3-pixel region centered at that pixel.

2.3.7 *Other features investigated for this study*

Several other features were tested but were not used due to their computational complexity or the lack of reliable computational infrastructure. For example, empirical mode decomposition is a decomposition method like the wavelet transform. Unfortunately, it takes considerable time to run when the method was tested on a 256-by-256 image. LBP is another method mostly used in

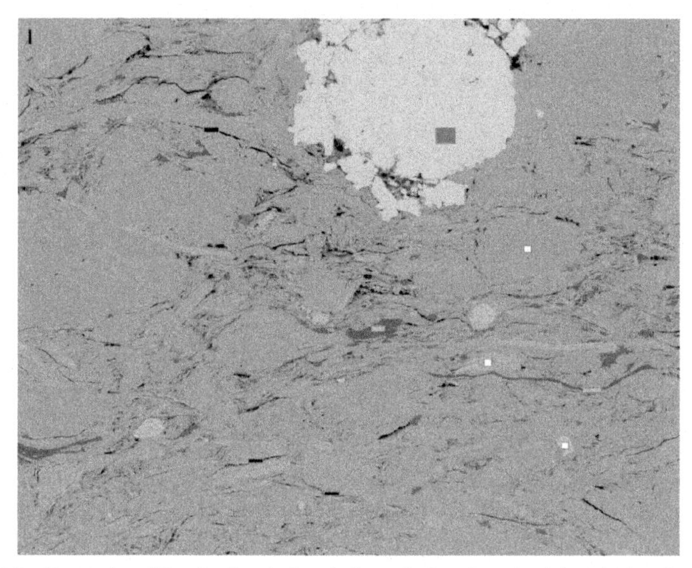

FIG. 10.5 Rectangles with red-colored edges indicate the location of training pixels, where green, gray, black, and blue colors represent the kerogen/organic, matrix, pore/crack, and pyrite components.

texture classification of regions instead of individual pixels. SIFT and speeded up robust features are two popular feature extraction methods; however, we did not implement them because their detectors and descriptors are primarily used for object matching and tracking, which are not suitable for our case.

2.4 Training dataset

Training dataset influences the learning and generalization of a machine learning model. Good training set selection will aid the learning process resulting in better segmentation performance while reducing time to train. To that end, the training samples need to be high quality, such that the pixels are precisely assigned a component type. For robust classification, it is recommended that the training dataset should be a balanced dataset, that is, an equal number of pixels be used for each component type. The rectangles with red-colored edges in Fig. 10.5 show the areas where the training pixels were selected. 705, 2074, 17,373, and 15,000 pixels were selected for pore/crack, organic/kerogen, rock matrix, and pyrite components, respectively. After selecting the training pixels, 16 aforementioned features were extracted for each training pixel, which constitutes the training dataset required to train the classifier. It is recommended to scale features before training the classifier because certain learning techniques, such as neural network and k-nearest neighbor (kNN), are sensitive to feature scaling. However, feature scaling is not performed in our segmentation method because random forest classifier can deal with unscaled data.

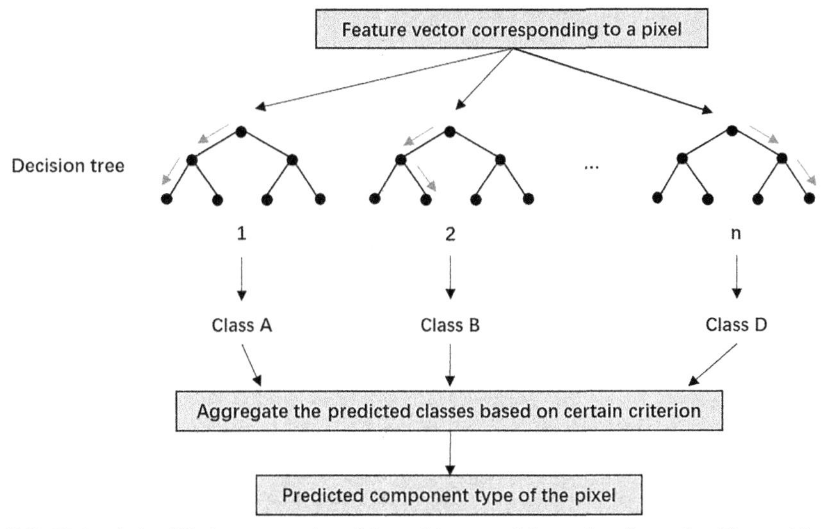

FIG. 10.6 A simplified representation of the architecture of the random forest classifier used for the proposed segmentation.

2.5 Classification of the feature vectors

Random forest model is widely used for classification. Random forest model is a bagging-type ensemble (collection) of decision trees that trains several trees in parallel and uses the majority decision of the trees as the final decision of the random forest model. Individual decision tree model is easy to interpret but the model is nonunique and exhibits high variance. On the other hand, random forest by combining hundreds of decision tree models reduces the variance and bias, which is hard to achieve due to the bias-variance threshold. Random forest (Fig. 10.6) trains several decision tree classifiers (in parallel) on various subsamples of the dataset (also referred as bootstrapping) and various subsamples of the available features. Random forest is an ensemble classifier based on bootstrap followed by aggregation (jointly referred as bagging). In practice, random forest classifier does not require much hyperparameter tuning or feature scaling. Consequently, random forest classifier is easy to develop, easy to implement, and generates robust classification. We notice that the use of random forest increases the reproducibility of the SEM-image segmentation. Hyperparameters of random forest need to be tuned to overcome the challenge of distinguishing pore/crack component from organic/kerogen component. Important hyperparameters include maximum depth of the trees that controls overfitting, number of subsamples of the original features used to build each decision tree, and the weight assigned to each class. Cross-validation pipeline should be used along with the hyperparameter optimization to ensure that the random forest is exposed to all the statistical distributions in the training dataset. In other words, hyperparameters should be determined by evaluating the average model performance on various splits of the training and testing dataset with different subsamples of the available features to ensure significant reduction in variance without increase in bias. It is important to ensure stratification when splitting the dataset into training and testing dataset or when splitting the training dataset for cross-validation. Stratification ensures that all splits have similar distribution of components as the parent dataset.

Apart from random forest, we tested the following classification techniques: kNN, logistic regression, linear support vector classifier (SVC), and multilayer perceptron (MLP). For each unsegmented pixel, kNN first finds k pixels from the training dataset, which have feature vectors that are closest to the feature vector of the unsegmented pixel. After that, the unsegmented pixel is assigned a component type that occurs the most among the k pixels. kNN requires careful selection of k, the number of neighbors, and generates inconsistent results when the dimensionality of the dataset increases. Linear SVC is a binary classifier that finds a boundary that best separates two classes, whereas logistic regression is a binary classifier that finds a boundary by identifying a log-likelihood distribution that best represents the distribution of the data for the two classes. Linear SVC and logistic regression require careful selection of hyperparameters: alpha and C that govern the nature of boundary and the penalty of misclassifying few data samples. Nonlinear SVC cannot be used for the proposed segmentation because it is inefficient for a large dataset with high-dimensional features. When using neural network model for classification, all features need to be properly scaled and require hyperparameter optimization with cross-validation to find the optimum values for the hyperparameters, such as regularization term (alpha), the number of hidden layers, and the number of neurons in each hidden layer. Unlike all these classification methods, random forest classifier is invariant to the scaling of data and requires little effort in tuning the hyperparameters while maintaining high reproducibility. Unlike linear SVC, random forest once trained is fast to deploy. Unlike neural networks, random forest has much lower variance and does not overfit resulting in better generalization. Unlike logistic regression, random forest can handle nonlinear trends in the dataset. Unlike random forest and other classification methods, kNN is nonparametric method requiring the entire training dataset during the deployment. Based on our extensive study, the random forest classifier was the most accurate, reliable, and computationally inexpensive as compared with other classifiers for the desired segmentation.

3 Results and discussions

3.1 Four-component segmentation

The segmentation is trained to identify the four components, namely, pore/crack (black); kerogen/organic (green); pyrite (blue); and rock matrix comprising clay,

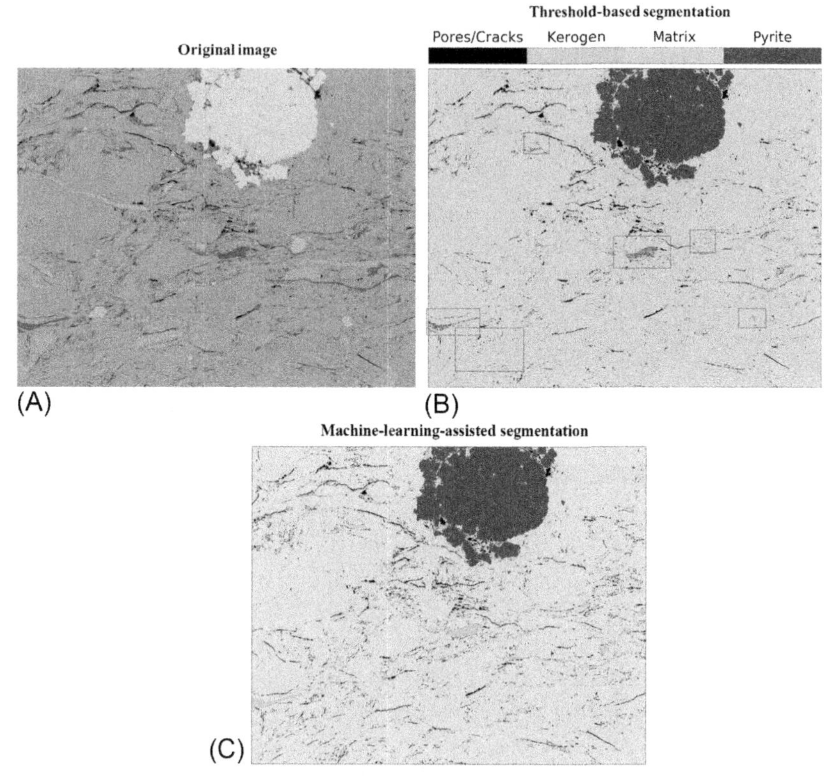

FIG. 10.7 Comparison of SEM-image segmentation generated by (B) threshold-based segmentation and that by the (C) proposed machine learning-assisted segmentation of (A) original image. Threshold-based segmentation performs poorly in regions indicated by the *red-edged boxes*.

quartz, and calcite (light gray). The segmentation methodology involves feature extraction followed by training a random forest classifier to assign a component type to each pixel. The proposed method performs better than other conventional methods, such as threshold-based segmentation (Fig. 10.7), object-based segmentation (Fig. 10.8), and ImageJ/Fiji segmentation (Fig. 10.9).

In the threshold-based method, the pixel intensity range needs to be determined for each component. In the SEM images, the intensity ranged from 0 to 255. In our segmentation task, pixels ranging between 0–80, 81–119, 120–190, and 190–255 are manually selected as pore/cracks, organic/kerogen, matrix, and pyrite components, respectively. A component type was then assigned to each pixel in the image by comparing the intensity of the pixel with the manually defined ranges. Fig. 10.7 compares threshold-based segmentation against the proposed machine-learning assisted segmentation. The threshold-based method fails for rectangular regions marked with red-colored edges in Fig. 10.7, for example, the method overpredicts pore/crack

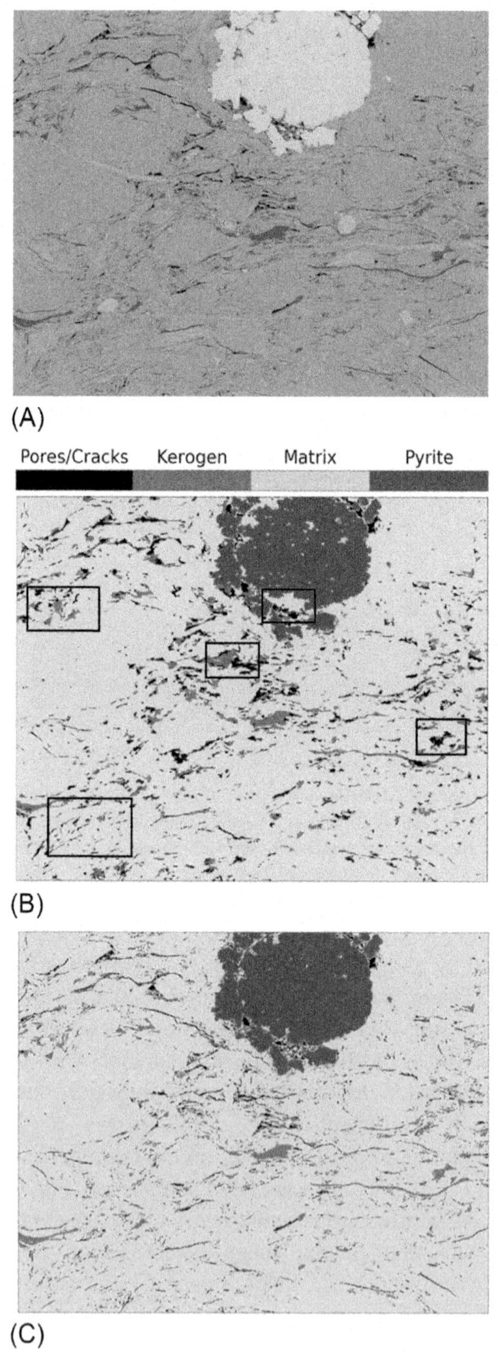

FIG. 10.8 Comparison of SEM-image segmentation generated by (B) object-based segmentation and that by the (C) proposed machine learning-assisted segmentation of (A) original image. Object-based segmentation performs poorly in regions indicated by the *red boxes*.

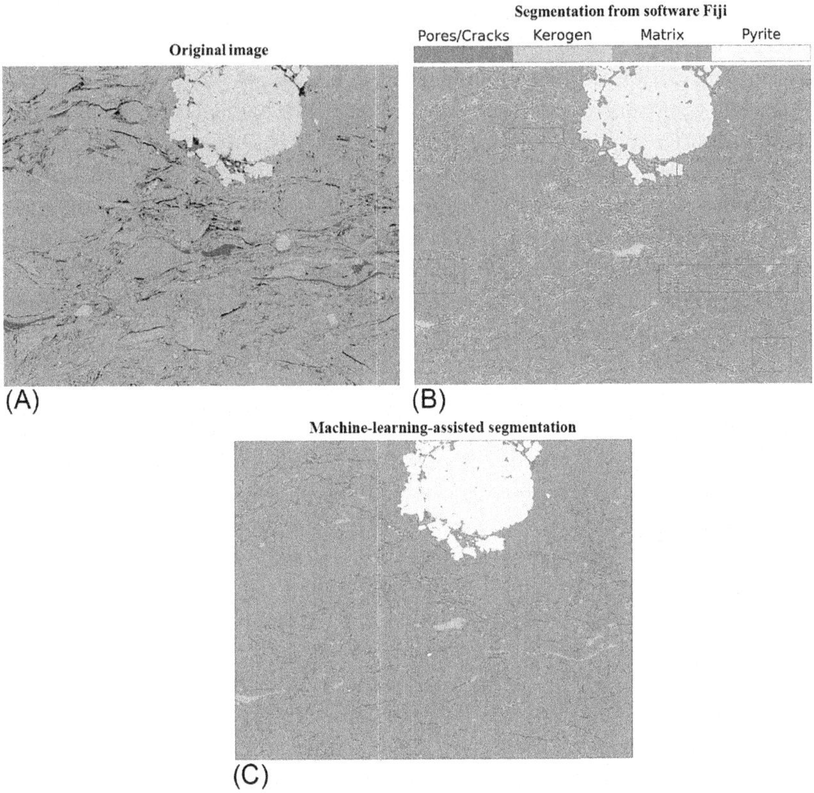

Original image

Segmentation from software Fiji

Pores/Cracks Kerogen Matrix Pyrite

(A) (B)

Machine-learning-assisted segmentation

(C)

FIG. 10.9 Comparison of SEM-image segmentation generated by (B) Fiji-based segmentation and that by the (C) proposed machine learning-assisted segmentation of (A) original image. Fiji-based segmentation performs poorly in regions indicated by the *red-edged boxes.*

by sprinkling pores all over the image, and the method fails to distinguish pore/crack from rock matrix. Pore/crack and organic/kerogen components tend to be misclassified in threshold-based segmentation.

Another popular method often used in segmentation task is the object-based segmentation. It involves object creation, feature extraction, and classification. It first partitions the image into a large number of primitive objects (regions) by a graph-based region comparison algorithm [21]. Then the local information features in a region, such as mean, median, minimum, maximum, skewness, and kurtosis of pixel intensities, are extracted. Next, these features vectors are fed into a trained classifier to assign the component types to the corresponding pixels. A significant disadvantage is that the object-based segmentation tends to be coarse (Fig. 10.8) because the oversegmentation function omits segments having the number of pixels lower than a certain threshold when generating those objects (segments). In Fig. 10.8, pores and cracks spread over a limited number of pixels cannot be identified using the oversegmentation method.

Robust segmentation requires that the classifier should not be very sensitive to the training set selection. Low sensitivity of the classifier to the training data ensures reproducible segmentation. An image processing package called Fiji is a popular open-source platform for biological image analysis. One of its plugin called the Waikato Environment for Knowledge Analysis (WEKA) can perform automated image segmentation [22]. Compared with our segmentation methodology, the performance of the Fiji segmentation method varies significantly when using different training sets. Also, as shown in Fig. 10.9, the Fiji segmentation method misclassifies the boundary regions around pore and crack as organic/kerogen matter and misclassifies regions around pyrite as pores and cracks.

3.2 Multilabel probability-based segmentation

Binary classification assigns one out of the two classes/labels to each sample (e.g., good or bad). Multiclass classification assigns one out of the many classes/labels to each sample (e.g., excellent, good, or bad). Multiclass and binary classifications assume that each sample can be assigned one and only one label (e.g., a laptop can be either Dell, HP, Microsoft, or Apple). By contrast, multilabel classification assigns more than one label to each sample, and sometimes the assignment of multiple label is accompanied by the probability of the sample to be one of the many classes (e.g., a day may be labeled as 10% windy, 40% sunny, and 50% rainy). For multiclass classification, we can implement a one-vs-one (OvO) strategy or one-vs-rest (OvR) strategy. Contrary to OvR, OvO requires more number of classifiers to be trained and is computationally slower approach but is not easily affected by data imbalance. Random forest, kNN, and neural network can perform binary, multiclass, or multilabel classification. For purpose of generating multilabel segmentation of an SEM image, we use random forest to generate four probability values that indicate the probability of a pixel to be one of the four rock components, namely pyrite, matrix, organic/kerogen, and pore/crack. In doing so, we can better assess the uncertainty in the component type assigned by the classifier. The probability values generated by the random forest describe the certainty in assigning the component types to each pixel. Therefore the proposed multilabel classification will aid multiclass classification to ascertain the segmentation results better.

Fig. 10.10 illustrates the probability distributions for the four components in one SEM image as identified by the multilabel classifier, such that red indicates high certainty and blue indicates low certainty. Classifier predictions and segmentation results for pixels located in the regions of low certainty, especially around the interface, have a low recall and low precision. Segmentation is also challenging for scattered/dispersed pores and organic matter in the matrix, which tend to be hard to differentiate. It is confirmed that for each component, regions having high prediction probability are always located in the inner region of that component, whereas uncertainty can be seen at boundary region. In practice, we set a threshold value of 0.7, such that a pixel

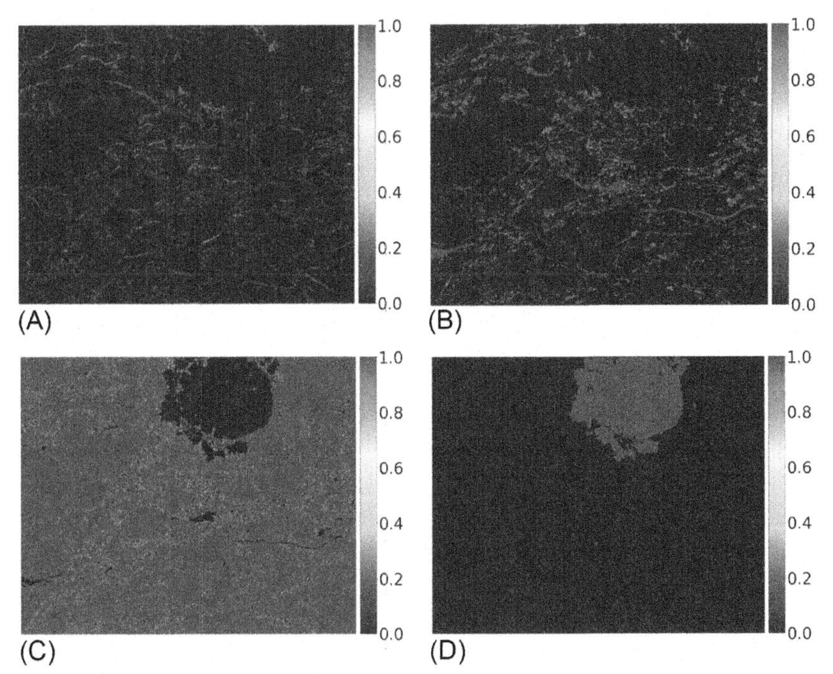

FIG. 10.10 Probabilities of a pixel to be (A) pore/crack, (B) organic/kerogen, (C) rock matrix, and (D) pyrite components as generated by the trained random forest classifier for purposes of multilabel classification. Each pixel is assigned four probabilities corresponding to the four components. Regions with probability <1 indicate the uncertainty in the assigned component-type.

is assigned one component type if the probability for the pixel to be that component is greater than 0.7. When none of the component types have a probability greater than the threshold of 0.7, we assign two labels (component types) to the pixel when the sum of probabilities to be those two components is higher than 0.7 and individually higher than 0.3. The tendency of misclassifying pixels near the interfaces of components was around 0.1%, which is calculated as the fraction of pixels for which the multilabel classification indicates that none of the four component types has a probability higher than 0.7.

3.3 Performance on testing dataset

The performance of the trained segmentation method is assessed on the test dataset. Compared with pixels in the inner region, the trained segmentation method has relatively poor performance for the pixels in the transition zone around the interfaces of the various components, as shown in Fig. 10.11. Test dataset was created with an emphasis on quantifying the performance of the segmentation method in the transition zones. Twenty percent Gaussian noise is intentionally added to the test dataset, to study the robustness of the segmentation method to noisy SEM image. The objective is to understand

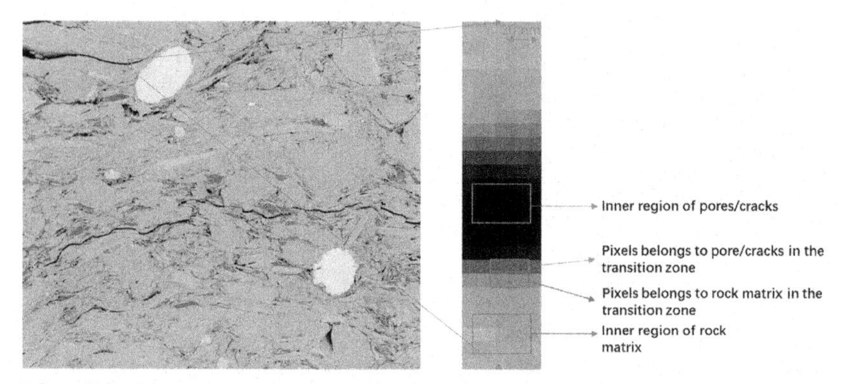

FIG. 10.11 Zoomed in visualization of the inner region (IR) and transition zone (TZ) around crack/pore and matrix interface. Interfaces exhibit grayscale transitions that are hard to segment.

the deterioration of the performance of the proposed segmentation method trained on clean, noise-free data when applied to noise-prone data. In the end, we also compare the performance of the proposed method with other segmentation alternatives on inner-region and transition-zone pixels with and without the 20% Gaussian noise.

Pixels are manually selected from the inner region and transition zones of the components to constitute the inner-region (IR) and transition-zone (TZ) test dataset, respectively. Manual selection of pixels from the inner region for building the test dataset is a straightforward task. However, the manual selection of pixels from the transition zone needs attention to details. As shown in Fig. 10.11, transition zone for the matrix and pore/crack interface is vague and may seem like organic/kerogen component. The locations where these selections of pixels are made are shown in Fig. 10.12. Also, the red rectangles in Fig. 10.12 represent the locations of test pixels, and the area of each rectangle approximates the number of pixels making up the test data (Table 10.1).

The performance on the test data is expressed in terms of F1 score, precision, and recall in Table 10.2. F1 score is the harmonic average of calculated precision and recall. Precision is the ratio of true positives to the summation of true positives and false positives. Recall (also referred to as sensitivity) is the ratio of true positives to the sum of true positives and false negatives. True positive is when the predicted class of a pixel is the true class of the pixel, whereas false positive is when a pixel is wrongly predicted to be the class of interest. The true negative is when a pixel is correctly predicted to be a class other than the class of interest, whereas false negative is when a pixel is wrongly predicted to be a class other than the class of interest.

The precision of the classifier specific to a component type is a measure of the reliability of the component type assigned by the classifier. Recall of the classifier specific to a component type is a measure of the classifier's ability

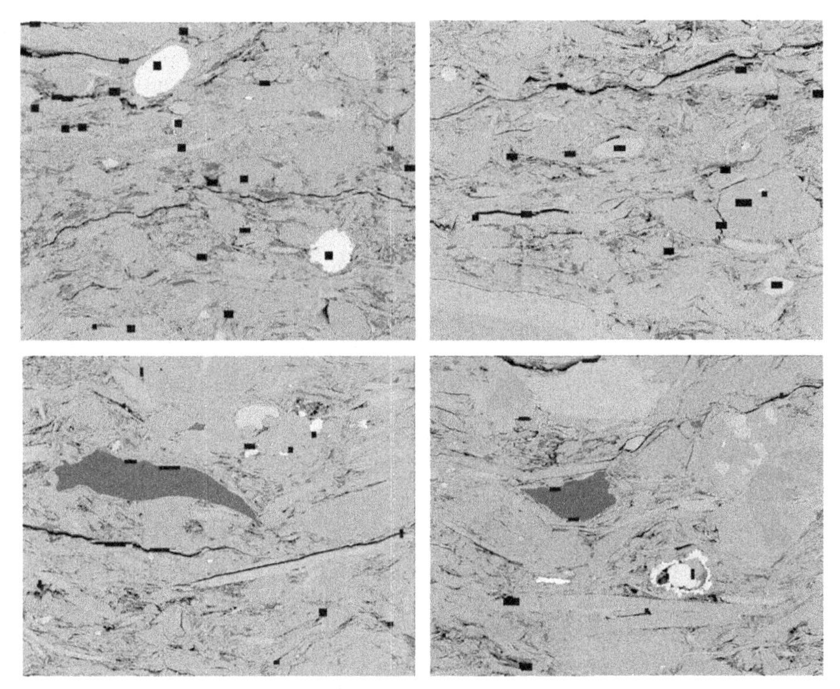

FIG. 10.12 Locations of test pixels, which were selected from both inner regions and transition zones of different images to effectively test the performance of the proposed segmentation methodology.

TABLE 10.1 Number of pixels in the inner regions and transition zones constituting the test dataset corresponding to the four components in the SEM image

	Number of pixels	
Components	**Inner region**	**Transition zone**
Pore and crack	2498	2623
Organic and kerogen	1977	4392
Matrix	2375	3623
Pyrite	1765	3010

to correctly assign the component type; in other words, it is the ability of the classifier to find the class of interest (similar to the sensitivity of the classifier to a certain class). For example, the scanners at the airport need high recall with respect to dangerous materials, but it is not crucial for the

TABLE 10.2 Performance of the image segmentation method on the test dataset without noise for the four rock components in the image, where IR and TZ stand for inner region and transition zone

	Precision		Recall		F1 score	
Components	IR	TZ	IR	TZ	IR	TZ
Pore and crack	1.00	0.93	1.00	0.97	1.00	0.95
Organic and kerogen	1.00	0.96	1.00	0.99	1.00	0.97
Matrix	1.00	0.79	1.00	0.90	1.00	0.84
Pyrite	1.00	1.00	1.00	0.74	1.00	0.85
Avg.	1.00	0.92	1.00	0.91	1.00	0.91

scanner to have high precision with respect to dangerous materials. F1 score, which is the harmonic average of precision and recall, ranges from 0 to 1, such that 0 indicates poor classification and 1 indicates robust performance.

For SEM image with and without 20% Gaussian noise, F1 scores are above 0.98 for all the four components in the inner region, as listed in Table 10.2 (without noise) and Table 10.3 (with noise). Majority pixels in these images can be correctly segmented, and the segmentation method has good tolerance to noise. Segmentation accuracy for the transition zone is lower than that for the inner region, especially for matrix and pyrite components. Matrix component in transition zone has low precision of 0.79 and high recall of 0.9, which indicates that pixels segmented as matrix component have higher

TABLE 10.3 Performance of the image segmentation method on the test dataset containing 20% Gaussian noise for the four rock components in the image, where IR and TZ stand for inner region and transition zone

	Precision		Recall		F1 score	
Components	IR	TZ	IR	TZ	IR	TZ
Pore and crack	1.00	0.91	0.98	0.96	0.99	0.93
Organic and kerogen	0.99	0.99	0.98	0.96	0.99	0.97
Matrix	0.97	0.64	1.00	0.89	0.98	0.75
Pyrite	1.00	1.00	0.99	0.65	1.00	0.79
Avg.	0.99	0.89	0.99	0.87	0.99	0.86

uncertainty and the segmentation method has good ability to correctly identify the actual matrix component. The exact opposite trend is exhibited for the pyrite component in the transition zone, which has a precision of 1 and recall of 0.74, which indicates that pyrite component is never assigned to any other component type but pyrite component tends to be wrongly labeled as other components.

With respect to the transition zone, the F1 scores for pore/crack and organic/kerogen components of noise-bearing test dataset (Table 10.3) are like those of noise-free test dataset (Table 10.2), which indicates that the machine-learning assisted segmentation method is reliable in differentiating the pore/crack from organic/kerogen when SEM image has lower acquisition quality. However, in the presence of noise, the method cannot reliably segment matrix and pyrite components in the transition zone, where the F1 score drops from 0.84 and 0.85 to 0.75 and 0.79, respectively. In the transition zone, the addition of noise to SEM image greatly deteriorates the precision for the matrix component and recall of the pyrite component to values close to 0.65. Organic/kerogen component has the best F1 score in the transition region.

Table 10.4 lists the performance of threshold-based segmentation method, which can be compared with Table 10.2 to gauge the robustness of the newly proposed segmentation method. The threshold-based method exhibits good performance in the inner region for all components except the pore/crack and organic/kerogen components. For transition zone, there is a significant drop in performance for the pore/crack and organic/kerogen components, whereas there is an increase in performance for the pyrite component primarily due to the improvement in recall. For both inner and transition zones, pore/crack exhibits lower precision, whereas organic/kerogen exhibits lower recall.

TABLE 10.4 Performance of thresholding-based segmentation method on the test dataset without noise for the four rock components in the image, where IR and TZ stand for inner region and transition zone

	Precision		Recall		F1 score	
Components	**IR**	**TZ**	**IR**	**TZ**	**IR**	**TZ**
Pore and crack	0.85	0.77	1.00	1.00	0.92	0.87
Organic and kerogen	0.99	0.89	0.78	0.85	0.87	0.87
Matrix	0.98	0.87	1.00	0.82	0.99	0.84
Pyrite	1.00	1.00	0.97	0.86	0.99	0.93
Avg.	0.95	0.88	0.94	0.87	0.94	0.87

TABLE 10.5 Performance of object-based segmentation method on the test dataset without noise for the four rock components in the image, where IR and TZ stand for inner region and transition zone

	Precision		Recall		F1 score	
Components	**IR**	**TZ**	**IR**	**TZ**	**IR**	**TZ**
Pore and crack	0.93	0.59	0.94	0.99	0.93	0.74
Organic and kerogen	0.97	0.89	0.91	0.71	0.94	0.79
Matrix	0.73	0.50	1.00	0.75	0.84	0.60
Pyrite	1.00	1.00	0.57	0.08	0.72	0.15
Avg.	0.90	0.75	0.87	0.64	0.87	0.59

For object-based segmentation of pyrite, the recall is very low, and the precision is high indicating the pixels from pyrite region are not reliably segmented and the predictions of pyrite as the component type have low uncertainty (Table 10.5). The object-based method is not as robust as the threshold-based method, especially for the pyrite and matrix components. Pyrite component has perfect precision for both inner and threshold regions. Matrix component has a perfect recall for only inner region.

3.4 Feature ranking using permutation importance

Sixteen features are used in the proposed machine-learning assisted segmentation methodology. Feature ranking identifies the most important features for the proposed segmentation of SEM image of shale samples. For feature ranking, we implement permutation importance, which replaces one feature at a time with random noise having mean and variance equal to those of the replaced feature. After the replacement of feature with noise, this ranking scheme measures the reduction in the classification score (in our case, we use the F1 score). Feature importance is directly proportional to the reduction in score, that is, an important feature when replaced by noise with similar statistical distribution will result in a large drop in the predictive performance. The permutation importance-based ranking of the 16 features for the proposed segmentation task from high importance to low importance is as follow: Gaussian blur, HL_{d2}, H_{xx}, pixel intensity, local minimum, HH_{d1}, Sobel operator, local mean, HL_{d1}, LH_{d2}, LH_{d1}, H_{xy}, DoG, local maximum, HH_{d2}, and H_{yy}. Permutation importance ranking indicates that when only the three top-ranked features (Gaussian blur, HL_{d2}, and H_{xx}) are used, the performance reduces only 10% of the performance achieved when using all the features, which is a reduction from 0.95 to 0.86 in the averaged F1 Score for the IR and TZ pixels.

3.5 Deployment of the segmentation method on new, unseen SEM images

After model evaluation and testing, the trained segmentation can be directly applied on SEM images of shale samples. For one SEM image of 2058-pixel by 2606-pixel in size, it takes 5 seconds for feature extraction followed by 30 seconds to perform the classification that determines the rock components in the image. Doubling of the image size will result in an increase in runtime by four times. Few segmented images shown in Fig. 10.13 clearly outline the excellent performance of the proposed segmentation methodology.

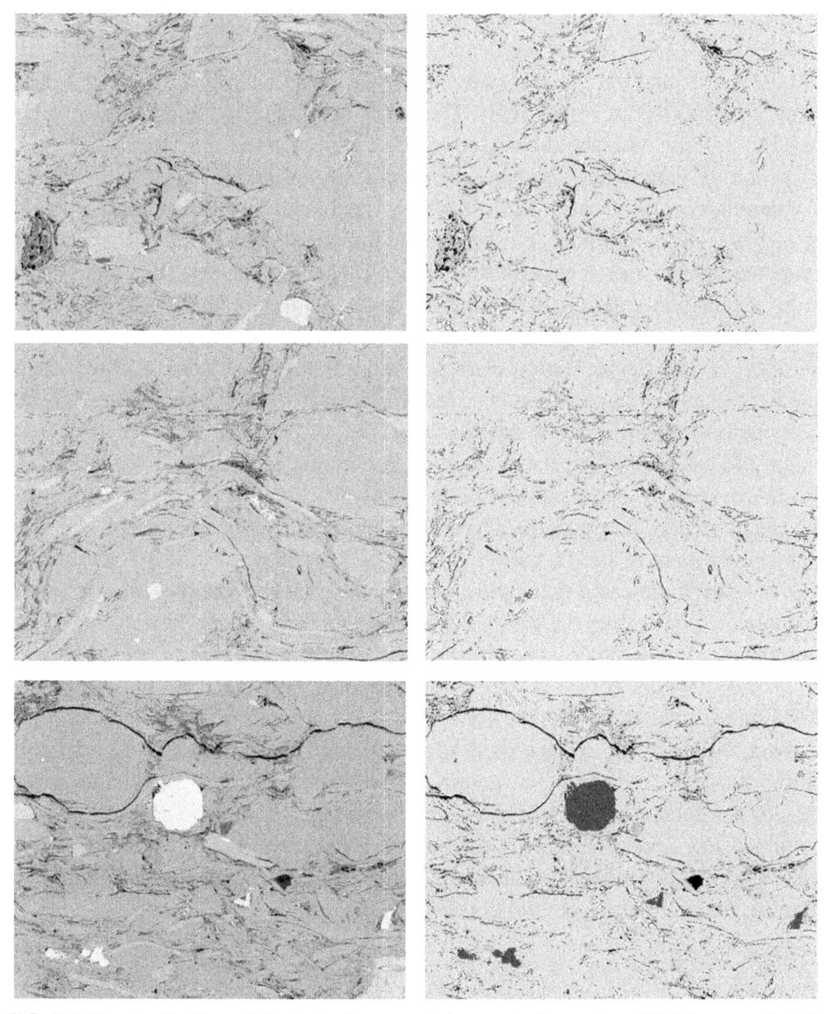

FIG. 10.13 Application of the trained segmentation method on other SEM images of shale samples. The segmented images exhibit good consistency when compared to the real images, where *green*, *blue*, *gray*, and *black* represent organic/kerogen, pyrite, matrix, and pore/crack components, respectively.

4 Conclusions

Machine learning-assisted segmentation workflow successfully located kerogen/organic, pore/crack, pyrite, and matrix components in SEM images of shale samples. This segmentation method involves two steps, feature extraction from SEM images followed by random forest classification of each pixel in the SEM image. The method was trained on 705, 15000, 17373, and 2074 pixels representing the four components, respectively. The trained method successfully segmented SEM images of size 2058 pixels by 2606 pixels. During the deployment phase on one SEM image, feature extraction takes 5 seconds, and the classification takes 30 seconds on an Intel Xeon CPU E5-1650 v3 at 3.5GHz, 32GB RAM desktop computer.

Average F1 scores of the segmentation of both inner and transition regions are 0.94, 0.97, 0.8, and 0.83 for (1) pore/crack, (2) organic/kerogen, (3) matrix, and (4) pyrite components, respectively. The newly developed method is shown to be superior to the threshold-based method, object-based method, and Fiji segmentation plugin. The tendency of misclassifying pixels near the interfaces of components was around 0.1%, which is calculated as the fraction of pixels for which the multilabel classification indicates that none of the four component types has a probability higher than 0.7. When SEM image has lower acquisition quality, the proposed segmentation method is reliable for differentiating only the pore/crack from organic/kerogen in both inner and transition zones. Segmentation accuracy for the transition zone is lower than that for the inner region, especially for matrix and pyrite components. Matrix component in transition zone has low precision of 0.79 and high recall of 0.9, which indicates that pixels segmented as matrix component have higher uncertainty and the segmentation method has good ability to correctly identify the actual matrix component. The exact opposite trend is exhibited for the pyrite component in the transition zone, which has a precision of 1 and recall of 0.74, which indicates that pyrite component is never assigned to any other component type but pyrite component tends to be wrongly labeled as other components. Organic/kerogen component has the best F1 score of 0.97, and the matrix component has the worst F1 score of 0.75 in the transition region.

In future investigations, the following tasks need to be accomplished: (1) Improve the capability of the method to segment seven components, namely, pyrite, kerogen and organic matter, clay, quartz, organic pore, inorganic pore, and cracks; (2) improve the segmentation performance for the pixels in the transition zone by improving feature extraction and the classification tasks; and (3) apply unsupervised learning and deep learning techniques to improve feature extraction and classification.

Acknowledgments

We thank the financial support from the American Chemical Society's Petroleum Research Foundation (ACS-PRF #59363-DNI Grant). We thank the Integrated Core Characterization

Center at the University of Oklahoma for providing us the SEM images for this study. We thank Dr. Mark Curtis and Jeremy Jernigen, who acquired high-quality SEM images and provided guidance on the dataset and the quality of segmentation. We also thank Dr. Carl Sondergeld and Dr. Chandra Rai, who lead the Integrated Core Characterization Center (IC3) as co-PIs, for their technical guidance and support for this work.

Conflict of Interest

The authors declare that they have no competing interests. The authors developed and tested the machine learning and feature extraction methods used in the study. The authors prepared the first complete draft of the chapter. SEM maps of shale samples were acquired at the Integrated Core Characterization Center (IC3) in the Mewbourne College of Earth and Energy of the University of Oklahoma.

References

[1] Daniel P, Raju R, Neelima G. Image segmentation by using histogram thresholding. Int J Comput Sci Eng Technol 2012;2(1):776–9.

[2] Grau V, Mewes AUJ, Alcaniz M. Improved watershed transform for medical image segmentation using prior information. IEEE Trans Med Imaging 2004;23(4):447–58.

[3] Li H, Misra S. Prediction of subsurface NMR T2 distributions in a shale petroleum system using variational autoencoder-based neural networks. IEEE Geosci Remote Sens Lett 2017;14(12):2395–7.

[4] He J, Misra S, Li H. Comparative study of shallow learning models for generating compressional and shear traveltime logs. Petrophysics 2018;59(06):826–40.

[5] Anemone R, Emerson C, Conroy G. Finding fossils in new ways: an artificial neural network approach to predicting the location of productive fossil localities. Evol Anthropol 2011; 20(5):169–80.

[6] Bauer T, Strauss P. A rule-based image analysis approach for calculating residues and vegetation cover under field conditions. Catena 2014;113:363–9.

[7] Li H, He J, Misra S. Data-driven in-situ geomechanical characterization in shale reservoirs. In: SPE annual technical conference and exhibition. Society of Petroleum Engineers; 2018.

[8] Wu C, Cheng H, Li S, et al. ApesNet: a pixel-wise efficient segmentation network for embedded devices. IET Cyber Phys Syst Theory Appl 2016;1(1):78–85.

[9] Li H, Misra S. Assessment of miscible light-hydrocarbon-injection recovery efficiency in Bakken shale formation using wireline-log-derived indices. Mar Pet Geol 2018;89:585–93.

[10] Shen S, Sandham W, Granat M, et al. MRI fuzzy segmentation of brain tissue using neighborhood attraction with neural-network optimization. IEEE Trans Inf Technol Biomed 2005; 9(3):459–67.

[11] Ong SH, Yeo NC, Lee KH, et al. Segmentation of color images using a two-stage self-organizing network. Image Vis Comput 2002;20(4):279–89.

[12] Jiang Y, Zhou Z. SOM ensemble-based image segmentation. Neural Process Lett 2004; 20(3):171–8.

[13] Tripathi D, Hathon L, Myers M. Exporting petrophysical properties of sandstones from thin section image analysis. In: SPWLA 59th annual logging symposium; 2018.

[14] Budennyy S, Pachezhertsev A, Bukharev A. Image processing and machine learning approaches for petrographic thin section analysis. In: SPE Russian petroleum technology conference; 2017.

[15] Rahimov K, AlSumaiti A, AlMarzouqi H, et al. Use of local binary pattern in texture classification of carbonate rock micro- CT images. In: SPE kingdom of Saudi Arabia annual technical symposium and exhibition; 2017.
[16] Asmussen P, Conrad O, Günther A, et al. Semi-automatic segmentation of petrographic thin section images using a "seeded-region growing algorithm" with an application to characterize wheathered subarkose sandstone. Comput Geosci 2015;83:89–99.
[17] Zhao Y, Jiang H, Li J, et al. Study on the classification and formation mechanism of microscopic remaining oil in high water cut stage based on machine learning. In: Abu Dhabi international petroleum exhibition & conference; 2017.
[18] Hughes A, Liu Z, Raftari M, Reeves ME. A workflow for characterizing nanoparticle monolayers for biosensors: machine learning on real and artificial SEM images. PeerJ 2014; 2:e671v2. https://doi.org/10.7287/peerj.preprints.671v2.
[19] Tang D, Spikes K. Segmentation of shale SEM images using machine learning. In: SEG international exposition and annual meeting; 2017.
[20] Tran HT, Jernigen JD, Curtis ME, Sondergeld CH, Rai CS. Investigating microstructural heterogeneity in organic shale via large-scale, high-resolution SEM imaging. In: Unconventional resources technology conference; 2017.
[21] Felzenszwalb PF, Huttenlocher DP. Efficient graph-based image segmentation. Int J Comput Vis 2004;59:167–81.
[22] Arganda-Carreras I, Kaynig V, Rueden C, et al. Trainable Weka segmentation: a machine learning tool for microscopy pixel classification. Bioinformatics 2017;33(15):2424–6.

Chapter 11

Generalization of machine learning assisted segmentation of scanning electron microscopy images of organic-rich shales

Siddharth Misra*, Eliza Ganguly* and Yaokun Wu[†,a]
*Harold Vance Department of Petroleum Engineering, Texas A&M University, College Station, TX, United States, †Texas A&M University, College Station, TX, United States

Chapter outline

1 Introduction

Microstructure of a material defines transport, chemical, geomechanical, and many other physical properties. For purposes of geomaterial characterization, subsurface geological samples are imaged and then analyzed to gain

[a] Formerly at the University of Oklahoma, Norman, OK, United States.

Machine Learning for Subsurface Characterization. https://doi.org/10.1016/B978-0-12-817736-5.00011-9

microstructural information [1]. Image segmentation is an essential step toward image analysis. Segmentation is the process of dividing an image into multiple superpixels and labeling the pixels constituting the superpixels on the basis of certain characteristics of the constituent pixels. The most popular segmentation methodologies include thresholding, region growing, statistical, and contour evolution methods. Thresholding is difficult to perform on complex heterogeneous samples, like shales, that has multiple components with overlapping pixel intensities. Region growing segmentation methods find ample application in medical imaging. However, it requires manual selection of seed points and does not provide an estimate of uncertainty.

In recent years, machine learning (ML) has been significantly contributing to image segmentation. ML has been implemented in several areas of geophysics, material characterization and petroleum engineering, e.g., automated detection of seismic faults and classification of well tops from labeled well logs. ML algorithms can learn from a large dataset of measurements to accomplish iterative, manual, time-consuming tasks in a much shorter duration as compared with physics-driven or human-led efforts while delivering relatively more accurate results by incorporating approaches that account for higher-order, higher-dimensional nonlinear trends.

Machine learning methods for image segmentation can be built using supervised or unsupervised learning techniques. In one ML approach referred to as supervised classification, ML algorithm processes the values of features and targets for thousands of samples to learn a function that can then be deployed on feature values of a new sample to predict its target values. Jobe et al. [2] presented two examples of image processing using machine learning techniques to predict geological and petrophysical properties of interest. In the first example, thin section images of carbonate rocks were classified by reservoir zone by using segmented pore geometries as input. Four machine learning methods, namely, distance to mean (DM), decision tree (DT), *k*-nearest neighbors (KNN), and multiclass support vector machines (MSVM), were used in the study, wherein DT and KNN methods proved to be most accurate in predicting the reservoir zones. In the second example, a convolutional neural network was implemented to assign 6 Dunham textures as labels to the carbonate thin sections. Andrew [3] compared the performance of machine learning-based image segmentation algorithm with two traditional segmentation methods on three sets of synthetic images. The machine learning-based classifier was found to have a relatively high tolerance to noise and a considerably low misclassification rate, thereby making it more efficient than the conventional segmentation methods.

In this chapter, the primary objective is to evaluate and improve the generalization of machine-learning assisted segmentation methodology for scanning electron microscopy (SEM) images. Generalization of a data-driven model is defined as the ability of the model to perform at high accuracy for new, unseen data without much sensitivity to the noise in training data. A well-generalized data-driven model will be effective across a range of data distribution beyond the limited data used for training and testing the model.

High generalization can be achieved by minimizing over-fitting on training data by using techniques like hyper-parameter tuning, cross-validation, and regularization. Performance of a model on the training data is generally quantified in terms of memorization error, whereas that on the testing data is quantified in terms of generalization error. Generalization of a model is improved by hyper-parameter tuning to reach to an optimal combination of hyper-parameters, such that the generalization error on testing data (or test fold) is at its lowest value and the memorization error on the training data (or training folds) is low and does not vary much with slight changes in hyper-parameter values. We first extract certain features from SEM images and then train random forest classifier to relate the feature values for each pixel to a specific component type of the pixel; thereby developing a methodology to divide SEM images of shale samples into four component types, namely, pore/crack, organic/kerogen, matrix, and pyrite. For purposes of evaluating the generalization capability, the data-driven segmentation model is trained on SEM images of shale samples from a specific formation, and then tested on SEM images of shale samples from a different formation. In our study, we use samples from Wolfcamp and Barnett shale formations.

The goals of this study are as follows:

- Evaluate the generalization capability of the ML-assisted segmentation method by using SEM images from Wolfcamp and Barnett Shale formations.
- Perform hyperparameter optimization/tuning to improve the generalization performance of the ML-assisted segmentation workflow.
- Investigate the robustness of the proposed segmentation method in the outer regions of components where the pixel intensities transition from one dominant component type to the other.
- Identify the important features influencing the generalization performance of the proposed SEM image segmentation methodology.

2 Method

This chapter deals with the validation of machine learning-assisted image segmentation workflow on high-resolution SEM images of shale samples. The segmentation workflow is elaborated in Fig. 11.1. The nanoscale attributes of shale samples are best captured by an SEM image. We aim at segmenting the pixels of the images into four component types, namely, pore/crack, organic/kerogen, matrix, and pyrite. The study makes use of two SEM maps: Map-1 from Wolfcamp Shale and Map-2 from Barnett Shale, which are continuous sequence of SEM images of organic-rich shale samples. These are referred to as Map-1 and Map-2 in the subsequent part of the chapter.

2.1 Image preprocessing

Map-1 covers an area of 260.6 μm by 2058 μm consisting of 26,060 × 205,800 pixels, and Map-2 covers an area of 164.64 μm by 182.42 μm containing

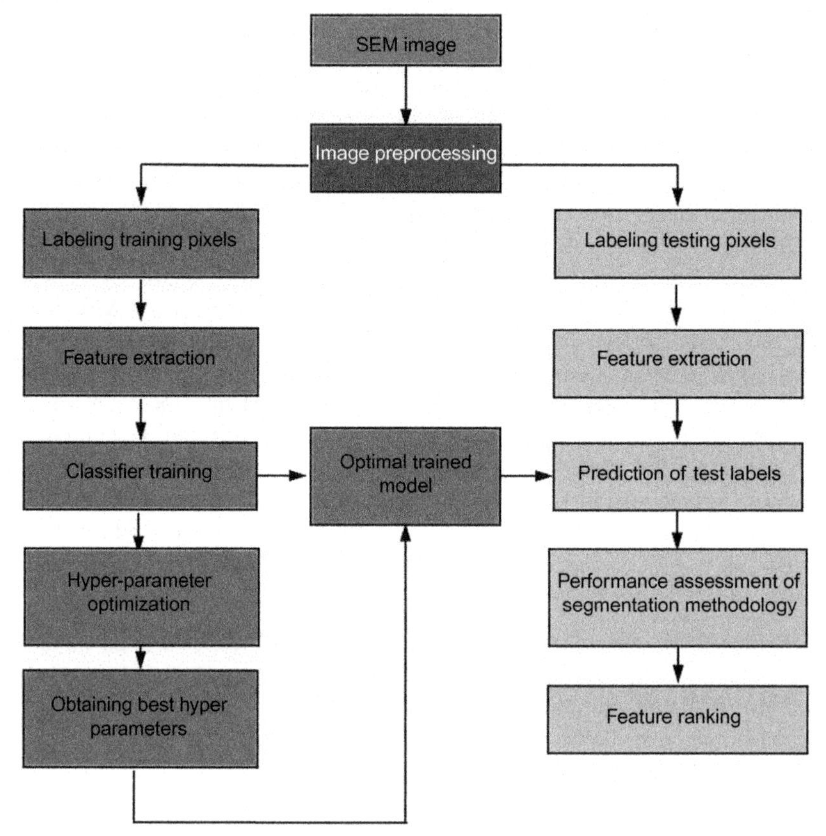

FIG. 11.1 Flowchart of the machine learning-assisted image segmentation methodology for SEM images of shale samples.

16,464 × 18,242 pixels. Map-1 and Map-2 were sliced into 1000 and 56 image slices, respectively, with no overlapping pixels. Image slices are numbered 1–1000 for Map-1 and 1–56 for Map-2, each consisting 2058 × 2606 pixels and covering an area of 20.58 μm by 26.06 μm.

Performance of a data-driven model improves when relevant features are used to build the model. For high-dimensional datasets, such as images and audio samples, feature extraction and feature engineering are required to facilitate the learning and generalization of data-driven models. Feature extraction was performed on the high-resolution SEM images for extracting 16 features used to train the classifier for image segmentation. Pixel intensity was used as the primary feature to differentiate between the four components. For each pixel, 15 additional features were computed by processing each pixel along with its neighboring pixels.

2.2 Segmentation performance evaluation

The performance of the proposed segmentation workflow is validated by performing the following tasks:

- Task 1: Training and testing on Map-1 (Model-1)
- Task 2: Training on Map-1 and testing on Map-2 (Model-1)
- Task 3: Training and testing on Map-2 (Model-2)
- Task 4: Training on Map-2 and testing on Map-1(Model-2)
- Task 5: Training on a combined dataset of Map-1 and Map-2 and testing on Map-1 (Model-3)
- Task 6: Training on a combined dataset of Map-1 and Map-2 and testing on Map-2 (Model-3)

In this chapter, the model trained on SEM images from Map-1 is referred as Model-1; the model trained on SEM images from Map-2 is referred as Model-2; and the model trained on SEM images from Map-1 and Map-2 is referred as Model-3. Tasks 2 and 4 are specially designed to evaluate the generalization performance of the ML-assisted segmentation method. The hypothesis is that a highly generalizable model will have high segmentation accuracy when applied on SEM images of shale samples from formations that are different from the formations from where SEM images are obtained for purposes of training. The proposed ML-assisted segmentation methodology will be considered to have high generalization capability when the data-driven model will exhibit high segmentation accuracy on the testing dataset in Tasks 2 and 4. Each model implements random forest classifier, which is a collection/ensemble of decision trees (DTs) that are independently trained in parallel by a technique involving bootstrapping followed by aggregation (referred as bagging). Bootstrapping involves random sampling of the training dataset with replacement. Multiple DTs are trained on separate bootstrapped samples using various subsets of available features. Finally, the predictions of multiple DTs are aggregated/averaged in order to reduce the overall variance of prediction. By integrating multiple decision trees, a random forest classifier achieves low bias and low variance [4].

2.3 Training dataset

Image Slice 90 is chosen from Map-1 to collect the training pixels for Model-1. From this slice, we select 705, 2074, 17,373, and 15,000 pixels corresponding to the pore/crack, organic/kerogen, matrix, and pyrite components, respectively, as the training pixels (Fig. 11.2). The training pixels are selected from easily identifiable regions of the image slice. For training of Model-2, we chose Image Slice 35 of Map-2. We selected 912, 8435, 5806, and 5387 pixels corresponding to the pore/crack, organic/kerogen, matrix, and pyrite components, respectively, as the training pixels (Fig. 11.3). For training of

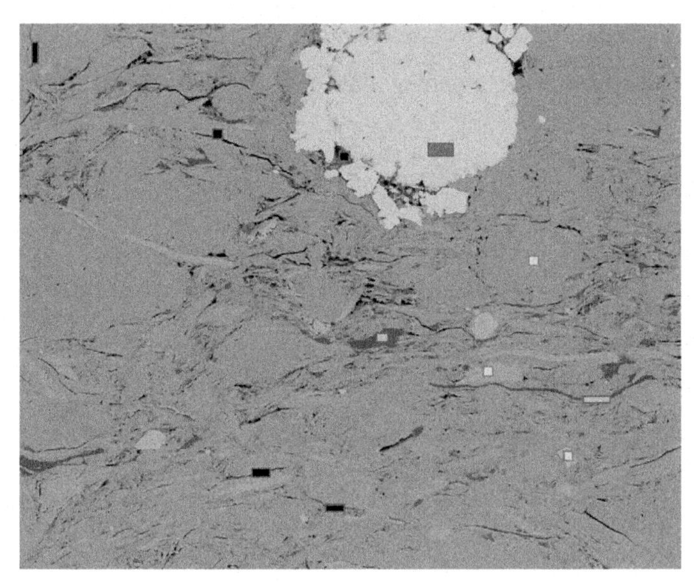

FIG. 11.2 Image Slice 90 of Map-1 and the training pixels, which constitute less than 0.7% of the total pixels in the Image Slice. Pixels corresponding to pore/crack, organic/kerogen, matrix, and pyrite components are represented by *black*, *green (light gray in print version)*, *gray*, and *blue (dark gray in print version)* rectangles, respectively.

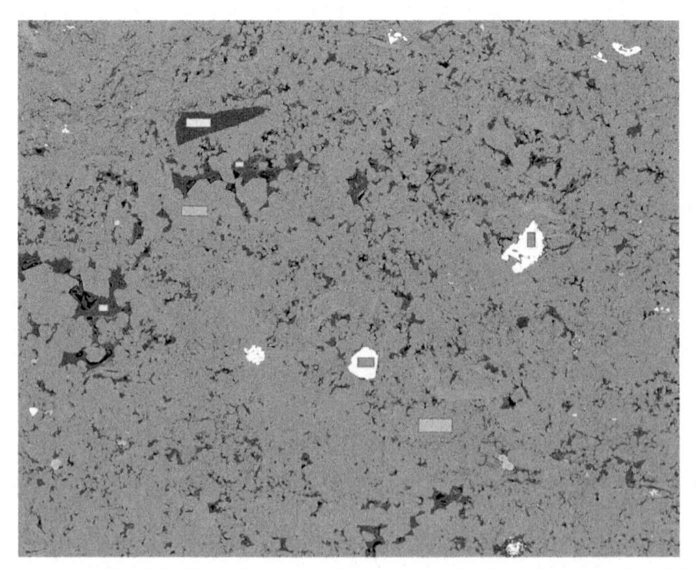

FIG. 11.3 Image Slice 35 of Map-2 and the training pixels, which constitute less than 0.5% of the total pixels in the Image Slice. Pixels corresponding to pore/crack, organic/kerogen, matrix, and pyrite phases are represented by *black*, *green (light gray in print version)*, *gray*, and *blue (dark gray in print version)* rectangles, respectively.

Model-3, the training dataset was created by combining the pixels from the Slice 90 of Map-1 and the Slice 35 of Map-2. A total of 1617, 10,509, 23,179, and 20,387 pixels belonging to pore/crack, organic/kerogen, matrix, and pyrite phase, respectively, were used to train Model-3. It is to be noted that only 35,512 pixels out of 26,060 × 205,800 pixels were used to train the Model-1, 20,540 pixels out of 16,464 × 18,242 pixels were selected to train the Model-2, and only 55,000 pixels out of the entire dataset were used to train the Model-3.

After selecting the training pixels, the training dataset is built by extracting 16 features for each of the 35,152 training pixels from Slice 90 and 20,540 training pixels from Slice 35. Features and feature extraction methods used in our study are as follows:

- Pixel intensity (one feature)
- Gaussian blur (one feature)
- Difference of Gaussian (one feature)
- Sobel edge detector (one feature)
- Hessian matrix (three features)
- Wavelet transform (six features)
- Local information (three features: minimum, maximum, and mean)

The 16 features extracted for each of the training pixels are fed into a random forest classifier to learn the function that maps each training pixel to its component type.

2.4 Hyperparameter optimization

Hyperparameters govern the learning process. Unlike parameters that are computed at the completion of the learning process, hyperparameters are defined by the users and are tuned to achieve optimal performance of the machine learning model. Hyperparameters are regulated to ensure low memorization error (performance on the training dataset) while ensuring lowest generalization error (performance on the testing dataset). In our study, hyperparameter optimization was conducted for each of the three random forest classifier models used in the three segmentation approaches, respectively. For purposes of hyperparameter optimization, we perform a grid search. The hyperparameters that were altered to obtain the best possible combination are as follows:

- N_estimators control the number of trees in a random forest. Higher number of trees reduce the model variance to enhance the generalization capability. A large number of trees can make the training process computationally slow: therefore it is necessary to find an optimum value for this parameter.

- Max_features represent the number of features taken into consideration while performing a split at a node. Overall, all features are used when training a deep tree; however, only a specific number of features (based on max_features) from the entire feature set are considered when splitting a node. Lower values of max_features ensures that the trees in the random forest are different from each other but does not guarantee maximum drop in entropy when splitting a node. When the algorithm cannot find a valid partition for a node, the algorithm will look for more features than max_features.
- Max_depth is the maximum depth of each tree in the random forest. For a higher value of max_depth, more splits are performed to finely partition the feature space, which may lead to overfitting.
- Min_samples_split is the minimum number of samples to be considered when splitting an internal node. With an increase in min_samples_split, the model gets constrained and cannot learn enough information about the data giving rise to underfitting.
- Min_samples_leaf is the minimum number of samples to be required at a leaf node. This hyperparameter has a similar effect as min_samples_split, where a larger value may cause underfitting.
- Bootstrap uses random sampling of data with replacement and helps to reduce variance and avoid overfitting of the model.

In the grid search approach, weighted F1 score was chosen as a basis of comparison to decide the best combination of hyperparameters. F1 score for each of the four components is the harmonic mean of precision and recall for the component when performing a binary classification (i.e., whether a pixel is the component of interest or not). Weighted F1 score is computed by weighing F1 score for each component by the corresponding support (the number of true instances for each component). Weighted F1 score accounts for class imbalance in the dataset; simple F1 score cannot be used as a reliable evaluation metric for imbalanced dataset. Class imbalance occurs when the numbers of samples of various component types are orders of magnitude different from each other. For dataset with class imbalance, stratification is performed during the train-test split and when splitting the training dataset during cross-validation to improve the generalization of the model. For developing the most generalizable Models 1 to 3 (described in Section 2.2), a total of 1512 combinations of hyperparameters were used to build 1512 models by learning from the training pixels, specific to each of the three models. This was followed by a threefold stratified cross-validation to identify the model that generates a low memorization error and the lowest generalization error. Table 11.1 lists the values of the hyperparameters that were found optimal for each of the three models, which are described in the Section 2.2.

TABLE 11.1 Optimal hyperparameters for the three random forest classification models implemented in the three segmentation approaches, respectively.

Hyperparameter	Model 1	Model 2	Model 3
Bootstrap	True	True	False
Max_depth	10	10	10
Min_samples_leaf	1	1	1
Min_samples_split	5	2	2
Max_features	1	Auto	1
N_estimators	333	200	200

2.5 Testing dataset

Two sets of image slices were chosen from Map-1 and Map-2 each to build the inner-region and outer-region testing datasets (Fig. 11.4). The inner-region pixels were selected from the Image Slices 13, 649, and 860, and the outer-region pixels were selected from the Image Slices 13, 203, 334, and 500 of Map-1. In case of Map-2, Image Slices 15, 26, and 28 were chosen to build the inner-region training dataset, while Image Slices 13, 17, and 26 were used to obtain the outer-region training pixels. These datasets were created by selecting pixels from the image slices corresponding to each of the four component types. Sixteen features were extracted from each of the testing pixels to create the testing dataset. Each of the three trained classifier models was then applied to the relevant testing datasets to evaluate the performance. Fig. 11.5 shows segmentation results for the Image Slice 35 of Map-2 to show the distribution of the four components as identified by the segmentation models.

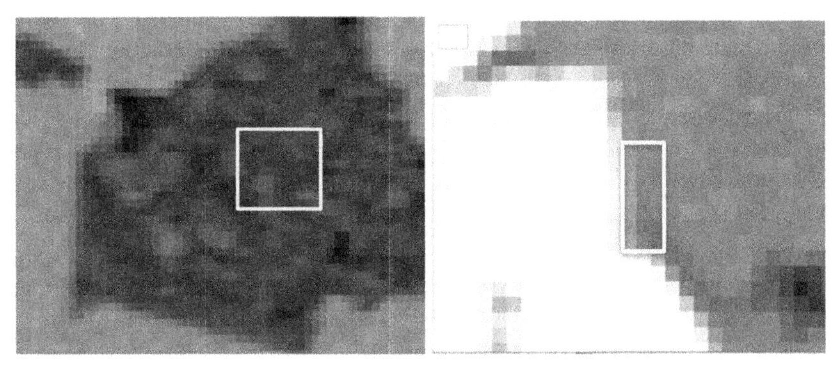

FIG. 11.4 Selection of pixels from the inner (left) and outer (right) regions of the Image Slice 35 of Map-2. These are highly zoomed-in 1-μm by 1-μm regions of the larger SEM image.

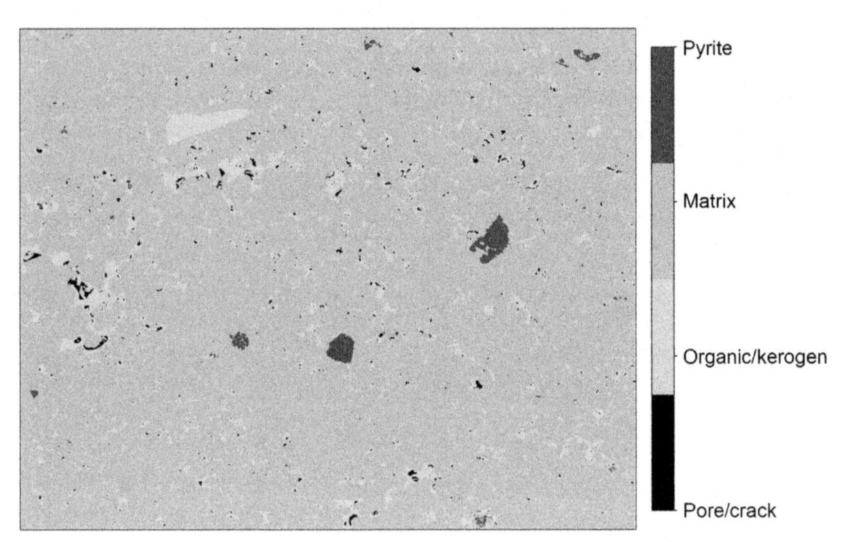

FIG. 11.5 Segmentation obtained when applying the second segmentation approach based on the Model-2 on the Image Slice 35 of Map-2.

The performance of the models (e.g., Table 11.2) is quantified in terms of F1 score. The F1 score is the harmonic mean of precision and recall. Precision is TP/(TP + FP), and recall is TP/(TP + FN), where TP, FP, and FN are true positives, false positives, and false negatives, respectively. True positive is when the predicted class of a pixel is the true class of the pixel, whereas false positive is when a pixel is wrongly predicted to be the class of interest. True negative is when a pixel is correctly predicted to be a class other than the class of interest, whereas false negative is when a pixel is wrongly predicted to be a class other than the class of interest. F1 score range from

TABLE 11.2 Classification report for Model-2 when applied on testing pixels from the inner regions of Slice 35 of Map-2 indicates excellent segmentation performance.

Components	Precision	Recall	F1 score	Support
Pore/crack	1.00	1.00	1.00	104
Organic/kerogen	1.00	1.00	1.00	476
Matrix	1.00	1.00	1.00	4500
Pyrites	1.00	1.00	1.00	91
Average/total	1.00	1.00	1.00	5171

0 to 1, such that zero indicates poor classification and one indicates robust performance. The precision of the classifier specific to a component type is a measure of the reliability of the component type assigned by the classifier. In other words, precision for component X can be defined as "when a model assigns a pixel to be component X, how often is the model correct?". Recall of the classifier specific to a component type is a measure of the classifier's ability to correctly assign the component type of interest; in other words, it is the ability of the classifier to find the class of interest (similar to the sensitivity of the classifier to a certain class). In other words, recall for component X is defined as "when a pixel is actually component X, how often does the model correctly assign that pixel as component X?". An example where recall is more important than precision: the scanners at the airport need high recall with respect to dangerous materials, but it is not crucial for the scanner to have high precision with respect to dangerous materials. An example where precision is more important than recall: when launching satellite into orbit, it is important to have a high precision with respect to whether it is a good day to launch or not, a low recall in this case does not matter much.

3 Results and discussion

In this chapter, we have applied a machine learning-assisted image segmentation workflow on SEM images of organic-rich shale samples from two different formations. The overall objective of this study is to understand the performance of the proposed segmentation workflow when it is trained on a slice of SEM map from one shale formation and then deployed on a slice of SEM map from an entirely different shale formation. The two formations differ in topology and distribution of the four components; in addition, the SEM maps vary in their pixel intensity because they were captured at different times using slightly different settings for the SEM acquisition. The study has been divided into six tasks to investigate the generalization potential of the proposed workflow. This numerical experiment tests the generalization of the machine learning assisted SEM image segmentation.

3.1 Task 1: Training and testing on Map-1

This section quantifies generalization of the ML assisted segmentation within a formation. Out of the 1000 slices of Map-1, Slice 90 is used to build the training dataset. The model trained on Map-1 is termed as Model-1. A total of 35,152 pixels are used to train the segmentation method on Map-1. Model-1 is tested on the inner-region pixels of Slices 13, 649, and 860 and outer-region pixels of Slices 13, 203, and 334 of the same map. For testing on inner-region pixels, 2916, 2819, 10,915, and 6197 pixels were selected from pore/crack,

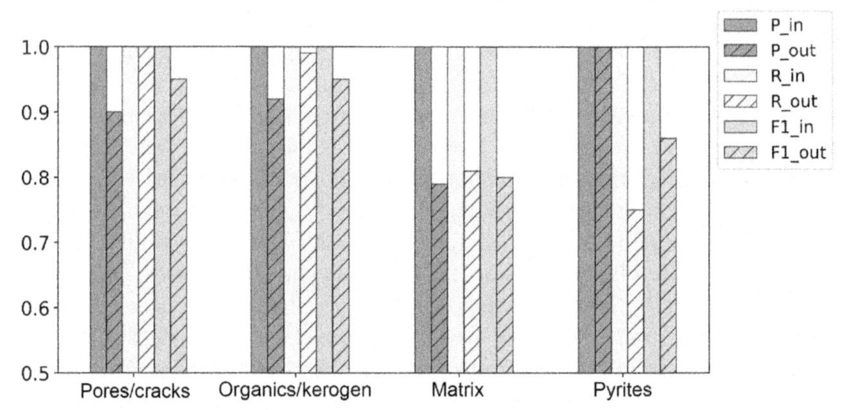

FIG. 11.6 Comparison of segmentation model performance (*P*, precision; *R*, recall; and *F1*, F1 score) on inner-region test pixels against that on outer-region test pixels of Map-1. The model was trained on training pixels from Slice 90 of Map-1. Model-1 exhibits good generalization within a formation for both inner and outer regions of pore/crack and organic/kerogen components.

organic/kerogen, matrix, and pyrite components, respectively, whereas for testing on outer-region pixels, 2623, 4392, 3623, and 3010 pixels from each component were chosen. The average F1 score was found to be 1.00 and 0.89 for the inner-region and outer-region testing pixels, respectively (Fig. 11.6). The method resulted in a perfect F1 score for the inner-region pixels but showed slightly lower performance for outer-region pixels, particularly for matrix and pyrite components. We observe a low precision and recall of 0.79 and 0.81, respectively, for the outer-region pixels of the matrix component. A low precision indicates that pixels belonging to other component types are being labeled as matrix, whereas low recall results when matrix is being classified as some other component by the model. For the outer region, we observe low precision with a high recall for the pore/crack and organic/kerogen components and an opposite behavior for the pyrite components. This suggests that boundary pixels of pyrites are being labeled as matrix, whereas boundary pixels of matrix are being misclassified as pore/crack or organic/kerogen. This observation is justified because in Map-1, the matrix shares considerable boundary area with all other components, including pore/cracks; consequently, matrix has both low recall and precision.

3.2 Task 2: Training on Map-1 and testing on Map-2

This section quantifies generalization of the ML assisted segmentation to a different formation. In this section, we study the performance of the model trained on Slice 90 of Map-1, that is, Model-1, when applied to testing pixels from the inner-region and outer-region of Map-2. Inner-region testing pixels were selected from Slices 15, 26, and 28 of Map-2, whereas the outer-region

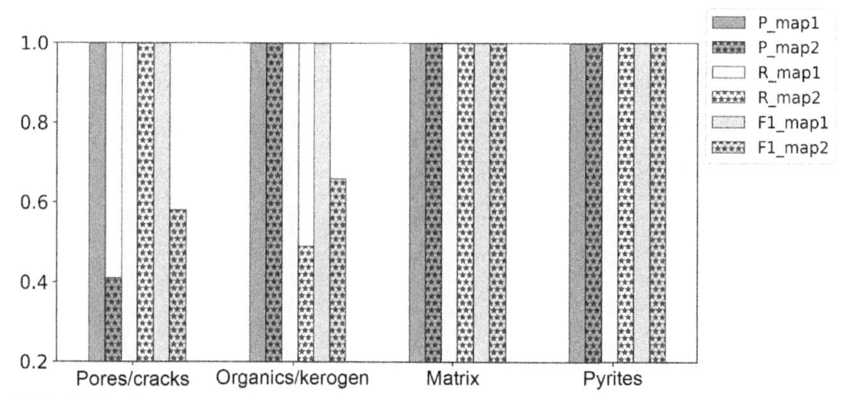

FIG. 11.7 Comparison of segmentation model performance (*P*, precision; *R*, recall; and *F1*, F1 score) on inner-region test pixels of Map-1 (Task 1) against those on inner-region test pixels of Map-2. The model was trained on training pixels from Slice 90 of Map-1. Model-1 exhibits excellent generalization to another formation for the inner regions of matrix and pyrite components.

testing pixels were selected from Slices 13, 17, and 26 of Map-2. In this section, we compare the performances of the same model (Model-1) on the inner-region testing pixels from the two maps (Fig. 11.7).

There are significant differences in the distributions/characteristics of pore/crack and organic/kerogen components between the two maps. Map-2 is dominated by the presence of pores embedded in organic/kerogen components, whereas Map-1 consists of both organic and inorganic pore systems (Figs. 11.2 and 11.3). In Map-1, the cracks are present in the form of thin strips, whereas Map-2 is characterized by clusters of black pixels representing the pores. As a result, a drop in the F1 score is observed for both the inner and outer-region pixels of the pore/crack and organic/kerogen components, when Model-1 is tested on Map-2 (Fig. 11.7). We tested the model on a comparable number of support pixels for each component (1447 pixels for pore/crack, 5263 pixels for organics/kerogen, 3701 pixels for matrix, and 2396 pixels for pyrite in the inner region) to obtain an unbiased performance. For the inner region the precision was 0.41 with a high recall for the pore/crack component, and the recall was 0.49 with high precision for the organic/kerogen component, which indicates a bad segmentation performance. As supported by the confusion matrix (Fig. 11.8), a large number of pixels (1615 pixels) belonging to the organic/kerogen in Map-2 are being classified as pore/crack by Model-1, thereby resulting in low precision for pore/crack (i.e. when the model predicts pore/crack, it is rarely correct) and low recall for organic/kerogen (i.e. when it's actually organic/kerogen, the model seldom correctly predicts). Matrix and pyrite components are robustly segmented both in terms of precision and recall. One explanation is that the difference in pixel intensities of pore/crack and organic/kerogen is much smaller than that between these components and the matrix or pyrite components.

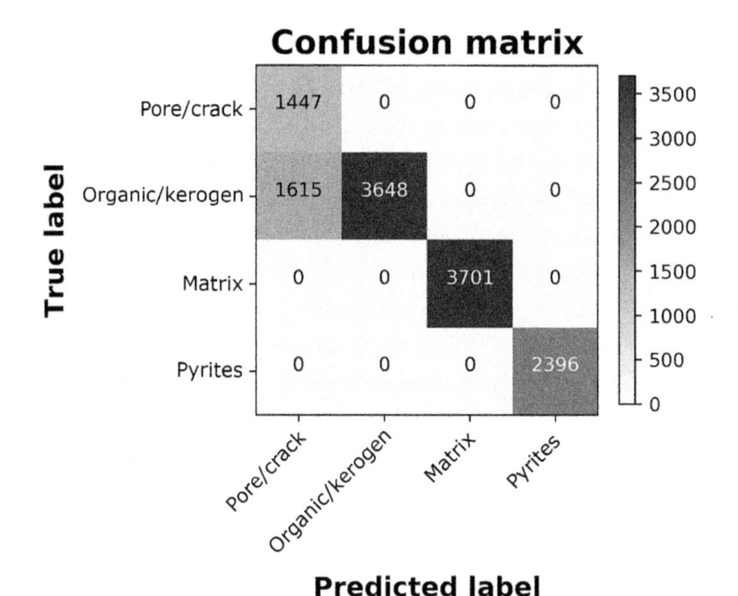

FIG. 11.8 Confusion matrix related to the segmentation performance of the model trained on Slice 90 of Map-1 when applied on the inner-region pixels of Map-2. 1615 out of 5263 organic/kerogen pixels got segmented as pore/crack pixel, resulting in a drop in precision of pyrite component and a drop in recall of organic/kerogen pixel.

In a confusion matrix, the diagonal elements represent the number of cases where the true label is same as the predicted label (i.e. true positives), whereas the off-diagonal elements show the number of cases where the components have been misclassified by the model (true negatives and false positives). Therefore the higher the diagonal values, the better the accuracy/recall of the model. In Fig. 11.8, for the matrix and pyrite components, the number of support pixels are equal to the number of diagonal elements, thereby proving that they have been correctly classified/segmented. But a significant number of support pixels (1615 out of 5263 pixels) in organic/kerogen phase has been classified as cracks, resulting in a low value of the F1 score for these two components.

For the outer region, Model-1 was tested on 395, 722, 693, and 2015 pixels corresponding to the pore/crack, organic/kerogen, matrix, and pyrite components, respectively, of Map-2. On an average, the model delivered a lower performance for the outer-region pixels, with F1 scores of 0.89 and 0.81 for Map-1 and Map-2, as compared with that of the inner-region pixels, with F1 scores of 1.00 and 0.82 for Map-1 and Map-2 (Fig. 11.9). This occurs since the model tends to misclassify the organic/kerogen pixels as pore/crack because the gray-scale intensities of the components have greater overlap in Map-2. For Map-1 (Fig. 11.9), we observe much lower precision

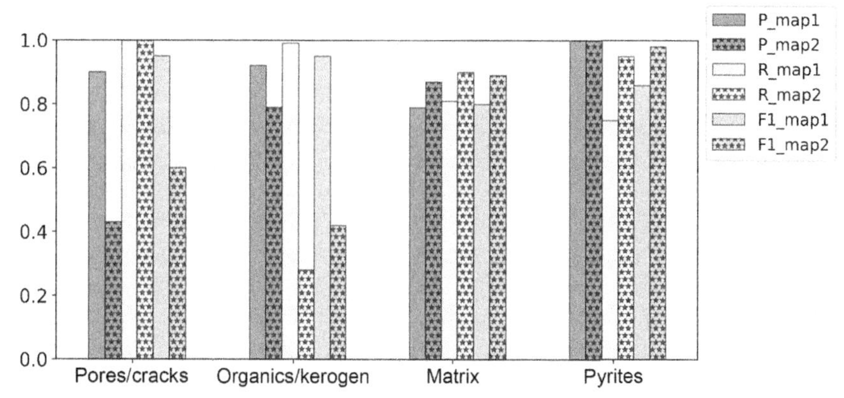

FIG. 11.9 Comparison of segmentation model performance (*P*, precision; *R*, recall; and *F1*, F1 score) on outer-region test pixels of Map-1 against those on outer-region test pixels of Map-2. The model was trained on training pixels from Slice 90 of Map-1. Model-1 exhibits good generalization to another formation for the outer regions of matrix and pyrite components.

for matrix and much lower recall for pyrite compared with others, suggesting that the pyrite pixels at the boundary of matrix and pyrite may have been classified as matrix. However, in Map-2, organic/kerogen exhibits very low recall indicating that Model-1 is not suitable for organic/kerogen detection. At the same time, the precision for pore/crack of Map-2 is very low, indicating a possibility that the organic/kerogen pixels at the interface of organic/kerogen and pore/crack are being segmented as pore/crack (Fig. 11.10). Interestingly, segmentation performance for matrix and pyrite components improve for Map-2 as compared with Map-1, primarily, due to the shaper contrast at the interfaces in Map-2.

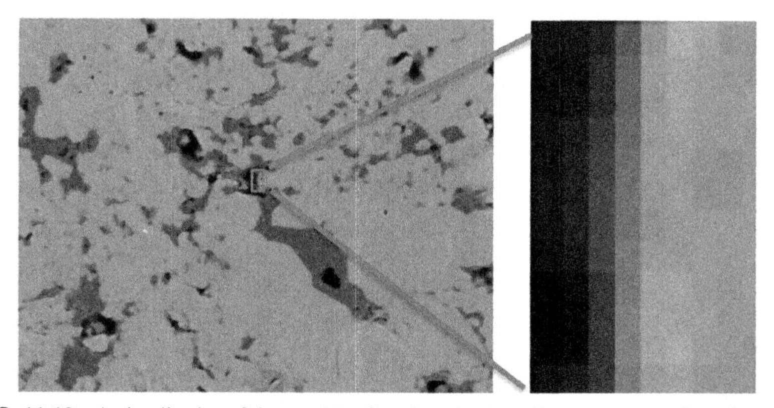

FIG. 11.10 A visualization of the transition from inner region of pore to outer region of pore to outer region of matrix to inner region of matrix, where the intermediate pixels can be misclassified as organic/kerogen due to intermediate intensities. Image belongs to Map-2.

3.3 Task 3: Training and testing on Map-2

This section quantifies generalization of the ML assisted segmentation within a formation. In the previous sections, we demonstrated the performance delivered by the machine learning-assisted segmentation workflow (Model-1) built on Map-1 from Wolfcamp formation. In this section, we will investigate the behavior of the machine learning-assisted segmentation workflow when it is trained on Map-2 from Barnett Shale formation. This model is being referred to as Model-2, which was trained on Slice 35 of Map-2. The training dataset from Map-2 comprises 912, 8435, 5806, and 5387 pixels corresponding to the pore/crack, organic/kerogen, matrix, and pyrite components, respectively. Model-2 was validated on the inner and outer regions of Map-2. The inner-region testing pixels were chosen from Slices 15, 26, and 28, whereas Slices 13, 17, and 26 were used to create the test dataset of outer-region pixels. Application of the model on these datasets results in an average F1 score of 1.00 for the inner-region pixels and that of 0.98 for the outer-region pixels (Fig. 11.11). F1 score was highest for pyrites from both inner and outer regions.

Near perfect segmentation was achieved using Model-1 and Model-2 on both training and testing datasets of inner-region pixels of Map-1 and Map-2 (Fig. 11.12). The off-diagonal values did not exceed 11 pixels for either case, which establishes the robustness of the proposed workflow on inner-region pixels despite the limited size of the training dataset.

3.4 Task 4: Training on Map-2 and testing on Map-1

This section quantifies generalization of the ML assisted segmentation to a different formation. In this section, Model-2 is being applied to inner and

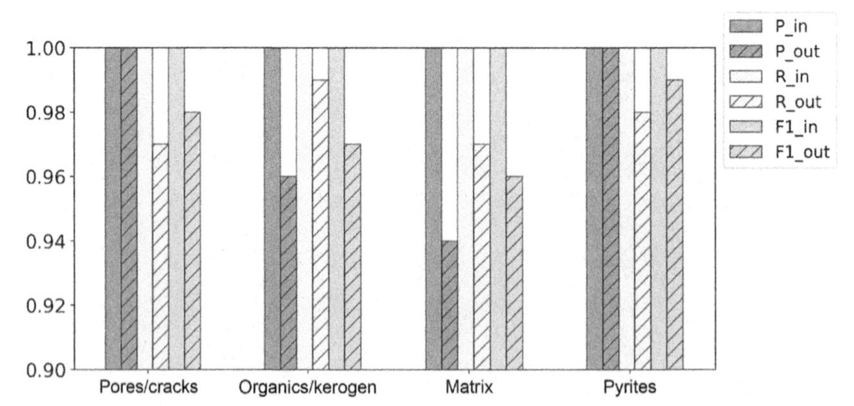

FIG. 11.11 Comparison of segmentation model performance (*P*, precision; *R*, recall; and *F1*, F1 score) for inner-region test pixels of Map-2 against that for outer-region test pixels of Map-2. The model was trained on training pixels from Slice 35 of Map-2. Model-2 exhibits good generalization within a formation for both inner and outer regions of all the four components.

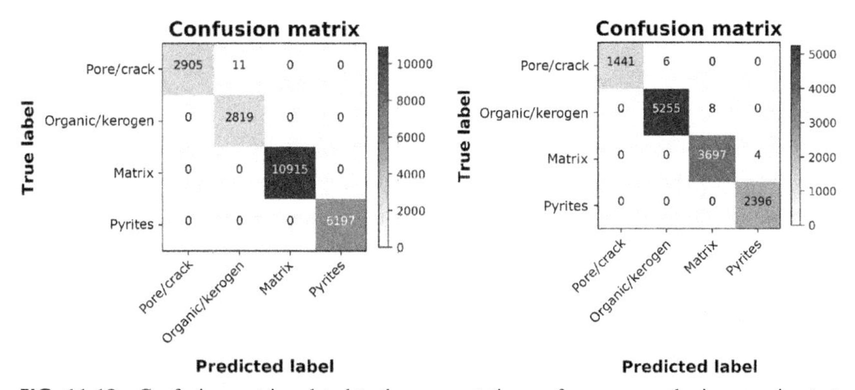

FIG. 11.12 Confusion matrix related to the segmentation performance on the inner-region test pixels for the model trained and tested on the inner regions of (left) Map-1 and (right) Map-2.

outer-region pixels of Map-1. The test dataset of the inner region consists of 8098, 7677, 8475, and 5965 pixels, while that of the outer region contains 2623, 4392, 3623, and 3010 pixels, respectively, from pore/crack, organic/kerogen, matrix, and pyrite components. Model-2 achieves an average F1 score of 0.89 for the inner-region testing pixels of Map-1 (Fig. 11.13), as compared with a perfect score of 1.00 when the Model-1 is applied on the same set of inner-region testing pixels of Map-1 in Task 1. But this score of 0.89 is higher than the average F1 score of 0.82 that was obtained by applying Model-1 on an inner-region testing set from Map-2 (Task 2). A similar result is observed in case of outer-region pixels having an average F1 score of 0.81 when Task 2 is performed and that of 0.91 in the case of Task 4. This suggests that Map-2 has more generalizable statistical features that

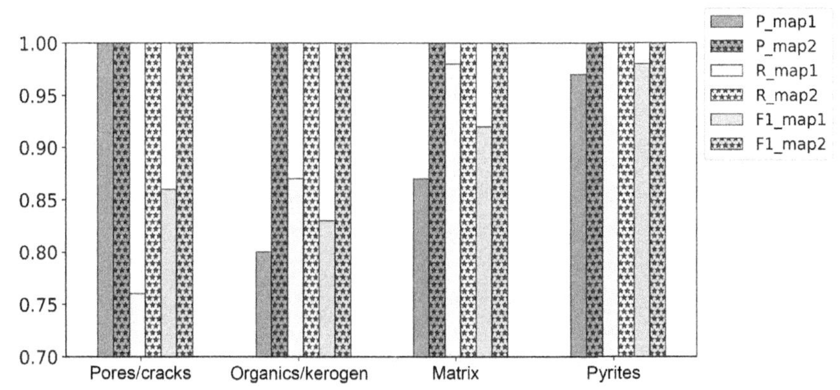

FIG. 11.13 Comparison of segmentation model performance (*P*, precision; *R*, recall; and *F1*, F1 score) for inner-region test pixels of Map-1 against that for inner-region test pixels of Map-2. The model was trained on training pixels from Slice 35 of Map-2. Model-2 exhibits good generalization to another formation for inner regions of matrix and pyrite components.

can serve to better train the segmentation model and the Map-1 has simpler features that are easier to segment. In case of the pore/crack and organic/kerogen components, Model-2 significantly outperforms for the inner-region of Map-2 as compared with the inner region of Map-1.

Model-2, when tested on Map-1, provides lower F1 scores for the pore/crack and organic/kerogen components due to the distinct distribution of these two components in the two maps. Map-1 has a dominant distribution of cracks, whereas Map-2 is mostly characterized by pores. The model, when trained on Map-2, can only identify black pixels present in a bulk as pores but fails to acknowledge strips of black pixels present in Map-1 as cracks. However, Model-2 shows a better accuracy in classifying the outer-region pixels of the matrix and pyrite components in Map-1 (Fig. 11.14), as compared with the performance achieved by Model-1 on Map-1. Model-1 has lower precision in comparison with Model-2 for the outer-region pore/crack component of Map-1, indicating that many outer-region pixels are incorrectly segmented as pore/crack by the Model-1 (Fig. 11.14). Model-2 has a lower recall in comparison with Model-1 for the outer-region pore/crack component of Map-1, indicating that many pore/crack pixels are being incorrectly segmented by the Model-2 (Fig. 11.14). Model-2 has a much higher recall in comparison with Model-1 for the outer-region pyrite component of Map-1, indicating that most of the outer-region pyrite pixels are being incorrectly segmented by the Model-1 (Fig. 11.14).

The figures in the succeeding text show the segmentation results on a slice of Map-1 when segmented by Model-1 (Fig. 11.15; left) and when segmented by Model-2 (Fig. 11.15; right). The *dark-blue (dark gray in the print version)* boxes indicate the portion of the pore/crack component in Map-1 that has been misclassified as organic component by Model-2, whereas the *dark-red*

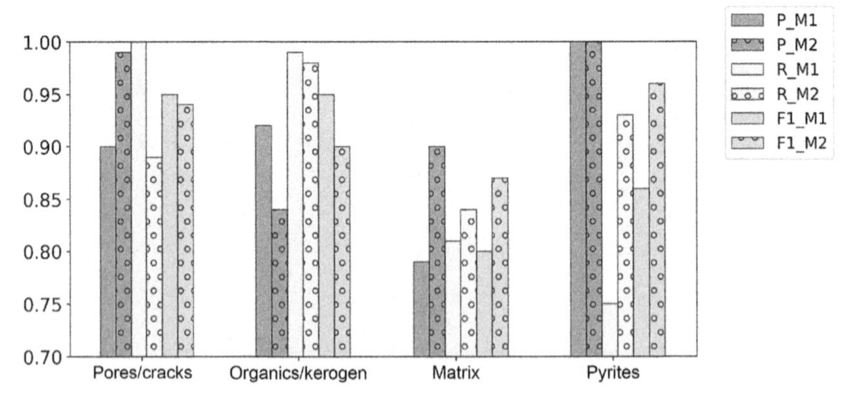

FIG. 11.14 Comparison of segmentation performances of Model-1 (M1) and Model-2 (M2) on the outer-region pixels of Map-1. Notably, for the outer-region pixels of matrix and pyrite components in Map-1, Model-2 exhibits better generalization as compared to Model-1. Model-1 exhibits good generalization for outer-region pixels of pores/cracks and organic/kerogen components. Model-2 exhibits good generalization for outer-region pixels of matrix and pyrite components.

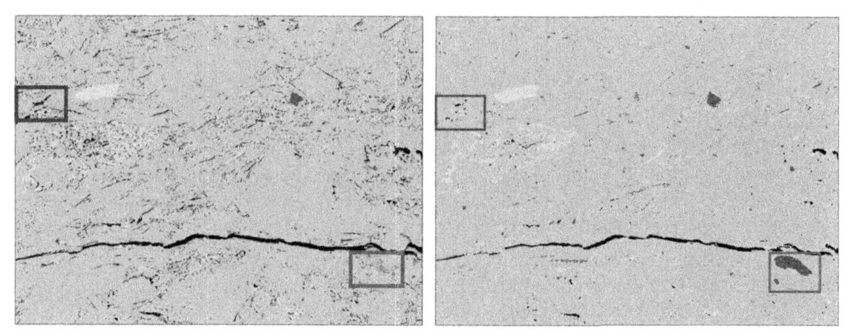

FIG. 11.15 Segmented image of Slice 20 of Map-1 generated by applying (left) Model-1 and (right) Model-2.

(light gray in the print version) boxes indicate the pixels belonging to pyrite components that have been misclassified as matrix by Model-1. A similar inference can be obtained by comparing the precision and recall values of these components listed in Fig. 11.14.

3.5 Task 5: Training on both Map-1 and Map-2 and testing on Map-1

We have observed from the aforementioned results that a model trained on a particular map performs well when tested on that map but may deteriorate in performance when applied to a map from a different formation. To optimize the performance, we train Model-3 on a combined dataset with training pixels from both Map-1 and Map-2. We used a total of 35,152 pixels from the Slice 90 of Map-1 and a total of 20,400 pixels from Slice 35 of Map-2. Sixteen features were extracted for each of the 55,552 pixels and fed to the random forest classifier. We studied the implementation of Model-3 individually on the inner and outer-region pixels of both Map-1 and Map-2. For testing Model-3 on the inner-region pixels of Map-1, we chose slices 13, 649, and 860 to create the testing dataset of inner-region pixels. A total of 2916, 2819, 10,915, and 6197 pixels were selected from pore/crack, organic/kerogen, matrix, and pyrite component types, respectively. Similarly, for testing on the outer-region pixels, Slices 13, 203, and 334 of Map-1 were used with 2747, 4842, 3951, and 3010 pixels corresponding to the four component types. Results for this task will be discussed together with Task 6 in the following section.

3.6 Task 6: Training on both Map-1 and Map-2 and testing on Map-2

For testing the newly trained Model-3 on Map-2, we applied it on the inner-region pixels of Slices 15, 26, and 28 and the outer-region pixels of Slices 13, 17, and 26. We used 1447, 5263, 3701, and 2396 inner-region pixels and

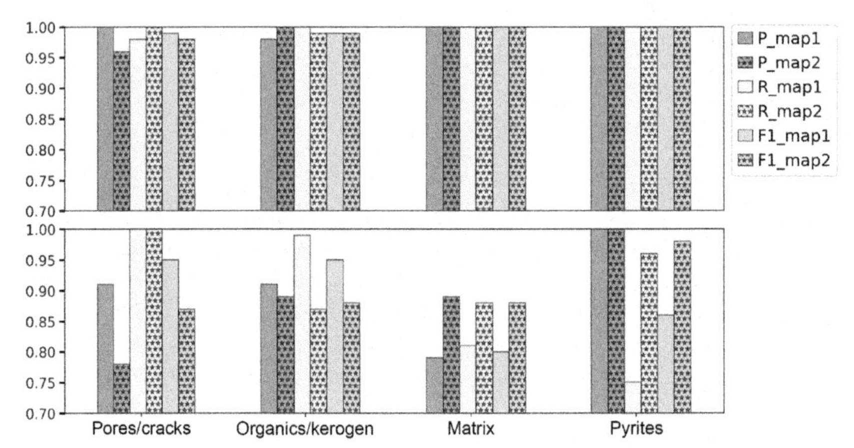

FIG. 11.16 Comparison of model performance for inner-region test pixels of Map-1 and Map-2 (top) against that for outer-region test pixels of Map-1 and Map-2 (bottom). Model 3 was trained on training pixels from Slice 90 of Map-1 and Slice 35 of Map-2. Model-3 exhibit good generalization for inner-region pixels of all components. Model-3 exhibits good generalization for outer-region pixels of pores/cracks and organic/kerogen components.

395, 722, 769, 2029 outer-region pixels as support for pore/crack, organic/kerogen, matrix, and pyrite components, respectively.

As expected, the inner regions show a much better classification with average F1 scores of 1.00 and 0.99 for Map-1 and Map-2, respectively. In contrast, the performance on the outer-region pixels is at average F1 scores of 0.89 and 0.93 for Map-1 and Map-2, respectively. In the first subplot of Fig. 11.16, we notice that the Model-3 trained on a combined dataset of Map-1 and Map-2 exhibits a perfect score of 1.00 in case of the matrix and pyrite components for the inner-region pixels of both Map-1 and Map-2. Outer-region matrix pixels (second subplot of Fig. 11.16) are difficult to correctly segment, and outer-region pixels of other components are being incorrectly predicted as matrix. Notably, in Map-1, outer-region pyrite pixels are most difficult to correctly segment, whereas in Map-2, the larger fraction of outer-region pixels tends to be incorrectly segmented as pore/crack pixels.

Model-3 is significantly better than Model-1 for inner and outer-region pixels of Map-2. Model-3 is significantly better than Model-2 for inner pixels of Map-1 (Fig. 11.17). The model trained on the combined dataset delivers a better performance than a model trained on one map and tested on a different one. This suggests that training the workflow with pixels from different maps results in a more generalized model, capable of segmenting SEM maps from different formations. Fig. 11.18 provides a visual description of the segmented images obtained using the three differently trained models. Model-2 trained on Map-2 (bottom left) exhibits good

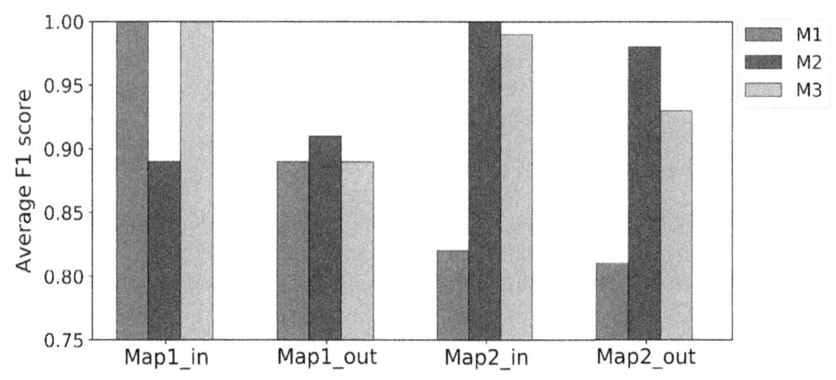

FIG. 11.17 Comparison of segmentation performance (in terms of average F1 score) of the three models on the inner and outer-region pixels of the two maps. M1, M2, and M3 represent Model-1, Model-2, and Model-3.

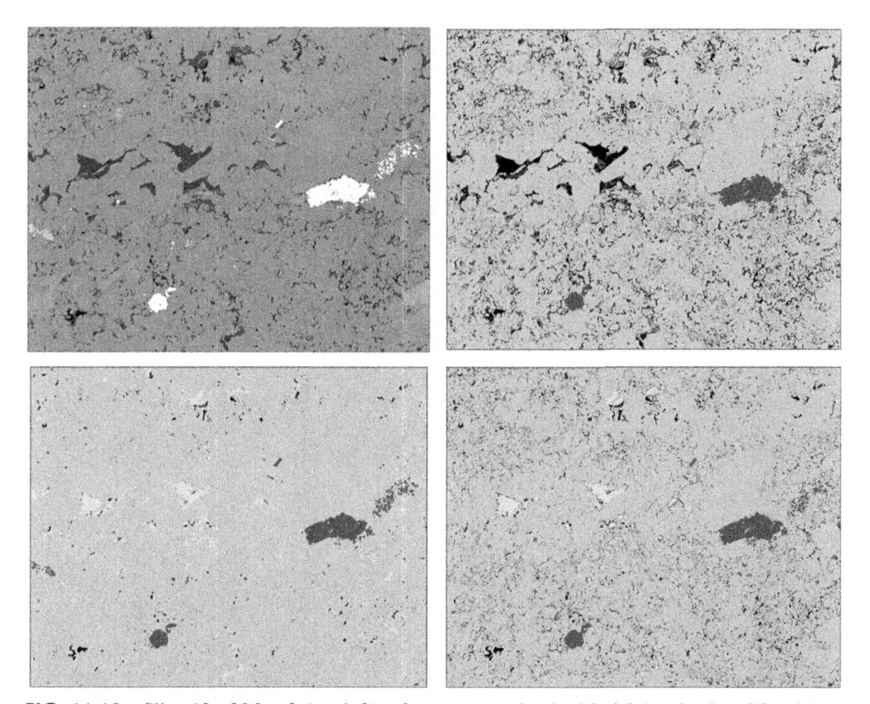

FIG. 11.18 Slice 10 of Map-2 (top left), after segmentation by Model-1 trained on Map-1 (top right), Model-2 trained on Map-2 (bottom left), and Model-3 trained on combined dataset of Map-1 and Map-2 (bottom right).

segmentation performance because the image comes from Map-2; however, Model-2 underpredicts pore space in Map-2. But in comparison with the performance of Model-1 (top right), where considerable number of pixels belonging to the organic/kerogen component were misclassified as pore/crack, we notice a significant improvement in the segmentation result by Model-3 (bottom right); thereby, proving that in this study, Model-3 delivers the best generalized segmentation results.

3.7 Feature ranking

Feature ranking is a crucial step required to understand the contribution of each feature to the overall performance of any supervised learning process. Feature ranking can assist the elimination of redundant features for purposes of enhancing the efficiency of the model and reducing the curse of dimensionality. Few popular feature ranking methods are (1) ranking based on feature importances computed using random forest, (2) permutation importance ranking, and (3) noise-based perturbation importance ranking. Permutation importance method is considered to be one of the best methods for feature ranking. Nonetheless, in this chapter, we use feature importances generated using random forest for ranking the 16 features used in the machine-learning assisted segmentation workflow. Feature importances in Fig. 11.19 are generated based on the average magnitude of reduction in impurity or increase in entropy when a feature and corresponding feature threshold are used to split the dataset for building the numerous decision trees comprising the random forest. Ranks are allocated to each feature depending on the mean decrease in impurity or the mean increase in entropy

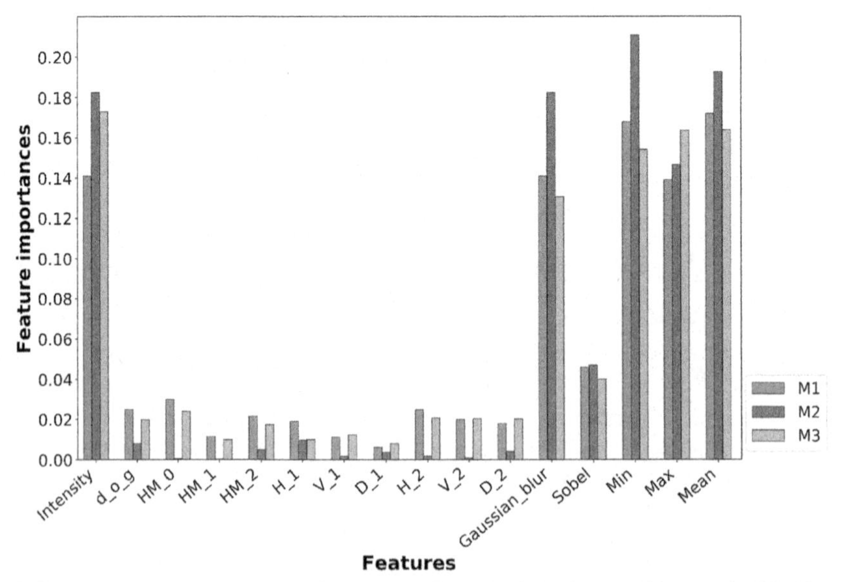

FIG. 11.19 Feature importances of 16 features for each of the three models, namely, Model-1 (M1), Model-2 (M2), and Model-3 (M3), as described in the previous sections.

averaged over the number of splits across all the trees in the random forest. Feature importance ranking method has few disadvantages, out of which the biggest challenge is when a feature is assigned a high importance, all the correlated/colinear features are automatically allocated low-importance values. It is evident in Fig. 11.19 that for all three models, the most dominant features are as follows:

- Pixel Intensity
- Gaussian blur
- Sobel operator
- Local minimum, maximum, and mean around the pixels

4 Conclusions

In this chapter, we investigated the generalization capability of the machine learning-assisted image segmentation of scanning electron microscopy (SEM) images of organic-rich shale samples from two shale formations, namely, Wolfcamp (Map-1) and Barnett (Map-2) shales. The segmentation workflow uses a random forest classifier to segment the images into four component types, namely, (1) pore/crack, (2) organic/kerogen, (3) matrix, and (4) pyrite. The two shale formations and the corresponding SEM maps differ in topology and distribution of the four components; in addition, the SEM maps vary in their pixel intensity because they were captured at different times using slightly different settings for the SEM acquisition. The segmentation workflow was rigorously tested on the inner-region and outer-region pixels of both maps in a series of six tasks. The deployment of the workflow takes an average of 50 seconds to segment one image slice of size 2058-by-2606 pixels using an Intel (R) Xeon (R) CPU E5-1603 @ 2080 GHz with 32 GB RAM.

The segmentation performance was quantified in terms of precision, recall, and F1 score. Machine learning-assisted segmentation method when trained and tested on SEM images from the same formation exhibits a performance of around 0.95 and 1 for the outer-region and inner-region pixels, respectively. Overall, for purposes of generalization, the outer-region and inner-region pixels of pore/crack and organic/kerogen are segmented at average F1 scores of 0.90 and 0.95, respectively. Models trained on Wolfcamp Shale SEM images can robustly detect matrix and pyrite in Barnett Shale SEM images but poorly perform for pore/crack and kerogen/organic components, where kerogen/organic pixels are wrongly segmented as pore/crack pixels. Models trained on Barnett Shale SEM images can robustly detect pyrite in Wolfcamp Shale SEM images but poorly perform for the remaining components, where pore/crack pixels are wrongly segmented as kerogen/organic pixels or matrix pixels and kerogen/organic pixels are wrongly segmented as matrix pixels. The Model trained on Barnett Shale cannot detect cracks in the Wolfcamp samples because the presence of cracks is very limited in the Barnett samples.

In terms of F1 score for segmenting the outer-region pixels, the model trained on Barnett performs poorly only for pyrite, whereas the model trained on Wolfcamp performs poorly for both pyrite and matrix components. In terms of F1 score for segmenting outer-region pixels, the model trained on both Wolfcamp- and Barnett Shale images perform better for pore/crack and kerogen/organic components of Wolfcamp Shale images as compared with Barnett Shale images, and perform better for matrix and pyrite components of Barnett Shale images as compared with Wolfcamp Shale images. Pixel intensity; Gaussian blur; Sobel operator; and local minimum, maximum, and mean around the pixels are the most important features for the proposed segmentation.

Acknowledgments

We thank the financial support from the American Chemical Society's Petroleum Research Foundation (ACS-PRF #59363-DNI Grant). We also thank The University of Oklahoma Research Council's Faculty Investment Program that supported few aspects of this project. We thank the Integrated Core Characterization Center at the University of Oklahoma for providing us the SEM images for this study. We thank the Unconventional Shale Gas Consortium for providing funding for acquiring the SEM images.

Declarations

The authors declare that they have no competing interests. EG, YW, and SM developed and tested the machine learning and feature extraction methods used in the study. EG and SM prepared the first complete draft of the chapter. MC and JJ acquired the SEM maps on shale samples at the IC3 Lab. MC and JJ provided guidance on the dataset and the quality of segmentation. CS and CR lead the IC3 Lab as co-PIs. MC, JJ, CS, and CR reviewed the first complete version of the chapter.

References

[1] Tang D, Spikes K. Segmentation of shale SEM images using machine learning. In: SEG Technical Program Expanded Abstracts 2017. Society of Exploration Geophysicists; 2017. p. 3898–902.

[2] Jobe T, Vital-Brazil E, Khaif M. Geological feature prediction using image-based machine learning. Petrophysics 2018;59(06):750–60.

[3] Andrew M. A quantified study of segmentation techniques on synthetic geological XRM and FIB-SEM images. Comput Geosci 2018;22(6):1503–12.

[4] Kim Y, Hardisty R, Torres E, Marfurt KJ. Seismic facies classification using random forest algorithm. In: SEG Technical Program Expanded Abstracts 2018. Society of Exploration Geophysicists; 2018. p. 2161–5.

Chapter 12

Characterization of subsurface hydrocarbon/water saturation by processing subsurface electromagnetic logs using a modified Levenberg-Marquardt algorithm

Siddharth Misra*, Pratiksha Tathed[†,a] and Yifu Han[‡,b]

Harold Vance Department of Petroleum Engineering, Texas A&M University, College Station, TX, United States, †BP, Houston, TX, United States, ‡Schlumberger Technology Company, Beijing, China

Chapter outline

---

a. Present address: BP, Houston, TX, United States.

b. Present address: Schlumberger, Beijing, China.

Machine Learning for Subsurface Characterization. https://doi.org/10.1016/B978-0-12-817736-5.00012-0

Symbols and abbreviations

α	geometrical fitting parameter assumed as 0.5
ε_b^*	bulk complex permittivity of the formation (F/m)
ε_m	matrix permittivity (F/m)
ε_o	hydrocarbon/oil permittivity
ε_w^*	complex permittivity of water
φ_t	total porosity of the formation
B	equivalent conductance of sodium clay exchange cations (mS/meq)
C_w	brine conductivity (S/m)
$Cond^{meas,\ 1\ GHz}$	measured electrical conductivity at 1 GHz
$Cond^{mod,\ 1\ GHz}$	modeled electrical conductivity at 1 GHz
CEC	cation exchange capacity
CRI	complex refractive index
ε_b	real value of water permittivity assumed as 80
ε_o	vacuum permittivity with universal value of 8.85×10^{-12} (F/m)
$Error_{CRI}$	error in permittivity and conductivity predictions based on CRI model
$Error_{SMD}$	error in permittivity and conductivity predictions based on SMD model
$Error_{Total}$	total error
$Error_{WS}$	error in resistivity prediction based on Waxman-Smits model
EM	electromagnetic
$F(p)$	cost function vector
GHz	gigahertz
IP	interfacial polarization
$J(p^k)$	Jacobian matrix computed at the *k*th iteration of the inversion
kHz	kilohertz
LMA	Levenberg-Marquardt algorithm
LR	Lichteneker-Rother model
m	cementation exponent
m'	rock textural parameter
MHz	megahertz
n	saturation exponent
$Perm^{meas,\ 1\ GHz}$	measured dielectric permittivity at 1 GHz
$Perm^{mod,\ 1\ GHz}$	modeled dielectric permittivity at 1 GHz
p^k	model parameter vector computed at the *k*th iteration of the inversion
Q_v	cation exchange capacity per unit pore volume (meq/cc)
R_t	true resistivity of the formation measured using galvanic resistivity tool at 1 kHz (ohm-m)

Res_{meas}	measured resistivity
Res_{mod}	modeled resistivity
RLA5	resistivity laterolog measured using galvanic resistivity tool at 1 kHz
S_w	water saturation
SMD	Stroud-Milton-De
TOC	total organic carbon
V_{py}	volume fraction of pyrite
WS	Waxman-Smits

1 Introduction

This chapter and Chapter 13 focus on the development and implementations of error minimization techniques that are crucial to machine learning and for building data-driven models. Most of the tasks in machine learning and data analysis need either deterministic or stochastic data-inversion (optimization) technique to quantify/assess the properties/behavior/trends of system/process. A system/process can be expressed as a mathematical model, physics-based model, empirical model, data-driven model, or numerical model. Measurements capture the response/behavior of the system/process of interest. Primary objective of data inversion is to process the measurements based on a forward model to find the model parameters that accurately determine the system/process of interest. Data inversion aims to minimize misfit/error between the measured data and modeled data computed using a mathematical model, physics-based model, empirical model, data-driven model, or numerical model. This is accomplished by systematically choosing model parameter values from within an allowed set of values (domain). The combination of values of model parameters that minimizes the misfit/error under certain constraints is referred as the optimal solution for the data-inversion problem. The optimal values of model parameters define the model that is most representative of the system/process that generated the measured data. In other words, a data-inversion problem finds a specific form of a forward model by identifying model parameter values that result in model predictions that best match the measured data. For example, seismogram data can be inverted to estimate various subsurface properties, such as acoustic velocity, formation and fluid densities, Poisson's ratio, formation compressibility, and shear rigidity. Data inversion is also referred as error minimization, optimization, or energy minimization. Data-misfit/error that needs to minimized is also referred as cost function, objective function or loss function, which is difference between model output/prediction and real observations/measurements. When objective function is not convex shaped for an inversion problem, there may be several local minima and it is hard to ascertain the optimal values of model parameters. An inversion problem may have none, one, or many objective functions. Moreover, the model parameters can be constrained or unconstrained depending on the nature of the problem. This chapter uses Levenberg-Marquardt algorithm

(LMA) for deterministic inversion of multifrequency electromagnetic measurements in shale formation to estimate water/hydrocarbon saturation and salinity under certain well-defined physics-based constraints.

1.1 Error-minimization methods

Error minimization techniques can categorized into two broad types: deterministic and stochastic techniques. Deterministic inversion methods are based on comparison of the model output with the measured data by continuously updating the model parameters to minimize the objective/loss function so as to find a single set of model parameters. In contrast, stochastic methods iterate and update model parameters based on a probability distribution to minimize the objective function to generate a distribution of alternate model parameters obeying the model constraints; in doing so, stochastic methods find all the possible nonunique solutions along with their likelihoods of occurrence. Stochastic inversion are suited when data contains noise, when the data has uncertainty (e.g., future data), and when data is generated by heterogeneous processes that are difficult to model using deterministic laws. Stochastic error-minimization methods are used for training data-driven models, like support vector machines, logistic regression and graphical models. In combination with the back-propagation algorithm, stochastic methods are popular algorithms for training neural network models. For instance, the training of neural network on a dataset involves updates of model parameters, weights and biases of neurons, which govern the model predictions of targets (Y) from the set of predictors/features (X) used in the model. The model parameters should be updated in the direction of optimal solution that minimizes the objective/loss function. During the error minimization, model parameters are updated by moving along a vector direction based on the first-order or second-order partial derivatives of the loss function with respect to model parameters. These derivatives can be expressed in matrix forms, known as Jacobian Matrix and Hessian Matrix, respectively. Calculation of the second-order partial derivatives tends to be more computationally expensive compared to the first order; nonetheless, the second-order partial derivative contain the information of the curvature of the error surface. Consideration of the curvature ensures that the second-order error minimization method does not converge slowly around saddle points. However, for faster computation of model parameters, first-order methods are more widely implemented.

1.2 Subsurface characterization problem

The Delaware basin forms the western subdivision of Permian Basin of west Texas and southeast New Mexico extending over 10,000 mile^2 [1]. In the Delaware Basin the Wolfcamp shale play forms one of the largest and complex unconventional reservoirs in the United States. Upper Wolfcamp formation is classified as a tight oil reservoir, with permeability usually in microdarcies and porosities ranging from 0.01 to 0.15 p.u. Upper Wolfcamp formation comprises sequences of carbonate, clastic sand, and shale laminations and beds [2]. Mineral constituents include varying amounts of quartz, calcite, dolomite, kerogen, illite, albite, and pyrite. This mix of minerals poses a major challenge when estimating porosity, water saturation, and net pay [3]. The presence of low values of porosity, interfacial polarization effects [4, 5], and large clay content and other factors in the Wolfcamp formation also affect resistivity interpretation and saturation [6]. Accurate evaluation of porosity is critical for the estimation of water saturation. Rosepiler [7] observed that errors in estimation of water saturation increased in low-porosity clay-rich formations. Water saturation estimates using Archie-type equations break down in organic-rich shales and tight hydrocarbon-bearing formations due to low porosity, increase in tortuosity, high connate-water salinity, interfacial polarization effects [4, 5], and large clay content.

Sarihi and Murillo [8] proposed a workflow to estimate water saturation using Waxman-Smits (WS) equation in tight-gas formations considering the conductivity and volume fraction of clay minerals. Their results on tight rock samples indicated a proportional relationship between clay content and the clay factor, which replaced BQ_v in the WS equation and was recommended for shale evaluation. Donadille et al. [9] addressed the limitations in determining high connate-water salinity with dielectric logs. At high salinity of about 70 ppk, the dielectric measurements lose sensitivity to salinity. Joint inversion of neutron sigma measurements and dielectric dispersion logs showed excellent sensitivity to high salinity values in Bakken shale formation. Chen and Heidari [10] proposed joint interpretation of dielectric and resistivity measurements that significantly improves water-filled porosity and hydrocarbon saturation assessment. They introduced analytical model combining conductivity and permittivity measurements for organic-rich source rocks with complex pore structure. They suggested that spatial distribution and tortuosity of water, kerogen, and pyrite networks significantly affect dielectric permittivity and electrical resistivity. Challenges with hydrocarbon saturation estimation in shale reservoirs due to dielectric effects have been reported by Misra et al. (2016) [11].

Han et al. [12] proposed a inversion methodology to process multifrequency EM log by combining (1) Lichtenecker-Rother model, (2) Stroud-Milton-De (SMD) model, and (3) PS model, a mechanistic pyrite-clay dispersion model, for the estimation of water saturation, formation water salinity, homogeneity index, and cementation index in clay-lean and clay-rich units of Bakken shale. They carried their interpretation for a 300-ft depth interval in Bakken Petroleum System and compared their estimates with Dean-Stark core water saturation, NMR interpretation, and service company's inversion results. Misra and Han [13] carried out joint inversion of conductivity and permittivity values obtained from EM induction at 26 kHz; EM propagation at 400 kHz and 2 MHz; and dielectric dispersion logs at 20, 100, 260 MHz, and 1 GHz to estimate the water saturation, bulk conductivity of brine, surface conductance of clay, and radius of spherical clay grains. This methodology has been extended to other shale reservoirs [11, 14].

Formation evaluation in conventional reservoirs generally involves the estimation of water saturation from the deep-sensing EM logs, such as induction log and laterolog, or high-resolution EM logs, such as EM propagation and dielectric dispersion logs. In unconventional reservoirs, water saturation estimation is difficult due to complex mineralogy, higher clay content, low porosity, textural features and high salinity. Interpretation with only 1-GHz dielectric permittivity log or with only laterolog or induction resistivity log or with only eight dielectric dispersion logs in the frequency range of 10 MHz–1 GHz is sensitive to model assumptions, noise in data, noise in model inputs, interfacial polarization mechanisms, and textural effects and has low sensitivity to certain petrophysical properties. These challenges can be addressed by performing a joint inversion of resistivity and dielectric dispersion logs using an integrated mechanistic model. In this chapter, eight dielectric dispersion logs and one laterolog resistivity logs are simultaneously processed to estimate the water saturation, brine conductivity, saturation exponent, and cementation exponent in a 520-ft depth interval of upper Wolfcamp shale.

2 Method

2.1 Relevant EM logging tools

Laterolog resistivity tool injects electric currents into geological formations and records the potential drop across a specific length along the open hole well. Laterolog measurements are related to the electrical resistivity of the formation. Laterolog tools are reliable in boreholes drilled with water-based muds. Laterolog tool has 2-ft resolution with 10, 20, 30, 60, and 90 in. of depths of investigation and operates at frequencies lower than 10 kHz. On the other hand, dielectric dispersion tool transmits electromagnetic (EM) waves and records the changes in amplitudes and the phases of the propagating wave, which are related to the dielectric permittivity and electrical conductivity of the formation and their dispersive behaviors. Dielectric dispersion tool has 1-in. vertical resolution and operates at multiple discrete frequencies in the range of 10 MHz–1 GHz.

2.2 Relevant log interpretation models

Interpretation of laterolog tool focuses on the petrophysical controls on charge transport through porous media, which can be formulated using mechanistic or empirical models similar to Archie's or WS type equations. Charge transport depends on pore connectivity, pore tortuosity, water saturation, wettability, water salinity, clays, and porosity. Interpretation of dielectric tool is based on the large contrast between permittivity of water and that of oil and rock minerals [9]. This interpretation primarily focuses on the petrophysical controls on charge storage, accumulation, and relaxation, thereby facilitating water volume estimation independent of water salinity. Dielectric behavior of a porous geomaterial is dependent on water saturation, salinity, frequency of measurement, bound water, polarization phenomena due to clays and conductive mineral, and pore connectivity/ tortuosity.

In this study, the joint inversion technique processes low-frequency laterolog measurements using WS model, the two 1-GHz conductivity and permittivity logs using complex-refractive index (CRI) model, and the three conductivity dispersion and three permittivity dispersion logs acquired in the range of 10 MHz–0.3 GHz using SMD model. Waxman and Smits [15] introduced a resistivity-based model for clay-rich geomaterials to estimate water saturation by accounting for additional conductivity due to the presence of clay with high cation exchange capacity (CEC). Waxman-Smits (WS) model is expressed as

$$\frac{1}{R_t} = \varphi_t^m S_w^n \left(C_w + \frac{B \cdot Q_v}{S_w} \right) \tag{12.1}$$

where R_t represents the true resistivity of the formation, φ_t is the total porosity, S_w is water saturation, C_w is the brine conductivity in S/m, Q_v is the cation exchange capacity (CEC) per unit pore volume in meq/cc, B is the equivalent conductance of sodium clay exchange cations in mS/meq that depends on the salinity of brine and temperature of formation, m is the cementation exponent, and n is the saturation exponent. Peeters and Holmes [16] showed that WS method is not very sensitive to variation of most of the parameters when B is fixed and reliable Q_v values are used. Based on the temperature of the formation and an estimate of salinity of brine provided by the operator company, B value was fixed to 13 mS/meq. For further simplification, Q_v was assumed to be 0.1 meq/cc based on the typical value suggested by the operator company. In future work, we will allow variable Q_v and B based on depth-wise clay types and their volume fractions. WS model focuses on dispersed shaly sand and does not account for the effects of shale laminations and thin beds.

CRI model is a commonly used single-frequency model for subsurface dielectric characterization at high frequencies close to 1 GHz. It is derived from the Lichteneker-Rother (LR) model, which can be expressed as [17],

$$\left(\varepsilon_b^*\right)^\alpha = (1 - \varphi_t)(\varepsilon_m)^\alpha + S_w \varphi_t \left(\varepsilon_w^*\right)^\alpha + (1 - S_w)\varphi_t (\varepsilon_o)^\alpha \tag{12.2}$$

where ε_b^* is the bulk complex permittivity of the formation, ε_m is the matrix permittivity, ε_w^* is the complex permittivity of water, ε_o is the hydrocarbon permittivity, φ_t is the total porosity, S_w is water saturation, and α is the geometrical fitting parameter ranging from -1 to 1. With $\alpha = 0.5$, Eq. (12.2) becomes the CRI model. In our work, one simplifying assumptions is

$\alpha = 0.5$ indicating a homogenous mixture without layering. CRI model does not consider the structure and spatial distribution of rock components [18].

Stroud et al. [19] introduced an analytical mixing formula to calculate complex dielectric permittivity and its dispersion in brine-saturated rocks as a function of total porosity, water saturation; dielectric permittivities of rock matrix, brine, and oil; rock textural parameter (m') that describes cementation/tortuosity; and brine conductivity. SMD model is expressed as

$$\varepsilon_b^* = (\varphi_t S_w)^{m'} \varepsilon_w^*(C_w) + \left[1 - (\varphi_t S_w)^{m'}\right] \varepsilon_{mo} - \varepsilon_{mo} \Gamma\left(\varphi_t S_w, m', \frac{\varepsilon_w^*(C_w)}{\varepsilon_{mo}}\right) \tag{12.3}$$

where

$$(\varepsilon_{mo})^\alpha = \frac{(1-\varphi_t)(\varepsilon_m)^\alpha + (1-S_w)\varphi_t(\varepsilon_o)^\alpha}{1-\varphi_t S_w} \tag{12.4}$$

$$\varepsilon_w^*(C_w) = \varepsilon_b + i\frac{iC_w}{\varepsilon_0 \omega} \tag{12.5}$$

where ε indicates relative permittivity, $\varepsilon_b = 80$, $\varepsilon_0 = 8.85 \times 10^{-12}$ F/m, and $\omega = 2\pi f$ with $f = 20$, 100, and 260 MHz. For purposes of simplifying the joint inversion of EM logs, we assume $m' = m$, that is, textural index in SMD model is equivalent to cementation exponent in WS model. There are few limitations of this approach. In strict terms, SMD model was not developed for clay- and pyrite-rich formations. Also, the integrated WS, CRI, and SMD model neglects the interfacial polarization (IP) effects of clays and conductive minerals in the shale formation.

2.3 Proposed deterministic inversion of EM logs

2.3.1 Introduction

An inversion technique (Fig. 12.1) is developed in this chapter to jointly process nine EM logs, namely one laterolog resistivity and eight dielectric dispersion

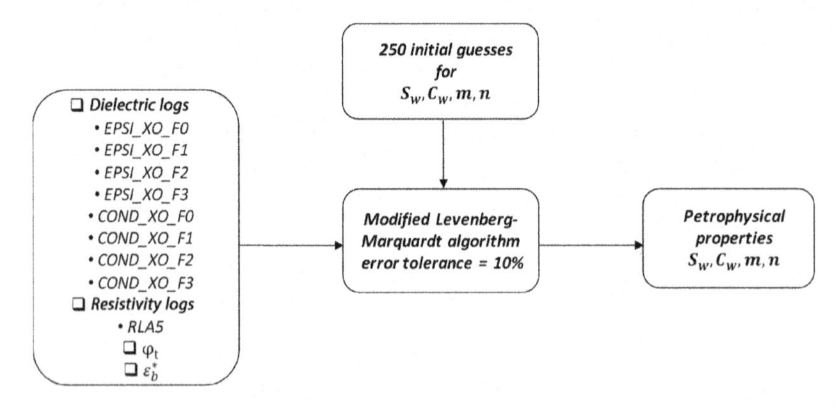

FIG. 12.1 Log processing workflow.

logs, for the simultaneous estimation of water saturation (S_w), brine conductivity (C_w), homogeneity index (m), and saturation exponent (n) in a shale formation with associated ranges of possible values of the estimates to account for the inherent uncertainty in the data-inversion procedure due to non-convex nature of the problem and inherent noise in the data acquired in the hostile borehole conditions. The estimates are generated at a vertical resolution of 1 inch. The joint inversion scheme is coupled to an integrated model, combining the WS, SMD, and CRI models, for iteratively finding the model parameters that lead to model predictions optimally fitting the nine measured EM logs. Unlike our objective of joint inversion of multifrequency EM log, Ramirez et al. [20] proposed a joint inversion of gamma ray, density, and resistivity logs of different depths of investigation at 1 frequency. Ramirez et al. [20] implemented Bayesian stochastic inversion to assess porosity, volumetric shale concentration and hydrocarbon saturation in thinly bedded formations.

Measured data vector $\boldsymbol{v}$, comprising 1 $Res^{meas,1\ \mathrm{kHz}}$, 4 $Perm^{meas,\ f_i}$, and 4 $Cond^{meas,f_i}$, is jointly inverted in the proposed log-processing scheme. $Res^{meas,\ 1\ \mathrm{kHz}}$ represents the true resistivity of the formation measured using the galvanic resistivity (laterolog) tool at 1 kHz. $Perm^{meas,\ f_i}$ and $Cond^{meas,\ f_i}$ represent effective permittivity and conductivity of formation measured at four distinct dielectric-dispersion log-acquisition frequencies of $f_2 < f_3 < f_4 < f_5$ in the range of 10 MHz to 1 GHz. The objective is to estimate petrophysical model parameter vector $\boldsymbol{p}$ comprising water saturation (S_w), formation water conductivity (C_w), homogeneity index (m), and saturation exponent (n). A joint inversion is implemented to process the measured data vector $\boldsymbol{v}$ using the integrated model, combining the WS, SMD, and CRI models, to estimate the model parameter vector $\boldsymbol{p}$. The joint inversion scheme is initiated using 250 random guesses of the vector $\boldsymbol{p}$, comprising S_w, C_w, m, and n.

Reliable estimates of petrophysical properties of the shale formation are such that the measured data vector $\boldsymbol{v}$, comprising one galvanic resistivity, four dielectric permittivity dispersions, and four electrical conductivity dispersions, are optimally fitted/matched by the modeled data vector $\boldsymbol{m}$ computed using the integrated CRI, SMD, and WS model. Various values of the vector $\boldsymbol{m}$ can be generated by varying the model parameter vector $\boldsymbol{p}$. Inversion method aims to bring $\boldsymbol{m}$ close to $\boldsymbol{v}$, such that an objective/loss function reaches a local optima under certain constraints on $\boldsymbol{p}$. To that end, the modified LMA inversion algorithm (explained in the next section) iteratively computes/updates the vector $\boldsymbol{p}$ for finding the best match (minimize error or minimize loss/objective function). This directs the convergence toward the optimum estimates for the unknown model parameters in vector $\boldsymbol{p}$. This best estimate of vector $\boldsymbol{p}$ generates best fit between the measured data vector $\boldsymbol{v}$ and modeled data vector $\boldsymbol{m}$, thus minimizing the data misfit. This inversion scheme requires minimization of cost/loss/objective function vector denoted as $\boldsymbol{F}(\boldsymbol{p})$. Individual components of the vector $\boldsymbol{F}(\boldsymbol{p}^k)$ at the kth iteration are denoted as $f_i(\boldsymbol{p}^k)$, where the superscript k is the iteration counter for the inversion scheme,

$\boldsymbol{p}^k$ is the model parameter vector computed at the kth iteration of the modified LMA-based inversion, and subscript i denotes one of the nine EM logs acquired at five log-acquisition frequencies. Each element $f_i(\boldsymbol{p}^k)$ in $\boldsymbol{F}(\boldsymbol{p}^k)$ is divided by the standard deviation of one corresponding log out of the nine EM logs for robust parameter estimation. Cost/loss/objective function vector at kth iteration $\boldsymbol{F}(\boldsymbol{p}^k)$ is expressed as

$$\boldsymbol{F}\left(\boldsymbol{p}^k\right) = \begin{bmatrix} \mathbf{f}_1(\mathbf{p}^k) \\ \mathbf{f}_2(\mathbf{p}^k) \\ \cdot \\ \cdot \\ \mathbf{f}_9(\mathbf{p}^k) \end{bmatrix}^{\mathrm{T}} \tag{12.6}$$

where $f_i(\boldsymbol{p}^k)$ is expressed as

$$\begin{aligned} f_1\left(\boldsymbol{p}^k\right) &= \left[Res^{mod,1\,\mathrm{kHz}}\left(\boldsymbol{p}^k\right) - Res^{meas,1\,\mathrm{kHz}}\right]^2 \text{ and} f_i\left(\boldsymbol{p}^k\right) \\ &= \left[Perm^{mod,f_i}\left(\boldsymbol{p}^k\right) - Perm^{meas,f_i}\right]^2 for\ i = 2,3,4,\ and\ 5\ and\ f_i\left(\boldsymbol{p}^k\right) \\ &= \left[Cond^{mod,f_i}\left(\boldsymbol{p}^k\right) - Cond^{meas,f_i}\right]^2 \text{ for } i = 6,7,8, \text{and } 9 \end{aligned} \tag{12.7}$$

The Jacobian matrix of $\boldsymbol{F}(\boldsymbol{p}^k)$ for the first-order error minimization using gradients with respect to model parameters is formulated as

$$J\left(p^k\right) = \begin{matrix} \dfrac{\partial f_1\left(\boldsymbol{p}^k\right)}{\partial p_1^k} & \cdots & \dfrac{\partial f_1\left(\boldsymbol{p}^k\right)}{\partial p_4^k} \\ \vdots & \ddots & \vdots \\ \dfrac{\partial f_9\left(\boldsymbol{p}^k\right)}{\partial p_1^k} & \cdots & \dfrac{\partial f_9\left(\boldsymbol{p}^k\right)}{\partial p_4^k} \end{matrix} \tag{12.8}$$

where $\boldsymbol{p}^k = [p_1^k; p_2^k; p_3^k; p_4^k]$.

We implemented the first-order central-difference formula to approximate the derivative of $f_i(\boldsymbol{p}^k)$ with respect to the jth unknown parameters formulated as

$$\frac{\partial f_i\left(p^k\right)}{\partial p_j^k} \approx \frac{f_i\left(p_j^k + \Delta p_j^k\right) - f_i\left(p_j^k - \Delta p_j^k\right)}{2\Delta p_j^k} \tag{12.9}$$

such that $\Delta p_j^k = 0.001 p_j^k$, $1 \leq j \leq 4$ representing the index for model parameters and $1 \leq i \leq 9$ representing the index for log-acquisition frequencies.

2.3.2 Modified Levenberg-Marquardt nonlinear inversion

The nonlinear inverse problem discussed in this chapter involves estimation of petrophysical parameters that generate least misfit between the integrated-model predictions and the measurements of multifrequency effective permittivity and conductivity. Least misfit is accomplished by minimizing the cost function described in Eq. (12.6). We modified the traditional Levenberg-Marquardt algorithm (LMA), for purposes of nonlinear inversion by introducing scaling matrix $\boldsymbol{W}_s$ into the mathematical formulation of the LMA inversion scheme. $\boldsymbol{W}_s$ is a 4×4 diagonal matrix that adaptively handles the non-singularity of the Jacobian matrix (described in Eq. 12.8) during the inversion. The modified LMA scheme is expressed as

$$\left[\left(\boldsymbol{J}(\boldsymbol{p}^k)\boldsymbol{W}_s\right)^T \cdot \left(\boldsymbol{J}(\boldsymbol{p}^k)\boldsymbol{W}_s\right) + \lambda\boldsymbol{I} + \alpha^2\boldsymbol{I}\right] \cdot \boldsymbol{W}_s^{-1} \cdot \Delta\boldsymbol{p}^k$$
$$= -\left(\boldsymbol{J}(\boldsymbol{p}^k)\boldsymbol{W}_s\right)^T \cdot \boldsymbol{F}(\boldsymbol{p}^k) - \alpha^2\boldsymbol{I} \cdot \boldsymbol{W}_s^{-1} \cdot \boldsymbol{p}^k \tag{12.10}$$

where $\boldsymbol{p}^k$ is the interpretation model parameter vector computed at the kth iteration of the inversion, $\boldsymbol{F}(\boldsymbol{p}^k)$ is the cost function vector, $\boldsymbol{J}(\boldsymbol{p}^k)$ is the Jacobian matrix of $\boldsymbol{F}(\boldsymbol{p}^k)$, $\Delta\boldsymbol{p}^k$ is the correction vector generated at the kth iteration that determines the direction toward the convergence of the parameter estimation (error minimization) process, λ is the damping parameter, α is the regularization parameter, $\boldsymbol{I}$ is an identity matrix, and T is matrix transpose operator. $\boldsymbol{W}_s$ is expressed as

$$\boldsymbol{W}_s = \begin{pmatrix} 10^{\alpha_1} & \cdots & 0 \\ \vdots & \ddots & \vdots \\ 0 & \cdots & 10^{\alpha_4} \end{pmatrix} \tag{12.11}$$

where the parameters ($\boldsymbol{\alpha}_1, \cdots, \boldsymbol{\alpha}_4$) in the diagonal scaling matrix denote the order of magnitude difference between the estimates of the four unknown model parameters. The parameters $\boldsymbol{\alpha}_1$, $\boldsymbol{\alpha}_2$, $\boldsymbol{\alpha}_3$, and $\boldsymbol{\alpha}_4$ correspond to the water saturation, formation water conductivity, homogeneity index, and saturation index, respectively, and their values are assumed to be 0, 2, 0, and 1, respectively.

Dielectric measurements at 1 GHz, which can be modeled using CRI model, lose sensitivity to water salinity for salinity above 60 kppm. Dielectric dispersion models, such as SMD and bimodal, lose their sensitivity at even lower water salinity compared with CRI model. All these models exhibit negligible sensitivity for porosity lower than 0.05 p.u. This loss of sensitivity of the forward models to water-filled porosity and salinity gives rise to the failure of traditional LMA-based inversion schemes. However, the modified LMA proposed in this chapter introduces a scaling matrix $\boldsymbol{W}_s$ to improve the inversion capabilities for high salinity and low porosity scenarios.

After convergence of the LMA-based inversion, a line search method is applied to find the smallest data misfit in the search space, which represents the optimal estimates for the unknown model parameters, namely S_w, C_w, m, and n, in the vector $\boldsymbol{p}$. The search for global minimum among the 250 inversion searches is affected by the noise in the data and uncertainties in the inputs to the three models, namely, CRI, SMD, and WS models, coupled with the inversion scheme. Therefore, we generate ranges (maximum and minimum) of possible values for the unknown model parameters rather than a single estimate. To that end, all possible values of the unknown model parameters, namely, S_w, C_w, m, and n, are searched around around the optimal estimate that lead to data misfit within the predefined fitting error that does not exceed 1.1 times the data misfit at the optimal estimate. This generates minimum and maximum values for the four inversion-derived estimates. The range of value indicates possible estimates of the four petrophysical properties of the shale formation. The total porosity used for water saturation estimation is computed using ELAN multimineral inversion of resistivity, gamma-ray, density, neutron, sonic, and elemental spectroscopy data [21]. The computations are based on a probabilistic approach combining the previously mentioned logs with additional constraints of total organic carbon (TOC), calcite transform, and pyrite transform.

2.4 Limitations and assumptions in the proposed multifrequency-EM log processing using modified Levenberg-Marquardt algorithm

The joint inversion assumes that the volumes sensed by the eight dielectric dispersion logs at the four frequencies and that sensed by the laterolog resistivity have similar petrophysical properties. This assumption requires the formation to be free from fractures and beds/laminations thinner than 2 ft. Vertical resolution refers to the thinnest bed the tool can accurately sense without the bed-boundary effects. Dielectric tool generally has about 2-in. vertical resolution and 3-in. depth of investigation, whereas laterolog resistivity generally has vertical resolution of 2 ft and depth of investigation of 5 ft. The pad-based microresistivity measurements in the 520-ft depth interval of the upper Wolfcamp shale indicate that there are few depths consisting of beds as thin as 0.1 ft. Another limitation involves the use of total porosity in the integrated model. The use of total porosity agrees with the physics of dielectric dispersion measurement; however, the use of total porosity will generate erroneous results for galvanic resistivity measurements when isolated pores are present in the shale formation.

3 Results and discussion

3.1 Sensitivity of the integrated WS, SMD, and CRI model

We couple the error minimization (inversion) workflow with the integrated WS, SMD, and CRI model to process the nine EM logs. Sensitivity of the integrated WS, SMD, and CRI model is compared with that of the integrated SMD and CRI model, required to process only the eight dielectric dispersion logs. In this comparison, we intend to test whether the joint inversion of one galvanic and eight dielectric dispersion measurements will provide better estimates as compared to the inversion of only the eight dielectric dispersion logs. We vary

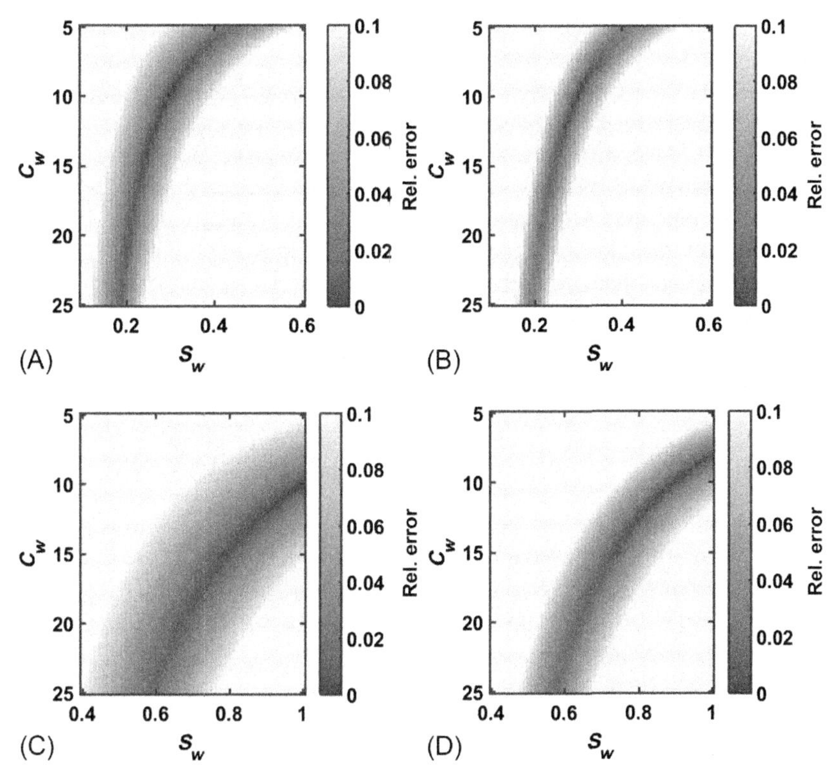

FIG. 12.2 Comparison of the sensitivity of (A and C) integrated CRI and SMD model used for the inversion of eight dielectric dispersion logs against that of (B and D) integrated CRI, SMD, and WS model used for the newly proposed joint inversion of one galvanic and eight dielectric dispersion logs to variations in S_w and C_w around the default parameters: $m = 2$, $n = 2$, $C_w = 10$ S/m, $\varphi_t = 0.02$, $BQ_v = 1.3$ mS/cc, and $S_w = 0.3$ s.u. and around the default parameters: $m = 2$, $n = 2$, $C_w = 10$ S/m, $\varphi_t = 0.02$, $BQ_v = 1.3$ mS/cc, and $S_w = 0.9$ s.u., respectively.

the brine conductivity in the range of 5–25 S/m and water saturation in the range of 0.1–1 saturation unit (s.u.) to generate four permittivity dispersion logs, four conductivity dispersion logs, and one resistivity log using the two integrated models (to be compared). When S_w and C_w are changed, the integrated model for the proposed joint inversion generates well-defined relative change in data misfit (relative error with respect to the global minimum). The contours of relative change in data misfit represent the efficacy of the error minimization during the joint inversion (Fig. 12.2B and D). In Fig. 12.2A and B, the global minimum is at $S_w = 0.3$ s.u. and $C_w = 10$ S/m because the model output variations are studied for variations in S_w and C_w around $m = 2$, $n = 2$, $C_w = 10$ S/m, $\varphi_t = 0.02$, and $S_w = 0.3$ s. u. The best set of local minima is identified as the black region in Fig. 12.2 indicating possible uncertainty in inversion-derived estimates because the minima are spread around the global minimum. It is evident from Fig. 12.2A and B that at low water saturation, the newly proposed joint inversion (as compared with inversion of dielectric dispersion logs using the integrated SMD and CRI model) shows sharper transition of relative errors around global minima; thereby, suggesting faster convergence to optimal solution. In Fig. 12.2C and D, the global minimum is at S_w= 0.9 s.u. and C_w = 10 S/m keeping all the other parameters equal to the previous case. At high water saturation, the transition of relative errors around global minimum are relatively uniform and flatter compared with the lower water saturation case. A comparison between Fig. 12.2A–D indicates that the proposed joint inversion can generate the desired estimates more reliably at low water saturations than at high water saturations; therefore, conducive for hydrocarbon estimation under uncertainty in brine conductivity.

For purposes of model sensitivity comparison critical to understanding the error minimization during the joint inversion, m and n are varied around a default parameter to generate Fig. 12.3. The integrated model for the proposed joint inversion with three models generates well-defined error contours around the global minimum that will better direct the error minimization process during the

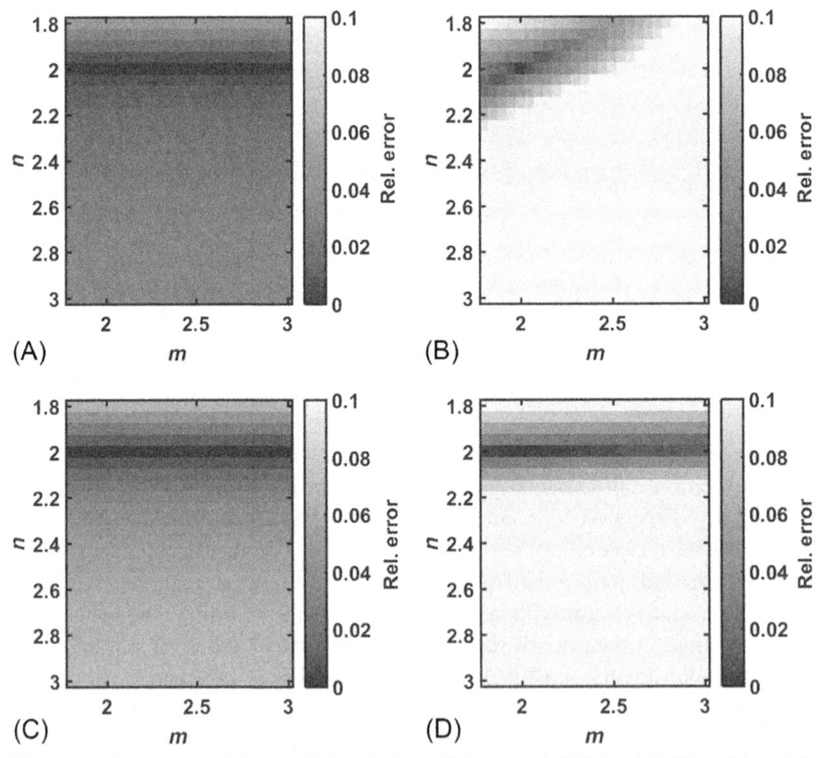

FIG. 12.3 Comparison of the sensitivity of (A and C) integrated CRIM and SMD model used for the inversion dielectric dispersion logs against that of (B and D) integrated CRIM, SMD, and WS model used for the newly proposed joint inversion of one galvanic and eight dielectric dispersion logs to variations in n and m around the default parameters: $m = 2$, $n = 2$, $C_w = 10$ S/m, $\varphi_t = 0.05$, $BQ_v = 1.3$ mS/cc, and $S_w = 0.3$ s. u. and around the default parameters: $m = 2$, $n = 2$, $C_w = 10$ S/m, $\varphi_t = 0.05$, $BQ_v = 1.3$ mS/cc, and $S_w = 0.9$ s. u., respectively.

joint inversion compared with those obtained using the integrated CRI and SMD model. A comparison between Fig. 12.3A and Fig. 12.3B indicates the substantial improvement when the integrated WS, SMD, and CRI model is implemented instead of the integrated SMD and CRI model. In Fig. 12.3A the spread of errors is very flat around the global minimum that indicates that the error minimization process will not have sufficient resolution to generate unique results. In Fig. 12.3A and B, the global minimum is at $m = 2$ and $n = 2$ for $S_w = 0.3$ s.u. Fig. 12.3C and D is generated for a higher saturation of $S_w = 0.9$ s.u., while keeping other parameters of the global minimum similar to the previous case. A comparison between Fig. 12.3B and D indicates that the proposed joint inversion can generate the desired estimates more reliably at low water saturations and the inversion-derived estimates will tend to be nonunique at high water saturations of around 0.9 s.u.; therefore, conducive for hydrocarbon estimation under uncertainty in m and n. This is similar to the observations for variations in S_w and C_w presented in the previous section. Similarly, in this case, with increase in water saturation, the sensitivity of the integrated WS, SMD, and CRI model to variation in m is significantly deteriorated while maintaining the sensitivity to variation in n. Deterioration of sensitivity of the integrated model to a specific model parameter indicates increased uncertainty in the corresponding estimates obtained from log processing.

After demonstrating that the newly proposed joint log processing will perform better compared with processing only the dielectric dispersion logs, we test the sensitivity of the integrated CRIM, SMD, and WS model to variations in n and C_w for various S_w. Sensitivity of the integrated model to the petrophysical parameters, in this case n and C_w, is inversely correlated to the uncertainty in the parameter estimation when performing the joint inversion because the model sensitivity directly affects the error minimization process. In Fig. 12.4, global minimum is at default parameters: $n = 2$ and $C_w = 10$ S/m, identified as the black region. Variations in n and C_w results in change in the relative error of the integrated model response with respect to the response obtained using the default parameters. It is evident from Fig. 12.4A–D that with increase in water saturation from 0.2 to 0.99 saturation units (s.u.), the sensitivity of the integrated model to n and C_w decreases. Consequently the uncertainty in the estimations of n and C_w should increase with increase in water saturation. At $S_w = 0.9$ the joint inversion should generate unreliable nonunique estimates (Fig. 12.4C). At high water saturation the sensitivity of the integrated model decreases more drastically for n in comparison with C_w. To a relatively limited extent, the proposed inversion should estimate C_w more reliably at higher water saturation.

Fig. 12.5 illustrates the sensitivity of the integrated model to variations in m and S_w for various S_w. Global minimum is at default parameters: $m = 2$ and $S_w = 0.9$ s.u., identified as the black region in Fig. 12.5. It is evident from Fig. 12.5A–C that with increase in n from 2 to 4, the sensitivity of the integrated model to m and S_w increases; consequently the uncertainty in the estimations of m and S_w should decrease. An increase in n relates to the

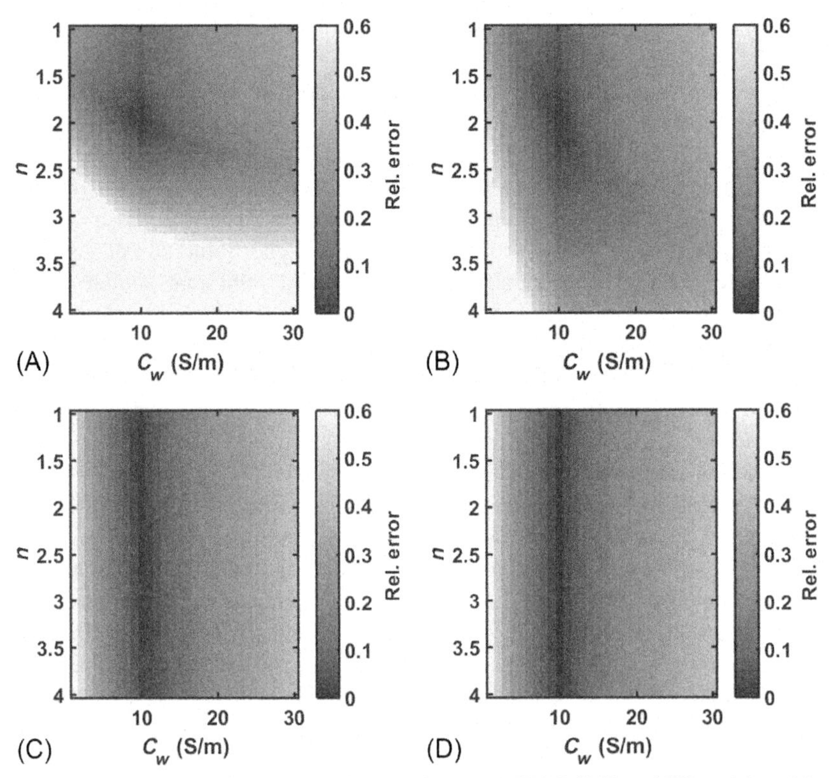

FIG. 12.4 Comparison of the sensitivities of the integrated CRIM, SMD, and WS model used for the newly proposed joint inversion of one galvanic and eight dielectric dispersion logs to variations in n and C_w for (A) $S_w = 0.2$, (B) $S_w = 0.5$, (C) $S_w = 0.9$, and (D) $S_w = 0.99$ around default parameters: $n = 2$, $m = 2$, $\varphi_t = 0.05$, $C_w = 10$ S/m, and $BQ_v = 1.3$ mS/cc.

increase in oil-wetness of the formation. The improvements in sensitivities for the two parameters with increase in n are relatively equal. With increase in C_w (Fig. 12.5D), there is a slight improvement in the model sensitivity to m and S_w, as observed between Fig. 12.5A and D. Therefore, the integrated model and the proposed inversion scheme is conducive for hydrocarbon estimation for scenarios of low water saturation, higher salinity, and high oil-wet conditions.

Errors in hydrocarbon saturation estimates from EM logs are largest for low-porosity, low-water-saturation, high-salinity formations [22]. The presence of pyrite in the formation leads to the underestimation of hydrocarbon saturation [23], especially when deriving them using dielectric dispersion logs. We analyze the effect of the presence of pyrite in low-porosity high-salinity formation on the proposed joint inversion in comparison with that on the inversion of dielectric dispersion log. To that end, we compare the sensitivity of the two integrated models, one for the proposed joint inversion and the other for the dielectric dispersion inversion, to the variations in S_w and volume fraction of pyrite (V_{py}) in the range of 0%–10%. In Figs. 12.6 and 12.7, the

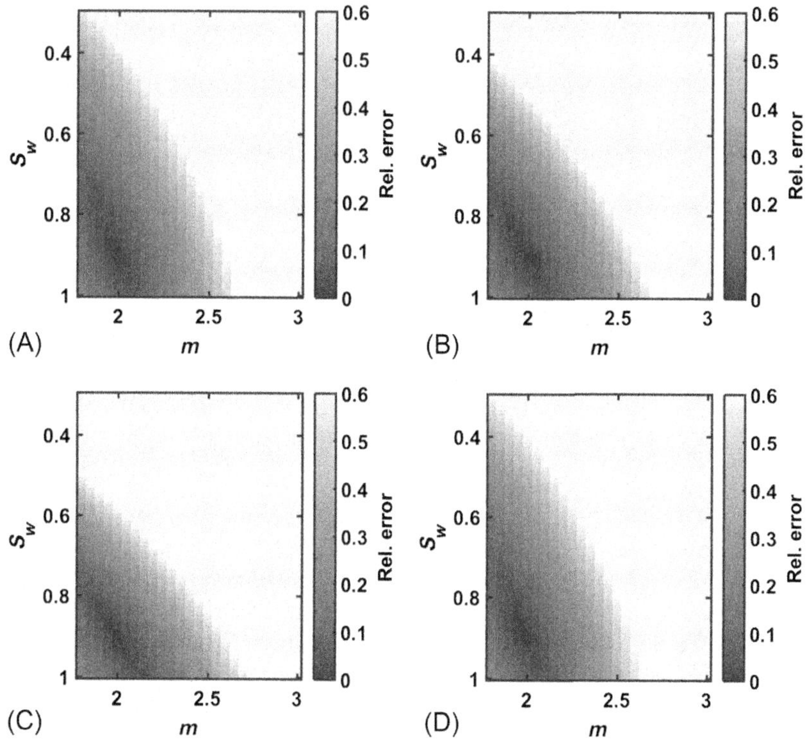

FIG. 12.5 Comparison of the sensitivities of the integrated CRIM, SMD, and WS model used for the newly proposed joint inversion of one galvanic and eight dielectric dispersion logs to variations in m and S_w for (A) $n = 2$, $C_w = 10$ S/m; (B) $n = 3$, $C_w = 10$ S/m; (C) $n = 4$, $C_w = 10$ S/m; and (D) $n = 2$, $C_w = 20$ S/m around the default parameters: $m = 2$, $\varphi_t = 0.05$, $S_w = 0.9$ s. u., and $BQ_v = 1.3$ mS/cc.

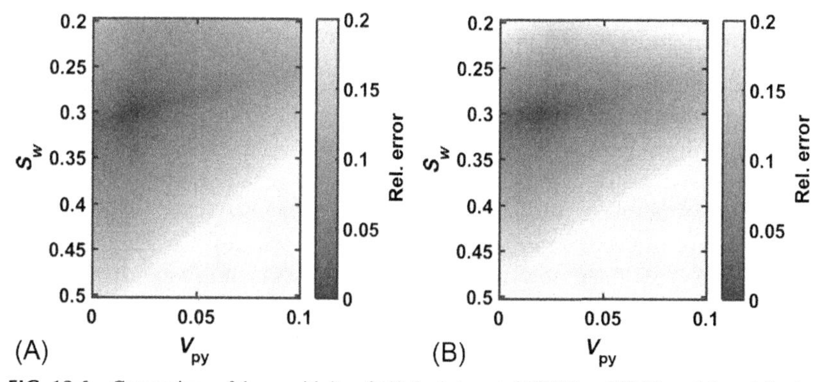

FIG. 12.6 Comparison of the sensitivity of (A) the integrated CRIM and SMD model used for the inversion of dielectric dispersion logs with that of (B) the integrated CRIM, SMD, and WS model used for the proposed joint inversion of one galvanic and eight dielectric dispersion logs to variations S_w and V_{py} around the default parameters: $n = 2$, $m = 2$, $\varphi_t = 0.02$, $C_w = 10$ S/m, $V_{py} = 0.02$, $BQ_v = 1.3$ mS/cc, and $S_w = 0.3$ s. u.

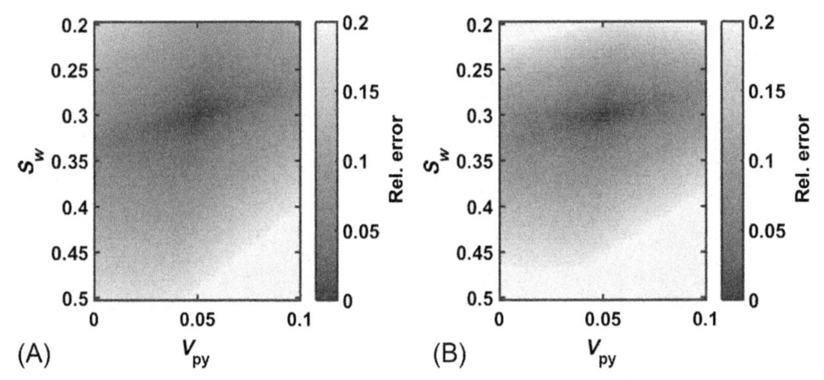

FIG. 12.7 Comparison of sensitivity of (A) the integrated CRIM and SMD model used for the inversion of dielectric dispersion logs with that of (B) the integrated CRIM, SMD, and WS model used for the proposed joint inversion of one galvanic and eight dielectric dispersion logs to the variations in S_w and V_{py} around default parameters: $n = 2$, $m = 2$, $\varphi_t = 0.02$, $C_w = 10$ S/m, $V_{py} = 0.05$, $BQ_v = 1.3$ mS/cc, and $S_w = 0.3$.

color of the map changes from yellow to dark blue indicating the magnitude of the errors. Comparison of Fig. 12.6A and B indicates that the integrated CRIM, SMD, and WS model used for the proposed joint inversion is more sensitive to variation in volume fraction of pyrite compared with the integrated CRIM and SMD model used for dielectric dispersion inversion. Consequently the proposed joint inversion is more suited in the presence of pyrite compared with the inversion of only the dielectric dispersion logs. Based on Figs. 12.6B and 12.7B, when initializing the inversion scheme, faster convergence can be achieved by initializing the inversion with a low value of water saturation. Comparison of Figs. 12.6B and 12.7B indicates that the sensitivity of the integrated model for joint inversion is relatively invariant to the change in V_{py}.

3.2 Application of the proposed log processing to synthetic data

The uncertainty in parameter estimation when using the proposed joint inversion scheme is studied in this section. 10% Gaussian noise was considered in the synthetic case. The porosity of the formation was selected to be 5% and brine conductivity as 10 S/m to create a synthetic model equivalent to the upper Wolfcamp formation. Values for other petrophysical properties are listed in Table 12.1. Using these properties, we generated one resistivity log at laterolog frequency and eight permittivity/conductivity dispersion logs at the dielectric tool frequencies. These simulated conductivity, permittivity, and resistivity logs were jointly inverted using the proposed joint inversion coupled with the integrated CRIM, SMD, and WS model to estimate C_w, S_w, n, and m of the synthetic formation using a scheme similar to that explained in Section 2.3 and used by Han and Misra (2017) [14]. Inversion-derived estimates of the unknown model parameters,

TABLE 12.1 Petrophysical parameters assumed for synthetic case study

Parameters	Values
Total porosity	5%
Volume fraction of clay	25%
Cation exchange capacity (CEC) per unit pore volume, Q_v	0.1 meq/cc
Equivalent conductance of sodium clay exchange cations, B	13 mS/meq
Relative permittivity of hydrocarbon	3.5
Relative permittivity of brine	80
Homogeneity index, α	0.5
Bulk conductivity of brine, C_w	10 S/m
Water saturation, S_w	30%
Cementation exponent, m	2
Saturation exponent, n	2

namely, C_w, S_w, n, and m, match with the assumed values listed in Table 12.1. Fig. 12.8 shows the histogram plot showing the frequency of occurrence of the four inversion-derived estimates for the 250 random initializations of the inversion scheme. For the synthetic case study, the inversion results show fast and robust convergence to desired values. Relative error in matching the synthetic data using the inversion-derived estimates is 0.01. The good agreement between the synthetic data and the integrated model predictions based on the inversion-derived estimates is illustrated in Fig. 12.9. For the 520-ft interval of upper Wolfcamp formation, no core data were available.

3.3 Application of the proposed log processing to the 9 EM logs acquired in the upper Wolfcamp formation

The joint inversion scheme was applied to dielectric dispersion and laterolog resistivity logs acquired in a well in upper Wolfcamp shale formation. Upper Wolfcamp comprises mostly of shale, varying from almost black to gray and greenish gray with several interbedded limestone and calcareous sandstone layers. A continuous estimation of S_w, C_w, m, and n was carried out across the 520-ft depth interval in upper Wolfcamp formation. Unlike conventional interpretation methods, 250 random initializations were performed for each depth to find the global minimum and the range of uncertainties in the estimates. Estimated time required for the proposed inversion method to process the logs acquired from 520-ft depth interval was about 20 hours on a

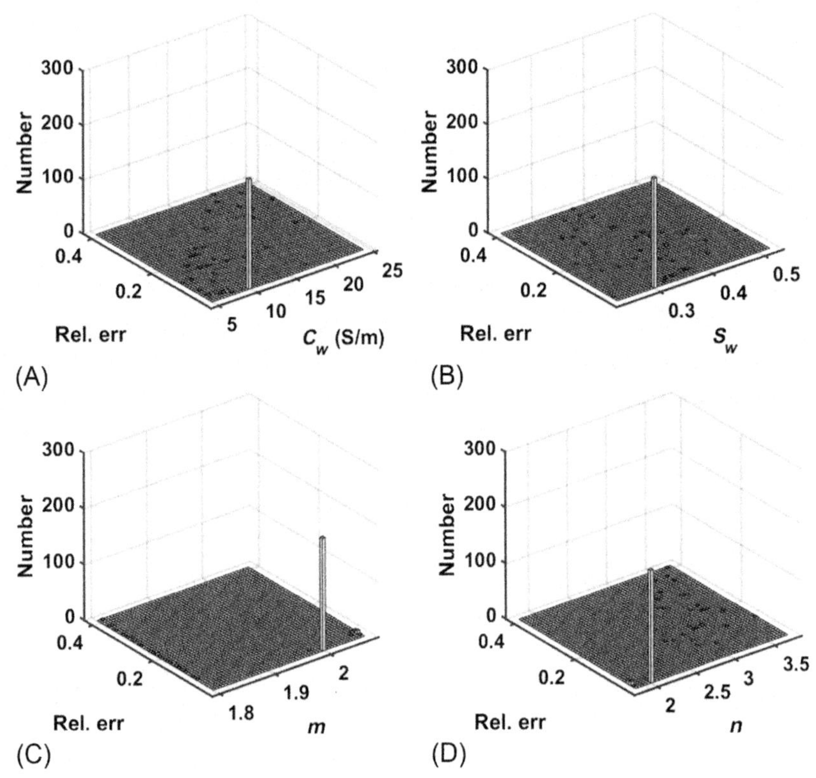

FIG. 12.8 Frequency of occurrence of the inversion-derived estimates of (A) $C_w = 10$ S/m, (B) $S_w = 0.3$, (C) $m = 2$, and (D) $n = 2$ with relative errors in matching the synthetic data for 250 random initializations of the proposed joint inversion of synthetic data, shown in Fig. 12.9.

single traditional laptop with 16GB RAM. The robustness of the inversion results is presented for the depth XX82 ft in upper Wolfcamp formation in Fig. 12.10. S_w, C_w, and m estimates have lower uncertainties and better convergences than n estimates. At the depth X82 ft, the ranges of estimates are 0.99–1 for S_w, 22–25 S/m for C_w, 1.87–1.92 for m, and 1.88–3.99 for n. Fig. 12.11 compares the effective permittivity, conductivity, and resistivity logs measured at depth XX82 ft with the integrated model predictions based on the inversion-derived estimates. Table 12.2 lists the values of other parameters assumed for this inversion.

3.4 Petrophysical interpretation of the inversion-derived estimates in the upper Wolfcamp formation

Figs. 12.12–12.14 contain logs of inversion-derived estimates in the upper Wolfcamp formation. Fig. 12.12 represents top part of upper Wolfcamp formation from X060 to X100 ft. This portion mainly comprises of 60%–80% quartz, 10%–20% clay and calcite (Track 5), and negligible TOC (total organic

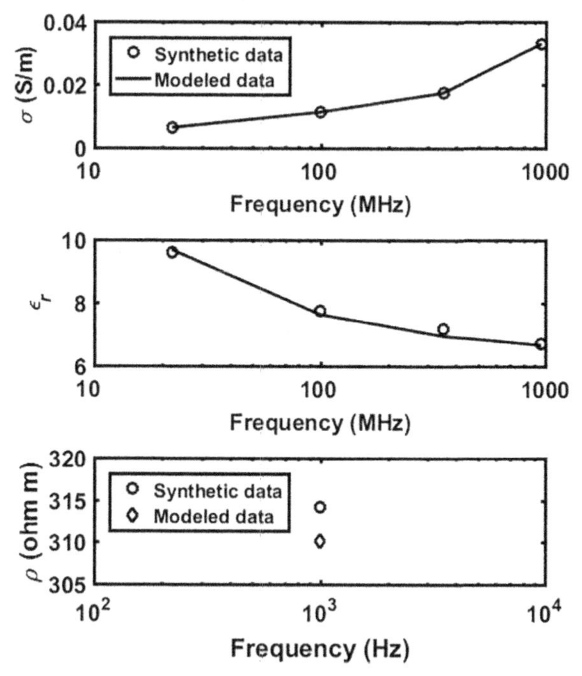

FIG. 12.9 Comparison of four effective conductivity values (top), four relative permittivity values (middle), and one resistivity value (bottom) for the synthetic case described in Table 12.1 against those predicted by the integrated model based on the inversion-derived estimates.

carbon). This interval of Wolfcamp formation has low resistivity on the laterologs (<10 Ω m). There are several laminations and beds exhibiting low porosity (Track 4). These low-porosity streaks are primarily calcite cementation (Track 5). Negligible effect of invasion was observed at different depths of investigation of laterolog resistivity measurements (Track 3). Therefore, we assume that the water saturations sensed by dielectric dispersion logs and resistivity log are equivalent in the absence of bed boundaries. With an increase in depth deeper into the well, the upper Wolfcamp formation mainly consists of tight carbonates (Fig. 12.13, middle XX30–XX60 ft) as calcite content increases (Fig. 12.13, Track 5). Formation microimages shows the presence of healed, resistive fractures and weak dissolution features in these tight carbonate formations. On triple combo logs, these intervals have low density (<2.55 g/cc) and elevated resistivity (>30 Ω m). These depths exhibit small pores and tight formations as observed from the NMR T2 distribution (Track 13, Fig. 12.13). With further increase in depth (Fig. 12.14), there are thin laminations and interbeds of calcite and clays. TOC ranges from 2% to 3%. There are several streaks of high-resistivity, low-porosity formations with high calcite deposition. These depths exhibit a wide range of pore size distribution as exhibited in Track 13 of Fig. 12.14.

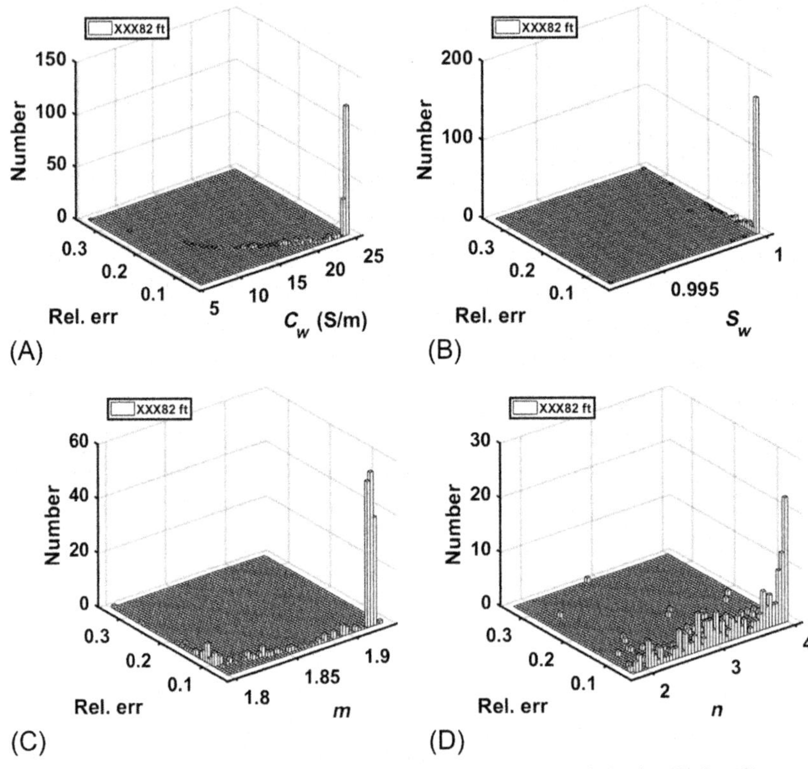

FIG. 12.10 Number of occurrences of inversion-derived estimates of (A) C_w, (B) S_w, (C) m, and (D) n with corresponding relative errors in matching the nine EM logs at depth XX82 ft for 250 random initializations of the proposed joint inversion of the nine EM logs at depth XX82 ft, shown in Fig. 12.11.

TABLE 12.2 Petrophysical parameters assumed for inverting the nine EM logs acquired at the depth XX82 ft

Parameters	Values
Total porosity	8.9%
Volume fraction of clay	9.9%
Cation exchange capacity (CEC) per unit pore volume, Q_v	0.1 meq/cc
Equivalent conductance of sodium clay exchange cations, B	13 mS/meq
Relative permittivity of hydrocarbon	3.5
Relative permittivity of brine	80
Homogeneity index, α	0.5
Bulk conductivity of brine, C_w	25 S/m
Water saturation, S_w	100%
Cementation exponent, m	1.9
Saturation exponent, n	3.99

Inversion-derived estimates are shown in **bold** font.

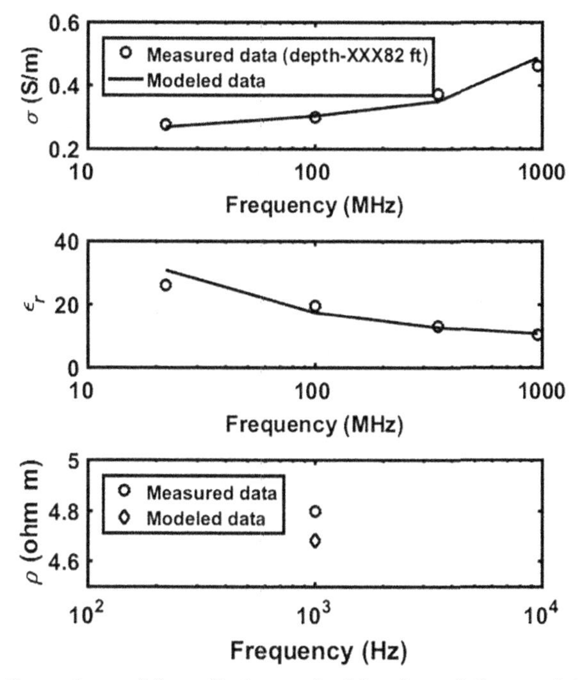

FIG. 12.11 Comparisons of four effective conductivity, four relative permittivity, and one resistivity logs measured at depth XX82 ft against those predicted using the integrated model based on the inversion-derived estimates.

In the depth interval from X060 to X110 ft, S_w estimates obtained using the proposed inversion (Fig. 12.12, Track 7) are in good agreement with those estimated using the multimineral inversion performed by an O&G service company (Fig. 12.12). Inversion-derived S_w estimates are more than 20% higher than those estimated by the service company's multimineral inversion for depth intervals in middle XX30–XX60 ft (Fig. 12.13) and in bottom XX30–XX60 ft (Fig. 12.14). Cementation exponent estimates computed using the inversion matches with the estimates obtained by the service company, as shown in Track 8 of Fig. 12.12. Estimates of cementation exponent obtained using the proposed inversion exhibit higher variations with the increase in depth, as shown in Track 10 of Fig. 12.14 in comparison with that of Fig. 12.12, that correlates with variations in clay content and porosity. Notably, based on the sensitivity analysis presented in Fig. 12.4, uncertainty in saturation-exponent estimates increases with increase in water saturation. Inversion-derived estimates of water saturation are maximum at depth in top X098–X104 ft, and in that depth interval the saturation exponent is the most uncertain. Similar to the sensitivity analysis performed in the earlier sections, water saturation and cementation exponent estimates have higher certainty in comparison with saturation exponent and brine conductivity estimates, as shown in Figs. 12.12–12.14. In Fig. 12.13, at depth in middle XX20–XX60 ft,

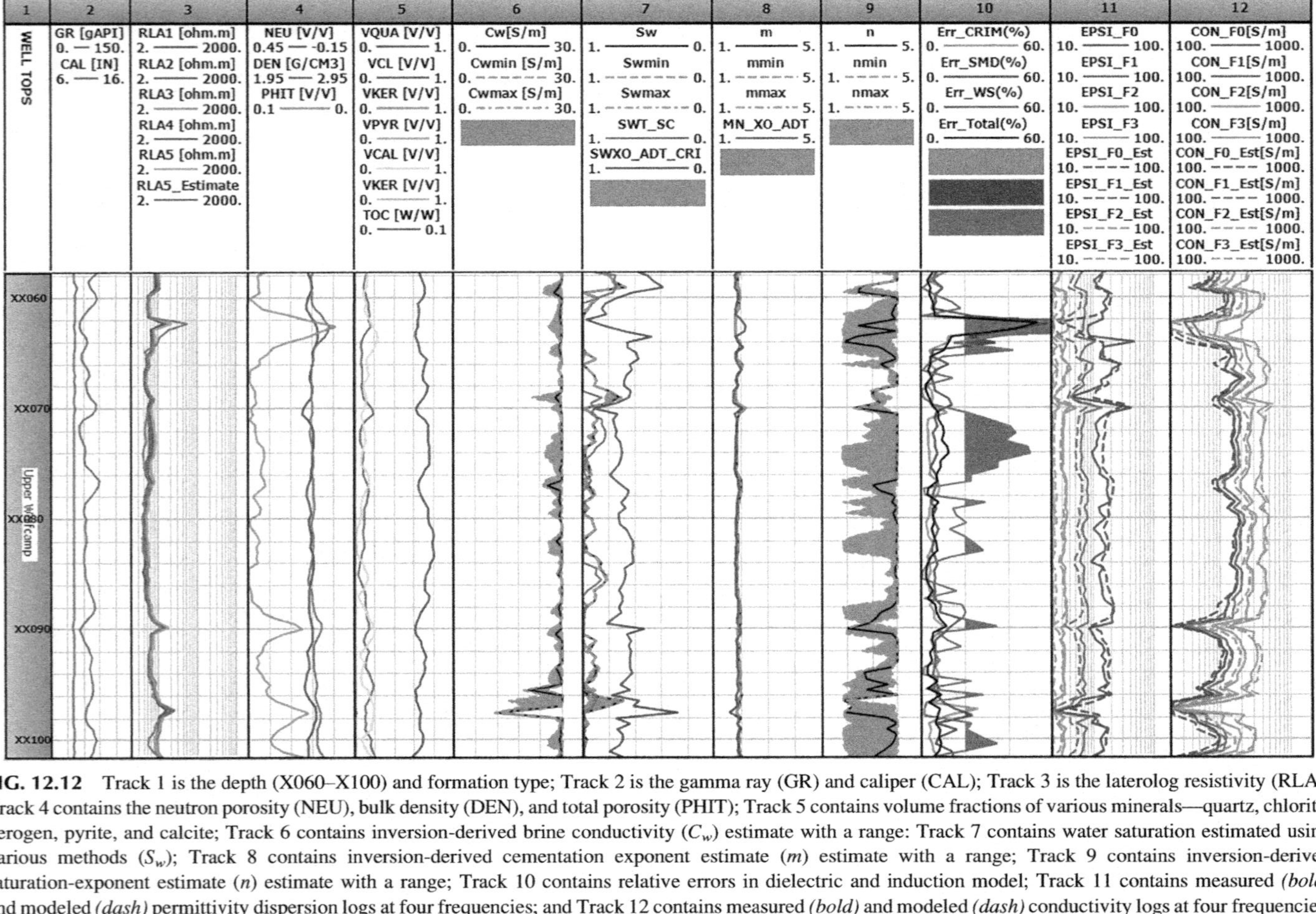

FIG. 12.12 Track 1 is the depth (X060–X100) and formation type; Track 2 is the gamma ray (GR) and caliper (CAL); Track 3 is the laterolog resistivity (RLA); Track 4 contains the neutron porosity (NEU), bulk density (DEN), and total porosity (PHIT); Track 5 contains volume fractions of various minerals—quartz, chlorite, kerogen, pyrite, and calcite; Track 6 contains inversion-derived brine conductivity (C_w) estimate with a range: Track 7 contains water saturation estimated using various methods (S_w); Track 8 contains inversion-derived cementation exponent estimate (m) estimate with a range; Track 9 contains inversion-derived saturation-exponent estimate (n) estimate with a range; Track 10 contains relative errors in dielectric and induction model; Track 11 contains measured *(bold)* and modeled *(dash)* permittivity dispersion logs at four frequencies; and Track 12 contains measured *(bold)* and modeled *(dash)* conductivity logs at four frequencies.

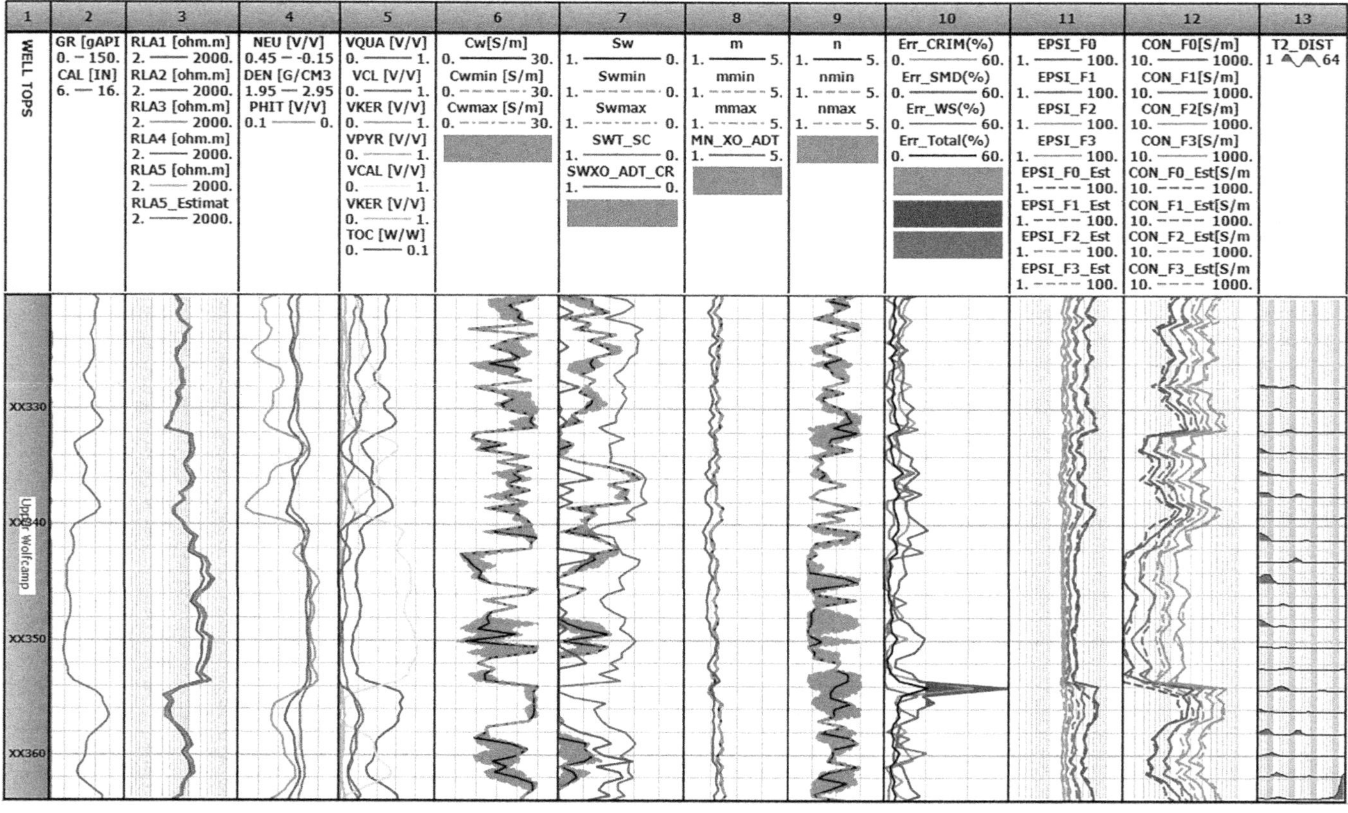

FIG. 12.13 Logs for formation depth from middle XX22 to XX60. The logs and the layout presented in this figure are similar to those in Fig. 12.12. Track 13 contains T2 NMR distribution.

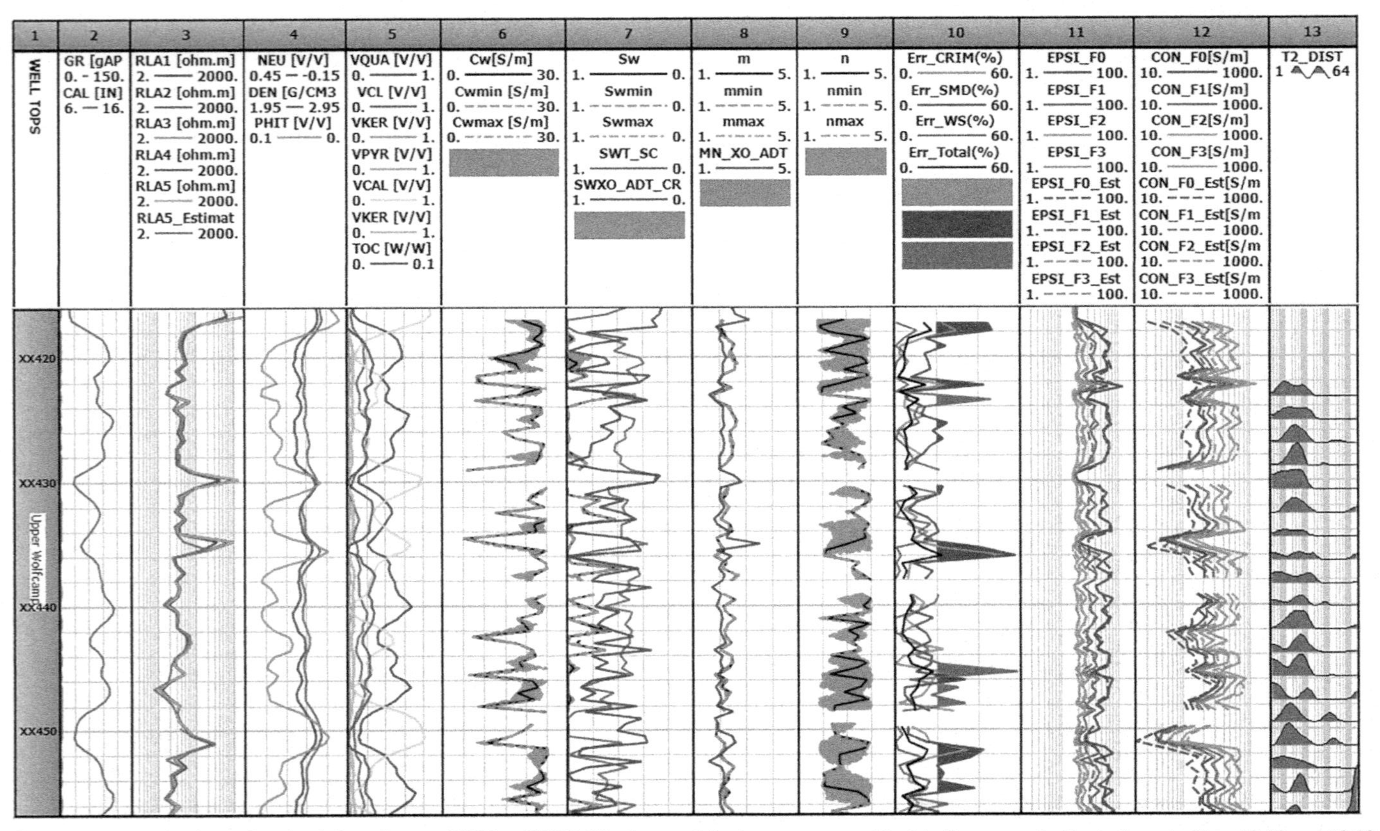

FIG. 12.14 Logs for formation depth from bottom XX12 to XX50. The logs and the layout presented in this figure are similar to those in Figs. 12.12 and 12.13.

high uncertainty observed in brine conductivity estimates is primarily because the integrated model is not able to account for the polarization mechanism that is resulting in large frequency dependence of the effective conductivity measured by the dielectric dispersion tool (Track 12). Uncertainty in C_w estimates in the formation shown in Fig. 12.12 is low that correlates with the low frequency dependence of the effective conductivity logs. Also, based on the sensitivity analysis, the proposed joint inversion shows unreliable estimates at high water saturation and low brine conductivity. From depth XX30–XX60 ft, uncertainty at high water saturations can be observed. The proposed inversion achieves high certainty for the estimates of water saturation, brine conductivity, and cementation exponent from depth X060–X100 ft. This is a consequence of low clay content and high porosity in that depth interval. Our estimates of brine conductivity and water saturation are uncertain in tight calcareous formations from XX30–XX60 ft. These unreliable estimates could be due to small pores and low-porosity beds in that depth interval.

3.5 Error analysis for the inversion-derived estimates in the upper Wolfcamp formation

Error in the integrated-model predictions is calculated as the difference between the modeled and measured dielectric dispersion and resistivity logs. The logs are modeled based on inversion-derived estimates obtained using the proposed inversion scheme coupled with the integrated WS, SMD, and CRI model. The total error averaged over the one resistivity, four conductivity, and four permittivity logs is reported as the mean relative error, which is expressed as

$$Error_{Total} = \left[\frac{Error_{SMD} + Error_{CRI} + Error_{WaxmanSmits}}{5}\right], \tag{12.12}$$

such that

$$Error_{CRI} = \frac{\left[\frac{abs\left(Perm^{meas,1\,GHz} - Perm^{mod,1\,GHz}\right)}{Perm^{meas,1\,GHz}} + \frac{abs\left(Cond^{meas,1\,GHz} - Cond^{mod,1\,GHz}\right)}{Cond^{meas,1\,GHz}}\right]}{2} \times 100\% \tag{12.13}$$

$$Error_{SMD} = \sum_{i=2}^{4} \frac{\left[\frac{abs\left(Perm^{meas,f_i} - Perm^{mod,f_i}\right)}{Perm^{meas,f_i}} + \frac{abs\left(Cond^{meas,f_i} - Cond^{mod,f_i}\right)}{Cond^{meas,f_i}}\right]}{6} \times 100\% \tag{12.14}$$

$$Error_{WS} = \frac{abs\left(Res_{mod,1\,kHz} - Res_{meas,1\,kHz}\right)}{Res_{meas,1\,kHz}} \times 100\% \tag{12.15}$$

Error in fitting the conductivity and permittivity logs acquired at 1 GHz using the CRI model predictions is referred as the $Error_{CRI}$. Error in matching the three conductivity and three permittivity dispersion logs acquired at 20, 100, and 260 MHz, respectively, using the SMD model predictions is referred as $Error_{SMD}$. Finally the error in fitting the resistivity laterolog using WS model predictions is referred as $Error_{WS}$.

Track 10 in Figs. 12.12–12.14 contain the various data misfits in matching the corresponding logs with CRI model, SMD model, and WS model predictions. In Fig. 12.12, the depth interval X060–X080 exhibits large error due to data misfit in modeling the laterolog resistivity, which on an average ranges around 40%. This interval exhibits low-resistivity and high-porosity formation with large volume fraction of quartz; these three characteristics correlate with zones where the $Error_{WS}$ is large. Volume of clay content is low in these formations and high $Error_{WS}$ may be a consequence of the assumed value of 1.3 for BQ_v for the entire formation. In Fig. 12.13, depth interval in middle XX43–XX54 ft shows highest interval of low porosity at an average of 2.5 porosity units. Error in fitting laterolog resistivity log ($Error_{WS}$) and that in fitting the highest-frequency dielectric log ($Error_{CRI}$) is lowest in this interval. Error in fitting the three lowest-frequency dielectric dispersion logs ($Error_{SMD}$) has the highest value in this interval. At the depth XX53.5 ft, $Error_{SMD}$ is about 60% along with $Error_{WS}$ attaining a large value of 50%, which might be a consequence of thin laminations and sand-shale bed-boundary effects at that depth. High errors are also observed at XX18, XX36, and XX52 ft in Fig. 12.14 due to the errors in SMD and CRIM models. These high errors could be because the integrated model is not able to account for the polarization mechanism that gives rise to a large frequency dependence of the effective conductivity measured by the dielectric dispersion tool.

4 Conclusions

We successfully processed eight dielectric permittivity and conductivity dispersion logs (10 MHz–1 GHz) and one laterolog resistivity (~1 kHz) acquired in 520-ft depth interval of the clay- and carbonate-rich upper Wolfcamp formation. In doing so, a continuous estimation of water saturation, brine conductivity, cementation exponent (textural index), and saturation exponent with ranges of possible values were obtained for the upper Wolfcamp formation. The proposed log-processing technique is robust compared with processing only the dielectric dispersion logs, especially in pyrite-rich, low-porosity, and hydrocarbon-bearing formations. These estimates have higher certainty and better convergence at lower water saturation, which is a desired feature of the processing technique. Compared with the estimates for saturation exponent and brine conductivity, water saturation and cementation exponent estimates exhibit higher certainty. Water saturation estimates obtained using the proposed inversion are in good agreement with those estimated using service company's multimineral inversion performed in the depth interval from X060 to X110 ft. Inversion-derived water saturation estimates are more than 20% higher than those estimated by the multimineral inversion for depth intervals

XX30–XX60 ft and XX30–X460 ft. Estimates of cementation exponent obtained using the proposed inversion exhibit higher variation with the increase in depth indicating an increase in heterogeneity/layering with depth. Hydrocarbon saturation in the interval XX34–XX46 ft is close to 30%. There are thin layers in the interval XX12–XX44 ft having hydrocarbon saturation close to 20%. Hydrocarbon-bearing formation has connate-water conductivity of 20 S/m. The saturation-exponent estimates are close to 2.8 in the hydrocarbon-bearing layers and tend to exhibit large uncertainty in the nonreservoir sections of the upper Wolfcamp shale.

References

[1] Engle MA, Reyes FR, Varonka MS, Orem WH, Ma L, Ianno AJ, Schell TM, Xu P, Carroll KC. Geochemistry of formation waters from the Wolfcamp and "Cline" shales: insights into brine origin, reservoir connectivity, and fluid flow in the Permian Basin, USA. Chem Geol 2016;425:76–92.

[2] Zhang XS, Wang HJ, Ma F, Sun XC, Zhang Y, Song ZH. Classification and characteristics of tight oil plays. Pet Sci 2016;13(1):18–33.

[3] Malik M, Schmidt C, Stockhausen EJ, Vrubel NK, Schwartz K. Integrated petrophysical evaluation of unconventional reservoirs in the Delaware basin. In: SPE annual technical conference and exhibition SPE 166264; 2013.

[4] Misra S, Torres-Verdín C, Revil A, Rasmus J, Homan D. Interfacial polarization of disseminated conductive minerals in absence of redox-active species—part 1: mechanistic model and validation. Geophysics 2016;81(2):E139–57.

[5] Misra S, Torres-Verdín C, Revil A, Rasmus J, Homan D. Interfacial polarization of disseminated conductive minerals in absence of redox-active species—part 2: effective electrical conductivity and dielectric permittivity. Geophysics 2016;81(2):E159–76.

[6] Little JD, Julander DR, Knauer LC, Aultman JT, Hemingway JL. Dielectric dispersion measurements in California heavy oil reservoirs. In: SPWLA 51st annual logging symposium. Society of Petrophysicists and Well-Log Analysts; 2010.

[7] Rosepiler MJ. Calculation and significance of water saturations in low porosity shaly gas sands. Oil Gas J 1981;79(28):180–7.

[8] Sarihi A, Murillo B. A method to compute water saturation in tight rocks accounting for conductivity of clay minerals. In: SPE annual technical conference and exhibition, SPE 177550; 2015.

[9] Donadille JM, Leech R, Pirie I. Water salinity determination over an extended salinity range using a joint interpretation of dielectric and neutron cross-section measurements. In: SPE annual technical conference and exhibition. Society of Petroleum Engineers; 2016.

[10] Chen H, Heidari Z. Assessment of hydrocarbon saturation in organic-rich source rocks using combined interpretation of dielectric and electrical resistivity measurements. In: SPE annual technical conference and exhibition. Society of Petroleum Engineers; 2014.

[11] Misra S, Lüling MG, Rasmus J, Homan DM, Barber TD. Dielectric effects in pyrite-rich clays on multifrequency induction logs and equivalent laboratory Core measurements. In: SPWLA 57th annual logging symposium. Society of Petrophysicists and Well-Log Analysts; 2016.

[12] Han Y, Misra S, Simpson G. Dielectric dispersion log interpretation in Bakken petroleum system. In: SPWLA 58th annual logging symposium. Society of Petrophysicists and Well-Log Analysts; 2017.

[13] Misra S, Han Y. Petrophysical interpretation of multi-frequency electromagnetic measurements in clay- and conductive-mineral-rich Mudrocks. American Association of Petroleum Geologists; 2016.

[14] Han Y, Misra S. Improved water-saturation estimates derived from inversion-based interpretation of broadband electromagnetic dispersion logs in organic-rich shale formations. In: SEG technical program expanded abstracts. Society of Exploration Geophysicists; 2017. p. 3458–62.

[15] Waxman MH, Smits LJM. Electrical conductivities in oil-bearing shaly sands. Soc Pet Eng J 1968;8(2):107–22.

[16] Peeters M, Holmes A. Review of existing shaly-sand models and introduction of a new method based on dry-clay parameters. Petrophysics 2014;55(6):543–53.

[17] Sabouroux P, Ba D. Epsimu, a tool for dielectric properties measurement of porous media: application in wet granular materials characterization. Prog Electromagn Res B 2011; 29:191–207.

[18] Brovelli A, Cassiani G. Effective permittivity of porous media: a critical analysis of the complex refractive index model. Geophys Prospect 2008;56(5):715–27.

[19] Stroud D, Milton GW, De BR. Analytical model for the di-electric response of brine-saturated rocks. Phys Rev B 1986;34:5145–53.

[20] Sanchez-Ramirez JA, Torres-Verdín C, Wang GL, Mendoza A, Wolf D, Liu Z, Schell G. Field examples of the combined petrophysical inversion of gamma-ray, density, and resistivity logs acquired in thinly-bedded clastic rock formations. Petrophysics 2010;51:247–63.

[21] Ferraris P, Borovskaya I, Ribeiro M. Advances in formation evaluation independent of conveyance method: state of the art logging while drilling & wireline petrophysical analysis in a carbonate reservoir offshore Brazil. In: SPWLA 53rd annual logging symposium. Society of Petrophysicists and Well-Log Analysts; 2012.

[22] Seleznev NV, Fellah K, Philips J, Zulkipli SN, Fournie B. Matrix permittivity measurements for rock powder. In: SPE annual technical conference and exhibition. Amsterdam, Netherland: Society of Petroleum Engineers; 2014.

[23] Clennell MB, Josh M, Esteban L, Piane CD, Schmid S, Verrall M, Hill D, Woods C, McMullan B. The influence of pyrite on rock electrical properties: a case study from NW Australian gas reservoirs. In: SPWLA 51st annual logging symposium. Society of Petrophysicists and Well-Log Analysts; 2010.

Chapter 13

Characterization of subsurface hydrocarbon/water saturation using Markov-chain Monte Carlo stochastic inversion of broadband electromagnetic logs

Siddharth Misra* and Yifu Han†,[a]
*Harold Vance Department of Petroleum Engineering, Texas A&M University, College Station, TX, United States, †Schlumberger Technology Company, Beijing, China

Chapter outline

[a] Present address: Schlumberger, Beijing, China.

Machine Learning for Subsurface Characterization. https://doi.org/10.1016/B978-0-12-817736-5.00013-2

Symbols and abbreviations

ε^*	bulk complex permittivity of the formation
ε_0	vacuum permittivity
ε_{rhy}	hydrocarbon relative permittivity
ε_{rc}	relative permittivity of surface charge-bearing particle (like clay)
ε_{ri}	relative permittivity of conductive particle (like pyrite)
ε_{rm}	matrix relative permittivity
ε_{rn}	relative permittivity of nonconductive particle (like sand)
ε_{rs}	relative permittivity of nonconductive particle (like sand)
ε_{rw}	relative permittivity of water
ε_w^*	water complex permittivity
λ_c	surface conductance of surface charge-bearing particle (like clay)
λ_s	surface conductance of nonconductive particle (like sand)
φ_t	total porosity of the formation
Γ	gamma function
ω	angular frequency
a	tortuosity factor
C_{sh}	low frequency conductivity in 100% clay-rich formation
C_w	connate water conductivity (S/m)
CEC	cation exchange capacity
CRI	complex refractive index
D_i	diffusion coefficient of conductive particle (like pyrite)
D_w	diffusion coefficient of water
EM	electromagnetic
GHz	gigahertz
IP	interfacial polarization
kHz	kilohertz
LR	Lichteneker-Rother model
m	cementation exponent
MCMC	Markov-chain Monte Carlo
MHz	megahertz
n	saturation exponent
PS	mechanistic model of frequency-dependent complex conductivity of a geomaterial containing clay particles and conductive mineral particles, such as pyrite; also, referred as the clay-pyrite interfacial-polarization (IP) model
r_c	radius of spherical surface charge-bearing particle (like clay)
r_i	radius of conductive particle (like pyrite)
r_s	radius of nonconductive particle (like sand)
R_t	formation resistivity
R_w	connate water resistivity (Ω m)
S	salinity (ppt)
S_i	bulk conductivity of conductive particle (like pyrite)
S_w	water saturation

SMD Stroud-Milton-De
T temperature (in °F)
TOC total organic carbon
V_c volumetric fraction of surface charge-bearing particle (like clay)
V_i volumetric fraction of conductive particle (like pyrite)
V_s volumetric fraction of nonconductive particle (like sand)
V_{sh} volumetric fraction of shale

1 Introduction

Chapters 12 and this chapter focus on the implementations of error minimization techniques that are crucial to machine learning and for building data-driven models. Most of the tasks in machine learning and data analysis need either deterministic or stochastic data inversion/optimization techniques to determine the underlying properties/behavior/trends of systems or processes. Data inversion aims to minimize misfit/error between the measured data and modeled data. This is accomplished by systematically choosing model parameter values from within an allowed set of values (domain). The combination of values of model parameters that minimizes the misfit/error under certain constraints is referred as the optimal solution for the data-inversion problem. The optimal values of model parameters define the model that is most representative of the system/process that generated the measured data. In other words, a data-inversion problem finds a specific form of a forward model by identifying model parameter values that result in model predictions that best match the measured data. For example, seismogram data can be inverted to estimate various subsurface properties, such as acoustic velocity, formation and fluid densities, Poisson's ratio, formation compressibility, and shear rigidity. Unlike the deterministic inversion method discussed in the previous chapter, stochastic inversion methods iterate and update model parameters based on a probability distribution to minimize the objective function to generate a distribution of alternate model parameters obeying the model constraints. In doing so, stochastic methods find all the possible nonunique solutions along with their likelihoods of occurrence that accounts for possible multimodality of the parameters' distribution. Stochastic inversion are suited when data contains noise, when the data has uncertainty (e.g., future data), and when data is generated by heterogeneous processes that are difficult to model using deterministic laws.

This chapter uses Markov-chain Monte Carlo-based stochastic inversion of multifrequency electromagnetic measurements in shale formations to reliably quantify the water/hydrocarbon saturation. Markov chain is a stochastic model that defines a sequence of events (random variables), wherein each event depends only on the state attained in the previous event. This requires the stochastic process to honor the Markov property, such that conditional probability distribution of future states of the process depends only upon the present state, not on the sequence of events that preceded it. Monte Carlo method is a computational technique that uses randomness to solve physical/mathematical problems with probabilistic interpretation by repeated random sampling and statistical analysis. An important assumption in Monte Carlo method is when samples are chosen at random, the samples tend to exhibit properties similar to those of the population from which they are drawn. Monte Carlo method is useful for obtaining numerical solutions to computationally expensive numerical problems, such as multidimensional numerical integration. Monte Carlo methods are useful for simulating phenomena with significant uncertainty in inputs and systems with a large number of coupled degrees of freedom. Markov Chain Monte-Carlo (MCMC) methods were created to address multidimensional problems better than simple Monte Carlo algorithms. MCMC is a technique to sample data from a large population exhibiting complex probability distributions using Markov chains. MCMC aims at sampling a subset such that the sample distribution matches the actual desired distributions. MCMC constructs Markov chains that draw samples which are progressively more likely realizations of the actual desired distributions. In doing so, the goal of MCMC is to estimate the posterior distribution indicating the parameter values that best honor the observed data (expressed in terms of the likelihood distribution) taking into account the prior distribution based on the beliefs about the parameters. MCMC assumes that the region of high probability tends to be "connected" without going through a low-probability region. When the number of dimensions rises, MCMC tends to suffer from the curse of dimensionality, which causes the regions of higher probability to stretch and get lost in an increasing volume of space. Due to the curse of dimensionality, the probability of rejection of the selected samples increases exponentially as a function of the number of dimensions. MCMC-based stochastic inversion is computationally-intensive method that exhibits computational time that is orders of magnitude more than its deterministic equivalent.

1.1 Saturation estimation in shale gas reservoir

This chapter uses Markov-chain Monte Carlo-based stochastic inversion of multifrequency electromagnetic (EM) measurements to estimate water saturation in a clay- and pyrite-bearing shale reservoir from lower Paleozoic basin. Such basins are considered prolific unconventional oil and gas resources [1]. The mineral constituents of the shale reservoir being studied in this chapter primarily include quartz, illite, chlorite, kaolinite, kerogen, calcite, and phosphate exhibiting high clay content, high TOC, low porosity, high tortuosity, and the presence of conductive minerals, such as pyrite [2]. In the presence of such subsurface near-wellbore reservoir petrophysical conditions, well-log-derived hydrocarbon/water saturation estimates obtained from the interpretation of various subsurface EM logs, such as induction resistivity log and dielectric dispersion log, and multimineral analysis solver are not consistent [3,4]. In addition, saturation estimates based on the interpretation of EM logs using conventional saturation models tend to break down in organic-rich shale gas formation because of the unaccounted interfacial polarization (IP) effects of clays and conductive minerals, such as pyrite [3,4,5].

1.2 Electromagnetic (EM) logging tools

Oil and gas operators commonly deploy only one type of EM logging tool for the subsurface near-wellbore reservoir characterization of saturations and certain characteristics of the pore space [6]. Wang and Poppitt [2] reported first ever deployment of three different EM logging tools in a well drilled in a shale reservoir resulting in first-ever acquisition of continuous broadband EM dispersion logs. The three EM logging tools were EM induction resistivity tool, EM propagation tool, and dielectric dispersion tool. The induction resistivity tool uses current flow in a coil to induce current flow in the formation under investigation and is designed primarily for formation conductivity measurements in wells drilled with oil-based mud. The induction resistivity tools have depth of investigation ranging from 10 to 90 in., with 1–4 ft vertical resolution. The induction resistivity tool is typically operated at frequency between 10 kHz and 60 kHz, and it was operated at 26 kHz in the well under investigation in this study, with depth of investigation around 36 in. [2]. EM propagation tool measures the attenuation, phase shift, and travel time for EM waves traveling through the formation at two receivers located in the tool at a certain spacing [6]. In this study, the EM propagation tool was operated at two discrete frequencies, 1 and 2 MHz with depth of investigation around 15 in. and 10 in., respectively [2]. The new-generation dielectric tools can measure continuous attenuation and phase shift at 1-in. vertical resolution and are suitable for low salinity formation, carbonate formation, and heavy oil reservoirs [6]. The new-generation dielectric tool operates at four distinct frequencies ranging from 10 MHz to 1 GHz, and it was operated at 20, 100, 260 MHz, and 1 GHz in this study. EM propagation tool and dielectric dispersion tool first measure the attenuation and phase shift of the propagating EM waves at multiple frequencies that are later transformed to conductivity and relative permittivity of formation [6].

1.3 Conventional EM-log-interpretation models

Archie's equation is an empirical electrical resistivity model to interpret subsurface resistivity logs acquired in clean sandstone formations at low EM frequencies [7,8,9]. Archie's equation doesn't account for the surface conductance of clay minerals and the interfacial polarization (IP) effects around the clays. Surface charge-bearing clay minerals in shaly sand formations can significantly change the bulk electrical resistivity; therefore, the Archie-derived water saturation in shaly-sand formations is an overestimation. Simandoux [10] proposed a shaly sand model that considers both the brine conductivity and the surface conductance of clay minerals. Simandoux equation was developed based on laboratory investigations on synthetic homogeneous mixtures of sand, conductive clay minerals, and water. Simandoux equation is expressed as

$$\frac{1}{R_t} = \frac{\varphi_t^{\,m}}{aR_w} S_w^n + \frac{V_{sh}}{R_{sh}} S_w^{n-1} \tag{13.1}$$

where R_t is formation resistivity (in Ωm), φ_t is total porosity, S_w is water saturation, R_w is connate water resistivity at formation temperature (in Ωm), V_{sh} is volumetric fraction of shale, R_{sh} is the resistivity of shale (in Ωm), m is cementation exponent ($m = 1.33$ in shaly sand [11]), n is saturation exponent, and a is tortuosity factor ($a = 1.65$ in shaly sand [11]). In the shale reservoir under investigation, Simandoux equation is used to process the EM induction resistivity log for estimating water saturation, which is then compared with those computed using the proposed stochastic inversion of broadband EM logs measured using the three EM logging tools.

CRI model is widely used to process complex permittivity logs at around 1 GHz, and it assumes that IP effects at grain surface are negligible around 1 GHz. CRI model is a specific case of the Lichtenecker-Rother (LR) model [12] expressed as

$$(\varepsilon^*)^\alpha = (1-\varphi_t)(\varepsilon_{rm})^\alpha + S_w\varphi_t(\varepsilon_w^*)^\alpha - (1-S_w)\varphi_t(\varepsilon_{rhy})^\alpha \tag{13.2}$$

where ε^* is bulk complex relative permittivity of geomaterials, ε_{rm} is matrix relative permittivity, ε_w^* is water complex relative permittivity, ε_{rhy} is relative permittivity of hydrocarbon, and α is a geometrical fitting parameter. With $\alpha = 0.5$, LR model reduces to CRI model. The tortuosity, rock fabric, pore-network, and minerals spatial distributions of geomaterials are not considered in LR and CRI model [13]. The water complex relative permittivity ε_w^* can be formulated as

$$\varepsilon_w^* = \varepsilon_{rw} + i\frac{C_w}{\varepsilon_0\omega} \tag{13.3}$$

where ε_0 is vacuum permittivity ($\varepsilon_0 = 8.85 \times 10^{-12}$ F/m), ω is angular frequency ($\omega = 2\pi f$, f is frequency), ε_{rw} is relative permittivity of water, and C_w is water conductivity (in S/m, $C_w = 1/R_w$). The relative permittivity of water ε_{rw} decreases with increasing temperature since thermal agitation reduces the overall water molecules alignment along the EM field direction [5] and is expressed as a function of temperature and salinity [3,14] as

$$\varepsilon_{rw}(0,T) = 94.88 - 0.2317T + 0.000217T^2 \tag{13.4}$$

$$\varepsilon_{rw}(S,T) = \left[\frac{1}{\varepsilon_{rw}(0,T)} + \frac{2.4372 \times S}{58.443(1000-S)}\right]^{-1} \tag{13.5}$$

where S is salinity (in ppt) and T is temperature (in °F). The water conductivity C_w can also be formulated as a function of temperature and salinity [3,14] expressed as

$$C_w(S,T) = \frac{T+7}{82}\left(0.0123 + \frac{3647.5}{(1000S)^{0.955}}\right)^{-1} \tag{13.6}$$

Stroud et al. [14] derived an analytical mixing law to calculate frequency-dependent complex dielectric permittivity of brine-saturated clean sandstone or carbonate. This mixing law is referred as SMD model and is a function of total porosity, water saturation, water salinity, relative permittivity of conductive and nonconductive phases, and rock textural parameter describing cementation/tortuosity [4,5]. SMD model is a commonly used spectral complex dielectric permittivity model and is applied to process the dielectric dispersion logs acquired around the frequency range of 10 MHz to 0.4 GHz [3,4,5]. SMD model is expressed as

$$\varepsilon^* = (\varphi_t S_w)^m \varepsilon_w^*(S,T) + [1-(\varphi_t S_w)^m]\varepsilon_{rn} - \varepsilon_{rn}\Gamma\left(\varphi_t S_w, m, \frac{\varepsilon_w^*(S,T)}{\varepsilon_{rn}}\right) \tag{13.7}$$

where Γ is gamma function and ε_{rn} is relative permittivity of nonconductive phase, which is expressed as

$$(\varepsilon_{rn})^{1/2} = \frac{(1-\varphi_t)(\varepsilon_{rm})^{1/2} + (1-S_w)\varphi_t(\varepsilon_{rhy})^{1/2}}{1-\varphi_t S_w} \tag{13.8}$$

1.4 Estimation of water/hydrocarbon saturation based on the inversion of electromagnetic (EM) logs

Estimation of water/hydrocarbon saturation is essential for subsurface characterization. Water saturation can be derived from the EM logs, such as induction resistivity log, EM propagation log, or dielectric dispersion log. Water saturation estimation is challenging in shale formations because of the complex rock fabric, high clay content, low porosity, high tortuosity, and the presence of conductive minerals, such as pyrite. Conventional EM-log interpretation models tend to break down in shale and other complex formations under these adverse petrophysical conditions [15]. Han et al. [3] reported that the water saturation estimates from various EM logs using various conventional interpretation models can vary up to 0.4 saturation unit for a single depth in Bakken petroleum system. Moreover, the water saturation measurements on core samples are not consistent with well-log-derived water saturation estimates for several depths [3]. In Bakken shale formation, Pirie et al. [16] also had similar findings upon comparing water saturation estimates derived from triaxial induction resistivity log, dielectric dispersion logs, NMR log, core-scale Dean-Stark laboratory measurements, and multimineral analysis solver.

Interpretation of only induction resistivity log or only dielectric dispersion logs is sensitive to model assumptions, noise in data, model inputs, and polarization phenomena, and is not sensitive to certain petrophysical properties [5]. Donadille et al. [17] mentioned the limitation of dielectric dispersion log interpretation in determining formation water saturation and connate water salinity by using CRIM or SMD model. The dielectric measurements lose sensitivity to salinity above 70 ppk [17]. The joint interpretation was then proposed for improved hydrocarbon saturation evaluation. Ramirez et al. [18] implemented a Bayesian inversion approach of joint interpretation of gamma-ray, density, and resistivity logs to estimate porosity, water saturation, and volumetric fraction of shale in thin-bed formations. Chen and Heidari [19] proposed a joint interpretation of dielectric and resistivity measurements that significantly improves water-filled porosity assessment. They suggested that the pore structure, spatial distribution and tortuosity of water, kerogen and pyrite networks should be taken into account for joint interpretation of electrical conductivity and dielectric measurements. Tathed et al. [5] proposed a joint log processing of resistivity and dielectric dispersion logs using an integrated mechanistic model combining Waxman-Smits, CRIM, and SMD model. They developed a Levenberg-Marquardt-based deterministic inversion method to simultaneously process dielectric dispersion log and laterolog resistivity log to estimate water saturation, connate-water conductivity, cementation exponent, and saturation exponent in a 520-ft depth interval in upper Wolfcamp shale formation. Das et al. [20] also attempted to quantify the rock textural parameters by using constrained rock physics template and fluid response modeling. Tathed et al. [4] then modified the previously proposed deterministic inversion method [5] to jointly process array induction resistivity log and dielectric dispersion log acquired in a 350-ft depth interval in one of the science wells intersecting the Bakken petroleum system. The estimated water saturation using the improved interpretation method was compared with those obtained from NMR log, Dean-Stark core measurements, and oil and gas service company's dielectric inversion.

More importantly, conventional EM logs interpretation models that ignore the IP effects of clay and conductive pyrite are not applicable in clay- and pyrite-rich formations. Misra et al. [21] mentioned the challenges with log-scale hydrocarbon saturation estimation in unconventional reservoirs due to dielectric effects. They developed a mechanistic electrochemical model to consider the interfacial polarization (IP) effects of clay and pyrite [22]. Han et al. [3] proposed a deterministic dielectric dispersion log inversion method by combining LR model, SMD model, and the mechanistic clay-pyrite interfacial-polarization (IP) model to estimate formation water saturation, connate water conductivity, homogeneity index, and cementation index in both clay-lean and clay-rich units of Bakken petroleum system. They compared their water saturation estimates with log-scale NMR interpretation and service company's inversion results and core-scale Dean-Stark core water saturation measurements for a 300-ft depth interval in Bakken formation.

The hydrocarbon saturation assessment in unconventional reservoirs can be significantly improved by jointly processing broadband EM dispersion logs using mechanistic models that accurately account for the effects of clays and conductive minerals. In this chapter, broadband EM dispersion logs acquired by EM induction resistivity tool, EM propagation tool, and dielectric dispersion tool are simultaneously processed to estimate water saturation, connate water conductivity, and surface conductance of clay in certain depth intervals of an organic-rich unconventional shale gas formation in a Lower Paleozoic basin. We develop a Markov-chain Monte Carlo-based (MCMC) stochastic inversion scheme coupled with a clay-pyrite interfacial-polarization model (also, referred as the PS model) to process the EM broadband dispersion logs. Traditional deterministic inversion approaches, such as Gauss-Newton method and Levenberg-Marquardt method, implement gradient-based least-squares optimization that are sensitive to the initial guesses. The proposed MCMC sampling-based stochastic method is a global approach to estimate unknown petrophysical parameters and can better quantify the uncertainty of the inversion-derived petrophysical estimates [23].

2 Method

2.1 Mechanistic clay-pyrite interfacial-polarization model

SMD model is not applicable in clay- and pyrite-rich formations, and most of conventional EM log interpretation models ignore the IP effects of clay and conductive pyrite [3]. Misra et al. [22] derived a mechanistic electrochemical model to accurately quantify the frequency-dependent relative permittivity and conductivity by considering the IP effects of clay and pyrite. Misra et al. [22] referred the mechanistic model as the PS model that quantifies the frequency-dependent complex conductivity of a geomaterial containing clay particles and conductive mineral particles, such as pyrite. In this chapter, PS model will be referred as the clay-pyrite interfacial-polarization model. The size distribution and shape of inclusions are also considered and evaluated in the clay-pyrite interfacial-polarization model [22], and we assume that both the clay and pyrite are spherical in geometry and of constant size in this study. The petrophysical parameters in the clay-pyrite interfacial-polarization model are listed in Table 13.1. Fig. 13.1A and B illustrates a representative perfectly polarized conductive spherical pyrite and a representative surface charge-bearing spherical clay particle, respectively, surrounded by brine (connate water) under externally applied EM field [22]. The mechanistic clay-pyrite dispersion model is calibrated to the Simandoux equation at low frequency (1 Hz) and to the CRI model at high frequency (1 GHz). The mechanistic clay-pyrite interfacial-polarization model assumes clay and pyrite grains are uniformly distributed in the volume of homogeneous and isotropic formation under investigation. However, Maxwell-Wagner polarization effects [24] and effective porosity, pore-network tortuosity, isolated pores, and the connectivity of clay and pyrite grains are not taken into account in the clay-pyrite interfacial-polarization model implemented in this study.

TABLE 13.1 Petrophysical parameters assumed for the Synthetic Layer 1 and Synthetic Layer 2. Synthetic Layer 1 identifies a clay rich ($V_c = 0.5$), low porosity ($\varphi_t = 0.08$), and water bearing ($S_w = 0.9$) formation. Synthetic Layer 2 identifies a clay rich ($V_c = 0.55$), pyrite rich ($V_i = 0.03$), low porosity ($\varphi_t = 0.09$), and hydrocarbon bearing ($S_w = 0.5$) formation

Parameters	Unit	Layer 1	Layer 2
Volumetric fraction of pyrite grains, V_i	%	0	3
Bulk conductivity of pyrite, S_i	S/m		1000
Relative permittivity of pyrite, ε_{ri}			30
Diffusion coefficient of pyrite, D_i	m^2/s		5×10^{-5}
Radius of pyrite grains, r_i	μm		10
Volumetric fraction of clay, V_c	%	50	55
Relative permittivity of clay, ε_{rc}		5.8	5.8
Surface conductance of clay, λ_c	S	3×10^{-7}	2×10^{-7}
Radius of spherical clay grains, r_c	μm	0.6	0.25
Low frequency conductivity in 100% clay-rich formation, C_{sh}	S/m	0.4	0.35
Volumetric fraction of sand (nonconductive grains), V_s	%	42	33
Relative permittivity of sand, ε_{rs}		4.65	4.65
Surface conductance of sand, λ_s	S	1×10^{-9}	1×10^{-9}
Radius of sand grains, r_s	μm	100	100
Total porosity of rock, φ_t	%	8	9
Bulk conductivity of water, C_w	S/m	3.5	1
Relative permittivity of water, ε_{rw}		56	57
Diffusion coefficient of water, D_w	m^2/s	1.5×10^{-9}	1.5×10^{-9}
Relative permittivity of hydrocarbon, ε_{rhy}		2	2
Water saturation, S_w	%	90	50
Tortuosity factor, a		1.65	1.65
Cementation exponent, m		1.33	1.33
Saturation exponent, n		2	2
Formation temperature, T	°F	200	200

The parameters in bold font represent those that will be estimated for synthetic layers using the proposed inversion methodology.

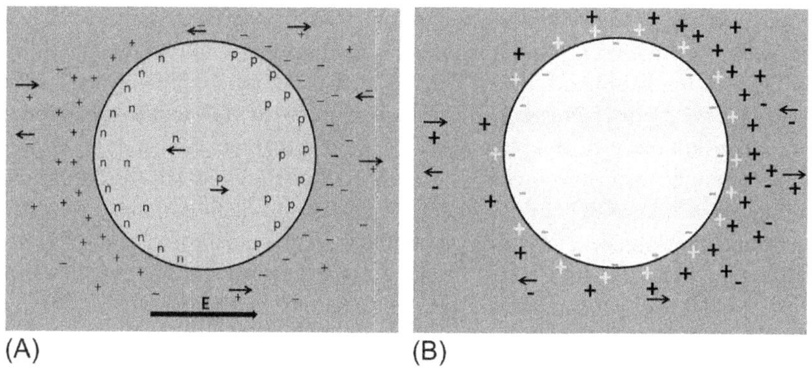

FIG. 13.1 (A) Representative volume comprising a perfectly polarized conductive spherical pyrite inclusion surrounded by conductive water and (B) representative volume comprising a surface charge-bearing spherical clay particle surrounded by conductive water [22].

2.2 Proposed inversion-based interpretation method

2.2.1 Bayesian formulation of the parameter estimation problem

Bayesian framework:

The MCMC sampling-based stochastic inversion method provides a global approach for estimating unknown parameters and a better way to quantify the uncertainty of unknown parameters as compared with the deterministic inversion method discussed in the previous chapter [23]. The MCMC sampling-based stochastic inversion method relies on Bayes' theorem expressed in Eq. 13.9. Bayesian formulation is used to update the probability for a hypothesis/event (in our case, model parameter) as evidence (in our case, broadband EM dispersion logs) is acquired. In the Bayesian framework, probabilities represent confidence levels for the occurrence of a certain event. We propose a Bayesian framework to estimate unknown petrophysical parameters in the clay-pyrite interfacial-polarization model representing a hydrocarbon-bearing shale formation that contains conductive mineral and clay grains. The broadband EM dispersion logs at seven log-acquisition frequencies constitute the input data vector in the Bayesian framework, and the unknown petrophysical model parameters are derived as the solution with associated probability distributions. If $\boldsymbol{d}$ denotes a data vector comprising broadband EM dispersion logs and $\boldsymbol{m}$ represents a petrophysical model parameters vector, then the Bayesian formulation for the posterior probability density function of model parameter $\boldsymbol{m}$ given data $\boldsymbol{d}$ and prior information $\boldsymbol{I}$, which is the prior knowledge of unknown petrophysical parameters (hypothesis or assumed model), is given by Bayes' theorem:

$$p(\boldsymbol{m}|\boldsymbol{d},\boldsymbol{I}) = \frac{p(\boldsymbol{m}|\boldsymbol{I})p(\boldsymbol{d}|\boldsymbol{m},\boldsymbol{I})}{p(\boldsymbol{d}|\boldsymbol{I})} \tag{13.9}$$

where $p(\boldsymbol{m}|\boldsymbol{d},\boldsymbol{I})$ is the posterior distribution of model parameter $\boldsymbol{m}$ given data $\boldsymbol{d}$ and prior information $\boldsymbol{I}$, $p(\boldsymbol{m}|\boldsymbol{I})$ is the prior distribution of model parameter $\boldsymbol{m}$, $p(\boldsymbol{d}|\boldsymbol{m},\boldsymbol{I})$ is the likelihood function, and $p(\boldsymbol{d}|\boldsymbol{I})$ is a normalizing constant, such that $p(\boldsymbol{d}|\boldsymbol{I}) = \int p(\boldsymbol{m}|\boldsymbol{I})p(\boldsymbol{d}|\boldsymbol{m},\boldsymbol{I})d\boldsymbol{m}$. The posterior $p(\boldsymbol{m}|\boldsymbol{d},\boldsymbol{I})$ is the joint probability distribution of model parameters of interest that we want to infer conditioned upon the measured data and an assumed model/hypothesis. According to the Bayesian framework (Equation 13.9), posterior probability density function represents the solution of the inverse problem and the denominator term is the evidence, a normalizing constant for the posterior. However, the integral operation to calculate the denominator term, i.e., normalizing constant $p(\boldsymbol{d}|\boldsymbol{I})$, over the entire model parameter space is computationally intensive. In the absence of computational resources to compute the normalizing constant, the posterior distribution $p(\boldsymbol{m} \mid \boldsymbol{d},\mathbf{I})$ of model parameter $\boldsymbol{m}$ can be considered to be proportional to likelihood function multiplied by the prior distribution. Bayes' theorem guarantees that MCMC sampling of the model space with the joint information given prior and likelihood can generate samples from a distribution that approximates the posterior distribution. Consequently, the use of MCMC sampling algorithms make it possible to obtain the posterior distribution of model parameter $\boldsymbol{m}$ without calculating the normalizing constant.

Prior models:

The prior distributions $p(\boldsymbol{m}|\boldsymbol{I})$ of petrophysical parameters of the clay-pyrite interfacial-polarization model can be obtained from prior knowledge of the formation under investigation. For example, the prior distribution of water saturation can be determined either from core-scale Dean-Stark and retort core measurements or log-scale EM logs, such as induction resistivity or dielectric dispersion logs interpretation [3,4,5] or reservoir-scale seismic interpretation [25]. The thermal neutron capture cross-sectional (sigma) subsurface measurement helps in determining the prior distribution of water conductivity or salinity [17]. The prior distribution of surface conductance of clay can be related to the cation exchange capacity (CEC) of clay and the prior distribution of radius of clay and pyrite can be determined from scanning electron microscope (SEM) or energy dispersive spectroscopy (EDS) images of rocks from the formation. However, the prior distribution of petrophysical parameters obtained from prior knowledge can be subjective and site specific. In this study, we assume each petrophysical parameter is independent of others, and the prior distributions are noninformative uniform distribution over the prior parameter ranges.

Likelihood function:

The likelihood function expresses the possibility of measuring a certain data vector $\boldsymbol{d}$ for a given set of model parameters $\boldsymbol{m}$ and prior information $\boldsymbol{I}$. Likelihood function quantifies how well a model $\boldsymbol{F}$ and the model parameter $\boldsymbol{m}$ can reproduced the measured/observed data $\boldsymbol{d}$. It can be expressed as a curve fitting function in the form of a normal distribution [26] expressed as

$$p(\boldsymbol{d}|\boldsymbol{m},\boldsymbol{I}) = \left[(2\pi)^N|\boldsymbol{C}_d|\right]^{-1/2} e^{\left[-\frac{1}{2}\left(\frac{\boldsymbol{d}^{meas}-\boldsymbol{d}(\boldsymbol{m})}{\boldsymbol{\sigma}}\right)^T \boldsymbol{C}_d^{-1}\left(\frac{\boldsymbol{d}^{meas}-\boldsymbol{d}(\boldsymbol{m})}{\boldsymbol{\sigma}}\right)\right]} \tag{13.10}$$

where $\boldsymbol{d}$ is the measured data vector that comprises the broadband EM dispersion logs (14 logs measured at 7 frequencies), $\boldsymbol{F}(\boldsymbol{m})$ is the modeled data vector computed using the mechanistic clay-pyrite IP model ($\boldsymbol{F}$) based on model parameter $\boldsymbol{m}$, $\boldsymbol{C}_d$ is the data covariance matrix that is a diagonal matrix with entries equal to the estimated data error variances for each of the broadband permittivity and conductivity logs, $\boldsymbol{\sigma}$ is the standard deviation vector for the broadband permittivity and conductivity logs, and N is the number of measurements ($N = 14$ for this study). There are 14 elements each in $\boldsymbol{F}(\boldsymbol{m})$ and $\boldsymbol{d}$. The normal distribution likelihood function assumes that the data error is Gaussian without any outlier. The likelihood function should be defined in the form of Laplace distribution with a broader tail if the data errors are expected to be non-Gaussian with some outliers [23]. For purposes of our study, $\boldsymbol{d}$ is the measured data vector comprising relative permittivity and conductivity measured at seven EM log-acquisition frequencies. The clay-pyrite interfacial-polarization model $\boldsymbol{F}$ is used to generate the modeled data vectors $\boldsymbol{Perm}(\boldsymbol{m})$ and $\boldsymbol{Cond}(\boldsymbol{m})$, comprising clay-pyrite interfacial-polarization model predictions of relative permittivity and conductivity, respectively, at the seven distinct log-acquisition frequencies, namely, $\omega_1, \omega_2, \ldots, \omega_7$, for a specific model parameter vector $\boldsymbol{m}$. The modeled data vectors can be expressed as

$$\boldsymbol{Perm}(\boldsymbol{m}) = [Perm(\boldsymbol{m}, \omega_1), Perm(\boldsymbol{m}, \omega_2), \ldots Perm(\boldsymbol{m}, \omega_7)], \tag{13.11}$$

and

$$\boldsymbol{Cond}(\boldsymbol{m}) = [Cond(\boldsymbol{m}, \omega_1), Cond(\boldsymbol{m}, \omega_2), \ldots Cond(\boldsymbol{m}, \omega_7)], \tag{13.12}$$

where $Perm(\boldsymbol{m}, \omega_i)$ and $Cond(\boldsymbol{m}, \omega_i)$ represent the modeled relative permittivity and conductivity calculated at specific log-acquisition angular frequency ω_i for a specific model parameter vector $\boldsymbol{m}$. The forward model predictions $\boldsymbol{F}(\boldsymbol{m})$ is thus expressed as a combination of the seven modeled relative permittivity $\boldsymbol{Perm}(\boldsymbol{m})$ and seven modeled conductivity measurements $\boldsymbol{Cond}(\boldsymbol{m})$ as

$$\boldsymbol{F}(\boldsymbol{m}) = [\boldsymbol{Perm}(\boldsymbol{m}), \boldsymbol{Cond}(\boldsymbol{m})]^T \tag{13.13}$$

2.2.2 *Metropolis-Hastings sampling algorithms*

Markov chain Monte Carlo (MCMC) helps in probabilistically solving an inverse problem (parameter estimation problem) by generating independent samples from a distribution approximating a target posterior distribution from which direct sampling is difficult. Target posterior distribution is a conditional probability density function (PDF) in the model parameter space describing the ability of different sets of model parameters to fit the measured data (likelihood). Various MCMC sampling algorithms have been used in Bayesian inference problems [27,28]. MCMC algorithms that implement transition kernels (proposal probability) and acceptance probabilities are called Metropolis-Hastings (MH) algorithm. The transition kernel defines a possible move of the Markov chain from the current model (state) $\boldsymbol{m}$ to a proposed model $\boldsymbol{m}'$. The proposed move of the Markov chain is accepted or rejected based on the acceptance probability. In other words, Metropolis-Hastings (MH) algorithm selects samples from a proposal distribution and then retains or ignores the selected samples according to an acceptance rule. Markov-chain Monte Carlo sampling through the Metropolis-Hastings algorithm demands a properly tuned choice of proposal distribution in order to achieve good efficiency. As more samples are drawn, MH algorithm produces distribution of values that more closely approximates the desired probability distribution. The MH sampling algorithm finds model parameters $\boldsymbol{m}$ from posterior distribution $p(\boldsymbol{m}\,|\,\boldsymbol{d},\boldsymbol{I})$ without explicitly calculating the posterior distribution. We apply the two-step Metropolis-Hastings sampling algorithm [29,30] to obtain a sequence of random samples to generate a desired Markov chain that can approximate the target posterior distribution. Step 1 uses a symmetric proposal distribution to propose a candidate next sample given the current sample, and Step 2 determines whether to accept or reject the candidate next sample based on an acceptance probability. In Step 1, a new model parameter vector $\boldsymbol{m}'$ is drawn from a proposal distribution, $q(\boldsymbol{m}'|\boldsymbol{m})$, given the current model parameter vector $\boldsymbol{m}$ in the Markov chain. In Step 2, the proposed model parameter $\boldsymbol{m}'$ is either accepted or rejected according to an acceptance probability [26]. A usual choice for the proposal distribution is a symmetric Gaussian distribution centered at the current sample, so that points closer to current sample are more likely to be the next samples. MH algorithm based MCMC sampling proceeds by randomly attempting to move (or halt) about the sample space with an initial tendency to generate correlated samples, which incorrectly reflects the target distribution. Such a sampling eventually converges to the desired target distribution; however, the initial samples tends to follow a very different distribution, especially when the starting points are in low-density regions. A burn-in period, involving throwing away of initial thousand samples, is required to reliably approximate the target posterior distribution. The sequence of model parameters thus obtained can then be used to approximate the posterior distribution. Removing samples from the burn-in period reduces the dependence on starting value by ensuring the stabilization of Markov chain. Autocorrelations due to Markov chain can be reduced through thinning.

Proposal distribution:

The efficiency of MCMC sampling algorithms strongly depends on the choice of proposal distribution $q(\boldsymbol{m}'|\boldsymbol{m})$. Computational costs increase with poor selection of the proposal distribution. When the proposed model parameters are too exploratory, most of the proposed samples for the next step are rejected for falling in the low probability regions. However, with limited exploration, the sampling cannot escape local minima because several accepted proposed samples are not far from the previous high probability regions [26]. Typically, a new model parameter $\boldsymbol{m}'$ is proposed based on multivariate Gaussian distribution that has the current model parameter $\boldsymbol{m}$ as its mean. The proposal distribution [26] is then expressed as

$$q(\boldsymbol{m}'|\boldsymbol{m}) = \left[(2\pi)^k |\boldsymbol{C}_k|\right]^{-1/2} e^{\left[-\frac{1}{2}\boldsymbol{m}^T \boldsymbol{C}_k^{-1} \boldsymbol{m}\right]} \tag{13.14}$$

where k is the dimension of model parameter space (i.e., number of unknown petrophysical parameters) and $\boldsymbol{C}_k$ is proposal covariance matrix, defined as the linearized estimation of the posterior model covariance [31,32] formulated as

$$\boldsymbol{C}_k \approx \left[\boldsymbol{J}^T \boldsymbol{C}_d^{-1} \boldsymbol{J}\right]^{-1} \tag{13.15}$$

where $\boldsymbol{J}$ is the Jacobian matrix of data misfit vector $[\boldsymbol{d} - \boldsymbol{F}(\boldsymbol{m})]$. In doing so, it is ensured that the unknown petrophysical parameters with high sensitivity have narrower search ranges and unknown petrophysical parameters with low sensitivity have wider search ranges over the entire model parameter space [26]. In our study, the number of unknown petrophysical parameters k is 3 for field application cases and 5 or 6 for synthetic cases. Moreover, k can be increased with increase in the number of frequencies at which conductivity and permittivity can be measured (in our cases, we have a total of 14 logs measured at 7 frequencies). In this work, we assume proposal distribution is multivariate Gaussian distribution, which is symmetric $\boldsymbol{q}(\mathbf{m} \mid \boldsymbol{m}') = \mathbf{q}(\boldsymbol{m}' \mid \mathbf{m})$.

Acceptance probability:

A starting model parameter vector $\boldsymbol{m}_0$ for a Markov chain is updated to the current model parameter $\boldsymbol{m}$ by Metropolis-Hastings sampling. At each step a new candidate sample $\boldsymbol{m}'$ is drawn from the proposal distribution, which can be referred as $q(\boldsymbol{m}'|\boldsymbol{m})$, the candidate sample is either accepted or rejected according to acceptance probability [33]. The acceptance probability [34] can be defined as

$$\alpha(\boldsymbol{m}|\boldsymbol{m}') = \min\left[1, \frac{p(\boldsymbol{m}'|\boldsymbol{d},\boldsymbol{I})\, q(\boldsymbol{m}|\boldsymbol{m}')}{p(\boldsymbol{m}|\boldsymbol{d},\boldsymbol{I})\, q(\boldsymbol{m}'|\boldsymbol{m})}\right] \tag{13.16}$$

This definition of acceptance probability α enables samples in the Markov chain to lie in higher probability regions of the model space while escaping local minima. The proposed model has higher chance to be accepted if the proposed model has higher posterior probability than current model [26]. The Markov chain always moves from $\boldsymbol{m}$ to the drawn candidate $\boldsymbol{m}'$ if the model acceptance probability $\boldsymbol{\alpha}(\mathbf{m} \mid \boldsymbol{m}')$ is higher than 1. When the model acceptance probability $\alpha(\boldsymbol{m} \mid \boldsymbol{m}')$ is smaller than 1, a random sample u is drawn from a uniform density between 0 and 1, and the drawn sample $\boldsymbol{m}'$ is accepted if $\alpha(\boldsymbol{m} \mid \boldsymbol{m}')$ is greater than u; or else it is rejected. In this work, we assume proposal distribution is multivariate Gaussian distribution, which is symmetric $q(\boldsymbol{m} \mid \boldsymbol{m}') = q(\boldsymbol{m}' \mid \boldsymbol{m})$. In doing so the acceptance probability reduces to the ratio of likelihood functions for the proposed model and current model [33].

2.2.3 Convergence monitoring

We run three Markov chains by starting from different sets of initial values covering the default range of each unknown petrophysical parameter, and the total number of iterations is T. The drawn samples in the first half of iterations ($0.5T$) depend on initial values, so the first half of iterations for each Markov chain are thrown away as they are in the burn-in period [23]. We calculate the scale-reduction score based on variance of each Markov chain sequence and the variance between three independent Markov chain sequence. The scale reduction score is used to monitor the convergence of three independent Markov chains to target distribution [35]. Markov chains are considered to be converged if the scale reduction score for each estimated parameter is lower than 1.2. The default maximum number of iterations in the proposed application of stochastic inversion-based log interpretation method for the shale gas formation under investigation is 50,000.

2.3 Limitations and assumptions in the proposed inversion-based interpretation method

The clay-pyrite interfacial-polarization model assumes the volume of formation under investigation is homogeneous and isotropic without Maxwell-Wagner polarization effects [24]. In addition, the proposed MCMC-based stochastic inversion assumes that the volumes of formation detected by EM induction resistivity, EM propagation, and dielectric dispersion logging tools at four frequencies have similar petrophysical properties and are free from laminations and fractures. In our study, MCMC-based inversion is applied to depth intervals with negligible mud-filtrate invasion. The clay-pyrite interfacial-polarization model, similar to certain log-scale or core-scale EM interpretation models such as CRI and SMD model, only considers total porosity. Effective porosity, pore-network tortuosity, isolated pores, and connectivity of clay and pyrite grains are not considered in our approach. The information of pore-network or clay- and pyrite-network can be obtained from pore-scale images of the rock or formation under investigation. Moreover, the advanced log-scale or core-scale EM interpretation model needs an upscaling method that includes the pore-scale network information for purposes of log-scale or core-scale modeling, which is beyond the scope of this chapter. We also assume that some petrophysical parameters such as cementation exponent and saturation exponent are constant in continuous depth intervals for the new inversion-based interpretation method, which should be different when lithology changes or there is presence of lamination and fractures.

3 Results and discussion

3.1 Application of the MCMC-based stochastic inversion to synthetic layers

For purposes of testing the newly developed MCMC-based stochastic inversion scheme coupled with the clay-pyrite interfacial-polarization model, we process synthetic EM broadband dispersion logs generated for two distinct synthetic layers very similar to the predominant clay- and pyrite-rich zones in the shale gas formation under investigation. Synthetic Layer 1 identifies a clay rich ($V_c = 0.5$), low porosity ($\varphi_t = 0.08$), and water-bearing ($S_w = 0.9$) formation. Synthetic Layer 2 identifies a clay rich ($V_c = 0.55$), pyrite rich ($V_i = 0.03$), low porosity ($\varphi_t = 0.09$), and hydrocarbon-bearing ($S_w = 0.5$) formation. Table 13.1 lists the petrophysical parameters assumed for the two synthetic clay- and pyrite-rich layers. The synthetic EM broadband dispersion logs are the modeled responses for the two synthetic layers computed using the clay-pyrite interfacial-polarization model for the assumed petrophysical parameters listed in Table 13.1, and 5% Gaussian noise is added to the synthetic EM broadband dispersion logs.

The high volume fraction of clay minerals, low porosity, and presence of conductive pyrite significantly influence the relative permittivity and conductivity of the formation under investigation [6,21,36,37]. Fig. 13.2 shows the sensitivity of relative permittivity and conductivity computed using the clay-pyrite interfacial-polarization model to volumetric fraction of pyrite (V_i) and volumetric fraction of clay (V_c). We use the clay- and pyrite-rich Synthetic Layer 2 for the sensitivity analysis, such that V_i varies from 0 to 0.1 for V_c of 0.35, 0.45, and 0.55, respectively. The volumetric fraction of sand (nonconductive grains) changes with the alterations in clay and pyrite volumetric fractions, while other petrophysical parameters, including total porosity and bulk conductivity of water, are kept constant. Fig. 13.2A and B shows the variations in relative permittivity and conductivity, respectively, with the variations in clay and pyrite volumetric fractions at 1 MHz, which is close to the log-acquisition frequencies of logging-while-drilling (LWD) EM propagation tool. At the volumetric fraction of clay of 0.55 and frequency of 1 MHz, the relative permittivity increases

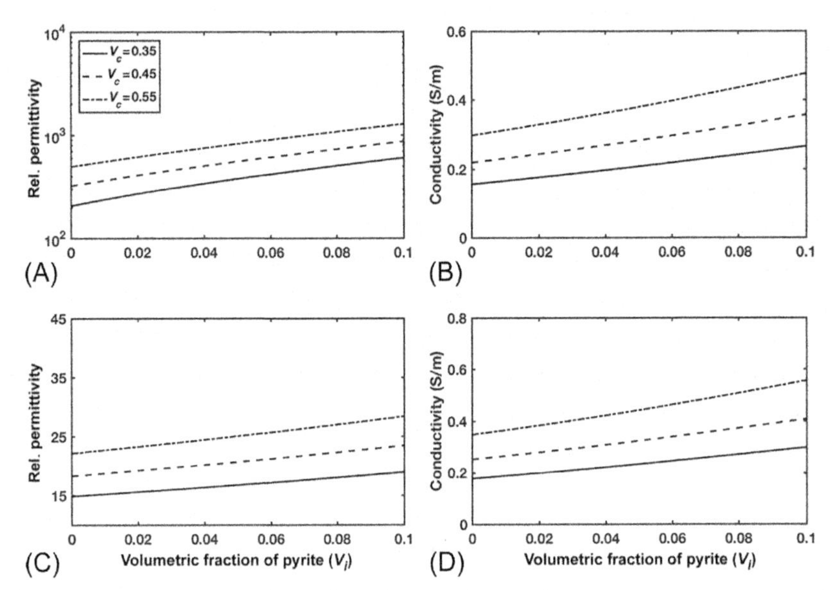

FIG. 13.2 Sensitivity of the (A and C) relative permittivity and (B and D) conductivity computed using the clay-pyrite interfacial-polarization model to volumetric fraction of pyrite (V_i) and volumetric fraction of clay at 1 MHz and at 100 MHz, respectively.

from 500 to 1250, and conductivity increases from 0.3 to 0.48 S/m with the increase in volumetric fraction of pyrite from 0 to 0.1. The relative permittivity equals to 670 and conductivity equals to 0.35 S/m at volumetric fraction of pyrite equaling to 0.03, which is equal to that in the Synthetic Layer 2. Fig. 13.2C and D shows the variations in relative permittivity and conductivity, respectively, with the variations in clay and pyrite volumetric fractions at 100 MHz, which is close to the log-acquisition frequencies of dielectric dispersion logging tool. When the volumetric fraction of clay equals to 0.55, the relative permittivity changes from 22.5 to 28.6, and conductivity changes from 0.35 to 0.56 S/m when increasing volume fraction of pyrite from 0 to 0.1. Relative permittivity equals 24.1, and conductivity equals 0.4 S/m at volumetric fraction of pyrite of 0.03 and frequency of 100 MHz. These results illustrate that volumetric fractions of clay and pyrite minerals strongly influence the relative permittivity and conductivity because of the frequency-dependent IP effects of clay and pyrite.

Table 13.2 shows the prior ranges and initial values of petrophysical parameters, which are inverted in the synthetic cases used in our study. The prior ranges of petrophysical parameters provide the range or boundary of each parameter in the model space. In Synthetic Layer 1, five petrophysical parameters are estimated using the proposed inversion scheme, whereas six petrophysical parameters are estimated for the Synthetic Layer 2. Figs. 13.3 and 13.5

TABLE 13.2 The default prior ranges and initial values of unknown petrophysical parameters assumed for the 3 Markov chains of the proposed stochastic inversion methodology

Model parameters	Prior ranges	Initial-1	Initial-2	Initial-3
S_w	(0.01,1)	0.4	0.6	0.8
C_w (S/m)	(0.01,20)	0.75	0.25	2.5
r_c (μm)	(0.1,10)	2	0.5	1.5
λ_c (S)	$(1 \times 10^{-9}, 1 \times 10^{-6})$	2×10^{-7}	1×10^{-8}	5×10^{-8}
C_{sh} (S/m)	(0.25,1)	0.25	0.5	0.75
r_i (μm)	(1100)	25	1	5

show the petrophysical parameters estimated at each iteration of the three Markov chains when inverting the synthetic broadband frequency EM measurements containing 5% Gaussian noise. The scale reduction score method applied to monitor convergence requires at least two Markov chains, and we run three Markov chains, similar to Chen et al. [23] for inverting spectral

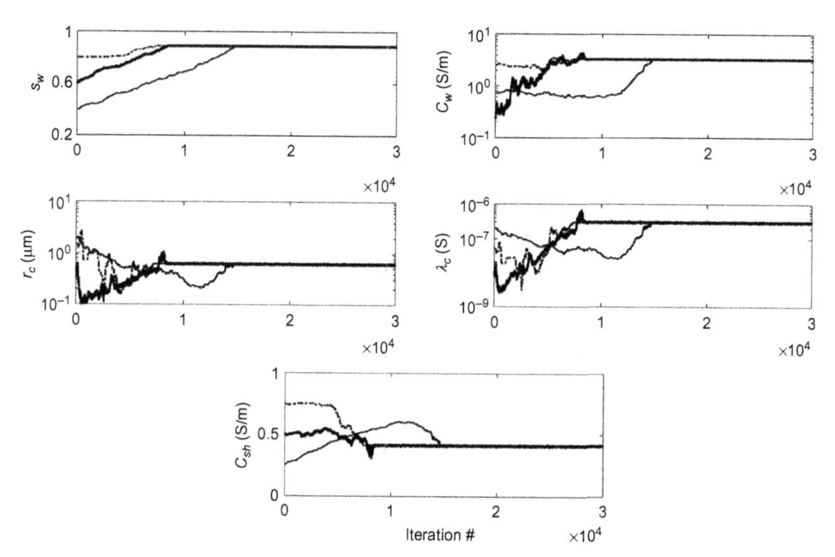

FIG. 13.3 Inversion-derived estimates of 5 unknown petrophysical parameters, namely, S_w, C_w, r_c, λ_c, and C_{sh}, for the Synthetic Layer 1 during the inversion of broadband frequency EM measurements containing 5% Gaussian noise for three Markov chains based on the Metropolis-Hastings sampling. The iteration processes of all parameters converge to stable value after 15,000 iterations.

induced-polarization data for Cole-Cole model parameters. The three Markov chains from very different sets of start points can help to detect possible local convergence [23]. The inversion process converges after 15,000 iterations for the Synthetic Layer 1, while it only takes 5000 iterations for the Synthetic Layer 2. Figs. 13.4 and 13.6 show the histograms of the estimated petrophysical parameters for the Synthetic Layers 1 and 2, respectively. All the samples in the second half of each Markov chain are drawn to generate the histograms in Figs. 13.4 and 13.5 for Synthetic Layers 1 and 2, respectively. Table 13.3 shows the real values and estimated 95% highest probability domains (HPD) for each petrophysical parameter. The estimated 95% HPD (Fig. 13.6) for all parameters are close to the real values, and the estimated water saturation, S_w, has only around 1% relative error in Synthetic Layer 1 and around 2.5% relative error for Synthetic Layer 2.

3.2 Application of the MCMC-based stochastic inversion to process broadband EM dispersion logs acquired in a shale gas formation

The well under investigation was drilled in an organic-rich shale gas formation. Oil-based mud was used to reduce the invasion of drilling fluid into formation, and the mud-filtrate invasion was negligible. EM induction resistivity, EM propagation, and dielectric dispersion logging tools were deployed over a depth interval of 1500 m in this well [2]. In doing so, for the very first time, continuous broadband EM dispersion measurements (comprising effective relative

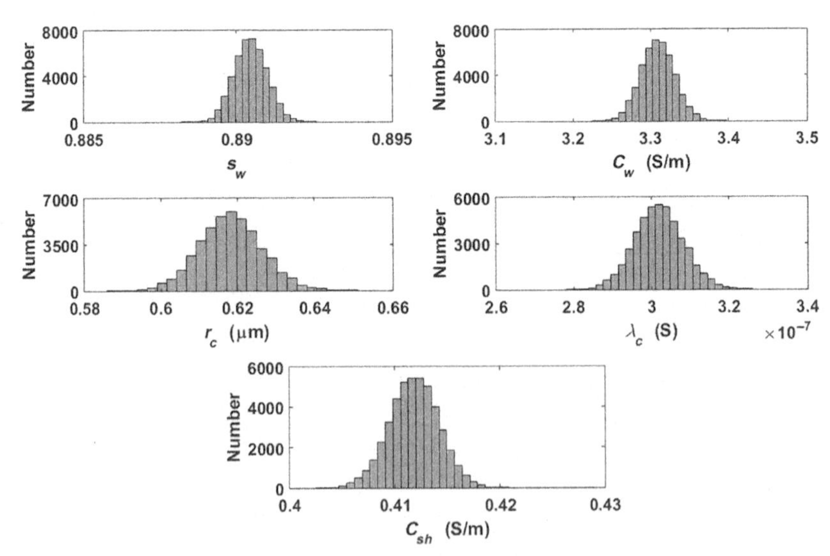

FIG. 13.4 Histograms of the five estimated petrophysical parameters, namely S_w, C_w, r_c, λ_c, and C_{sh}, for the Synthetic Layer 1. All the samples in the second half of each Markov chain are used to generate the histograms.

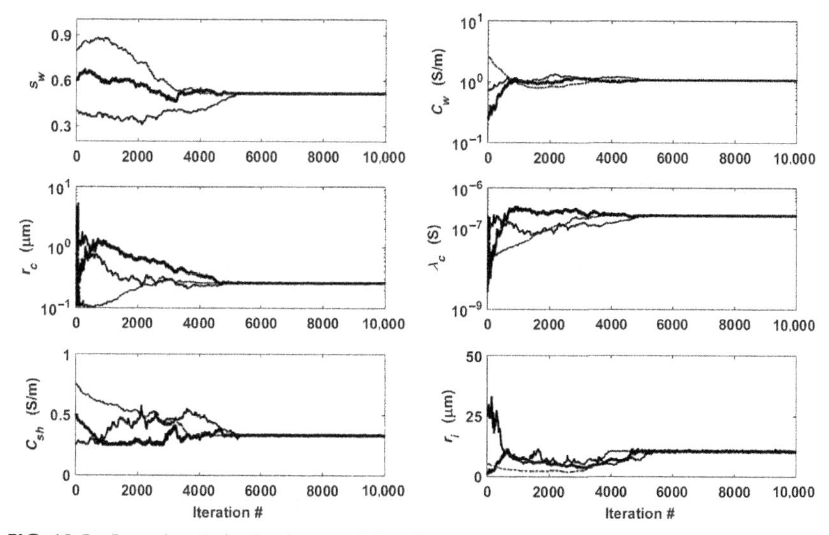

FIG. 13.5 Inversion-derived estimates of six unknown petrophysical parameters, namely, S_w, C_w, r_c, λ_c, C_{sh}, and r_i, for Synthetic Layer 2 during the inversion of broadband frequency EM measurements containing 5% Gaussian noise for three Markov chains based on the Metropolis-Hastings sampling. The iteration processes of all parameters converge to stable value after 5000 iterations.

TABLE 13.3 Estimated 95% HPD for the inversion-derived petrophysical parameters of Synthetic Layer 1 and Synthetic Layer 2

Model parameters	Layer 1 true values	95% HPD	Layer 2 true values	95% HPD
S_w	0.9	(0.889,0.892)	0.5	(0.511,0.516)
C_w (S/m)	3.5	(3.267,3.351)	1	(1.056,1.065)
r_c (μm)	0.6	(0.603,0.635)	0.25	(0.254,0.260)
λ_c (S)	3×10^{-7}	$(2.898 \times 10^{-7}, 3.149 \times 10^{-7})$	2×10^{-7}	$(2.076 \times 10^{-7}, 2.136 \times 10^{-7})$
C_{sh} (S/m)	0.4	(0.407,0.417)	0.35	(0.327,0.338)
r_i (μm)			10	(10.159,10.880)

permittivity and effective conductivity) at seven EM log-acquisition frequencies ranging from kHz to GHz were acquired for purposes of robust water saturation estimation. The newly developed stochastic inversion processed the broadband EM dispersion logs (7 effective relative permittivity and 7 effective conductivity) acquired in the shale gas well. In doing so, three unknown

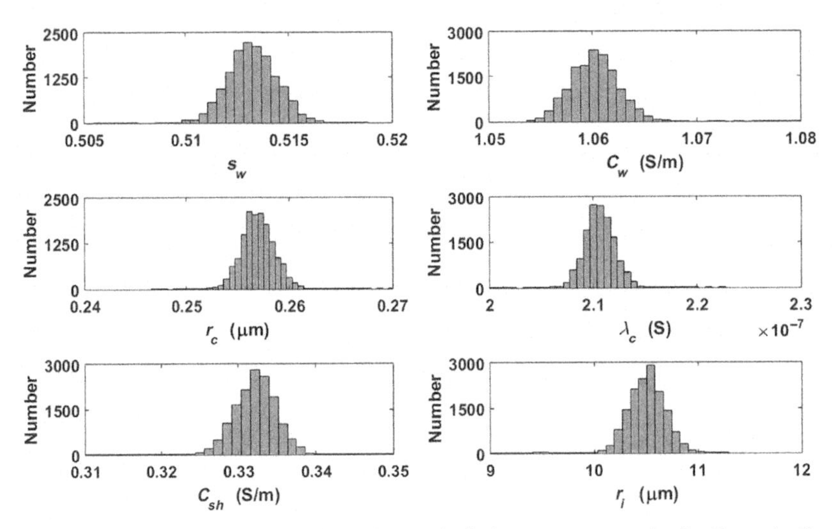

FIG. 13.6 Histograms of the six estimated petrophysical parameters, namely, S_w, C_w, r_c, λ_c, C_{sh}, and r_i, for the Synthetic Layer 2. All the samples in the second half of each Markov chain are used to generate the histograms.

petrophysical parameters, namely, water saturation S_w, water conductivity C_w, and surface conductance of clay λ_c, were continuously estimated along certain depth intervals. We assume that the average radius of clay r_c is 2 μm, average radius of pyrite grains r_i is 10 μm, and low-frequency conductivity of 100% clay-bearing zone C_{sh} is 0.25 S/m, which are typical values for organic-rich shale gas formations in that region. By assuming r_i, r_c, and C_{sh} to be constant, we mitigate the nonuniqueness problems of inversion results. The assumed values of tortuosity, cementation exponent, and saturation exponent are the same as those in Table 13.1 for the two synthetic layers, which are typical values for shaly sand [11].

For assessing the performance of the stochastic inversion, we use data misfit between clay-dispersion model predictions for the inversion-derived estimates and the measured broadband EM dispersion logs. Data misfit at two specific depths are elaborated in the next paragraph. The first chosen depth is X066 m, which is clay-rich and pyrite-bearing zone with low total porosity ($\varphi_t = 0.05$). At depth X066 m, the volumetric fraction of clay V_c is 0.47 and that of pyrite V_i is 0.01. The second chosen depth is X151 m, which is clay-rich ($V_c = 0.57$) with higher total porosity ($\varphi_t = 0.09$) and no pyrite content. Fig. 13.7 shows histograms of inversion-derived estimates of 3 petrophysical parameters, namely, S_w, C_w, and λ_c, for the clay-rich and pyrite-bearing depth at X066 m. All the samples in the second half of each Markov chain are drawn to generate those histograms. Figs. 13.8 and 13.9 show the comparisons of the modeled relative permittivity and conductivity dispersions computed using the

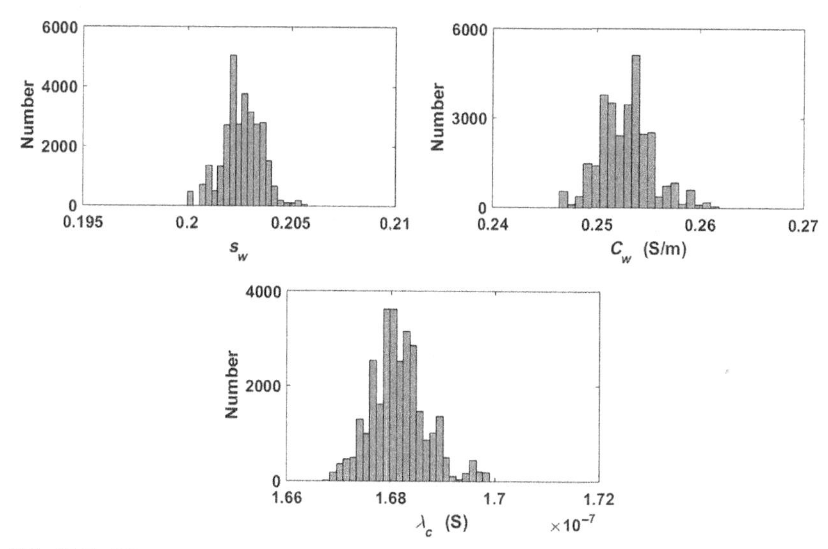

FIG. 13.7 Histograms of estimated petrophysical parameters, S_w, C_w, and λ_c in the clay-rich and pyrite-bearing depth of X066 m.

clay-pyrite interfacial-polarization model based on the inversion-derived estimates against the measured broadband relative permittivity and conductivity logs at depth X066 m and depth X151 m, respectively. The average relative errors between measured and modeled relative permittivity and conductivity are 0.25 for the depth X066 m (Fig. 13.8) and 0.13 for the depth X151 m (Fig. 13.9). Fig. 13.8 indicates that the modeled and measured relative permittivity and conductivity at depth X066 m exhibit good agreement except for relative permittivity at the EM propagation log-acquisition frequencies (1 and 2 MHz). On the other hand, at depth X151 m, the modeled and measured relative permittivity and conductivity show good agreement over the entire frequency range. Table 13.4 lists the estimated 95% HPD and medians of S_w, C_w, and λ_c. The median of estimated s_w for depth X066 m is 0.203 indicating hydrocarbon-bearing zone, and that for the depth X151 m is 1 indicating water-bearing zone. Both the depths have similar surface conductance of clay and brine conductivity of X151 m is higher than that of X066 m.

SMD model can process dielectric dispersion logs acquired in clean sandstone formation for frequencies 20 MHz–1 GHz. Unlike SMD model, the clay-pyrite interfacial-polarization model (PS model) considers the interfacial polarization arising from grains with surface charges or those that have conductive/semimetallic properties. Consequently, the clay-pyrite interfacial-polarization model is more suitable for shales containing uniformly distributed clay and pyrite grains as compared with the SMD model. For demonstrating the robustness and physical effectiveness of the clay-pyrite interfacial-polarization

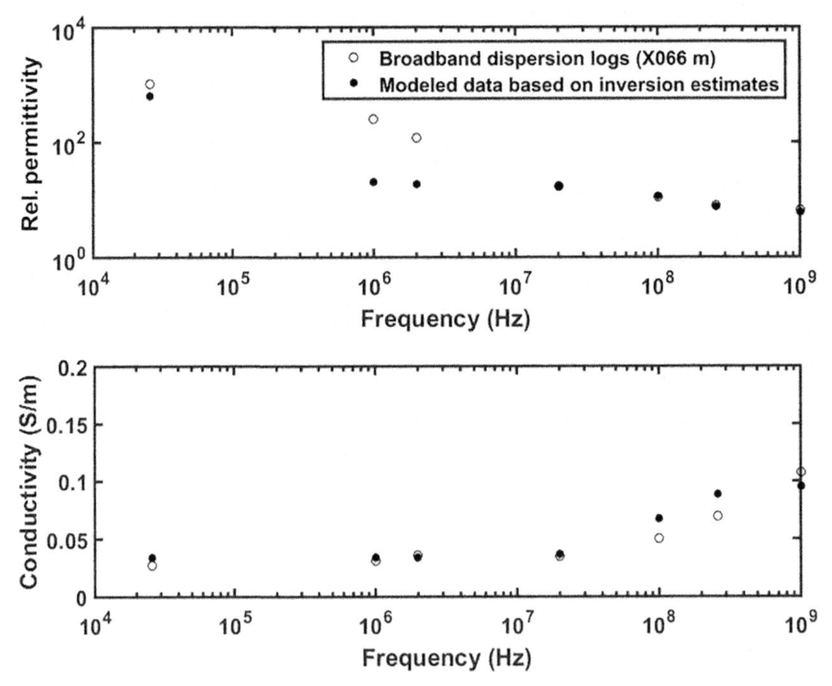

FIG. 13.8 Comparisons of the modeled and measured relative permittivity and conductivity at depth X066 m. Model predictions were generated using the clay-pyrite interfacial-polarization model based on the inversion-derived estimates of S_w, C_w, and λ_c.

model, we compare modeled and measured relative permittivity and conductivity dispersions at depth X070 m for the dielectric-dispersion log-acquisition frequencies in Fig. 13.10. The depth X070 m is clay-rich ($V_c = 0.36$) with low total porosity ($\varphi_t = 0.04$). Conductivity and permittivity data modeled using the clay-pyrite interfacial-polarization model (PS model) and the MCMC-based stochastic inversion-derived estimates are compared with those modeled using the joint SMD and LR model and the Levenberg-Marquardt-based deterministic inversion-derived estimates. More description of the deterministic inversion-based interpretation method coupling the joint SMD and LR model is in previous chapter [3]. The data misfit in terms of average relative error is 0.10 when the modeled relative permittivity and conductivity are computed using the inversion-derived estimates obtained with the clay-pyrite interfacial-polarization model as the forward model coupled to the stochastic inversion scheme, whereas the average relative error is 0.21 when the modeled relative permittivity and conductivity are computed using the deterministic inversion-derived estimates obtained with the joint SMD and LR model as the forward model. Implementation of clay-pyrite interfacial-polarization model reduces the average relative error in fitting the dielectric dispersion logs at depth X070 m by 52%. The median

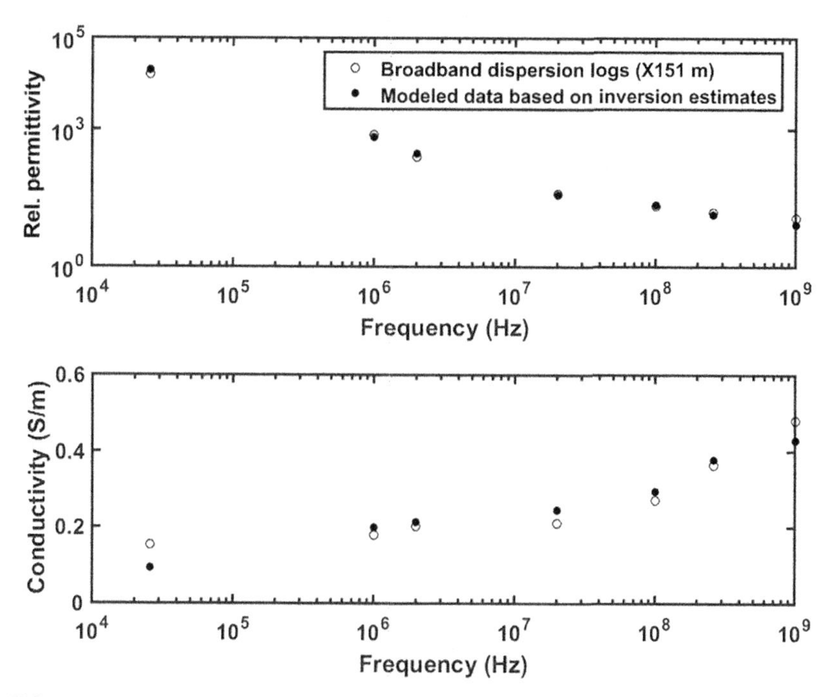

FIG. 13.9 Comparisons of the modeled and measured relative permittivity and conductivity at depth X151 m. Model predictions were generated using the clay-pyrite interfacial-polarization model based on the inversion-derived estimates of S_w, C_w, and λ_c.

value of the S_w estimates obtained when using the stochastic inversion coupled with the clay-pyrite interfacial-polarization model is around 0.39, which indicates the presence of hydrocarbon. On the other hand, the S_w estimated computed using the deterministic inversion coupled with the joint SMD and LR model is 1, which indicates water-filled zone. The results indicate that we would overestimate S_w by using the conventional EM logs interpretation models when we ignore the clay and pyrite IP effects, which is also reported in the publication by Han and Misra [3] for dielectric dispersion logs interpretation in Bakken petroleum system.

3.3 Petrophysical interpretation and log analysis of the inversion-derived estimates in the shale gas formation

The newly proposed stochastic inversion continuously processes the broadband EM dispersion logs acquired in the shale gas well for two depth intervals, namely, Formation Zone 1 (X058 m–X073 m) and Formation Zone 2 (X545 m–X555 m), which are clay- and pyrite-rich, homogenous, and isotropic zones. Continuous estimations of S_w, C_w, and λ_c were obtained across the two

TABLE 13.4 Estimated 95% HPD and medians of petrophysical parameters, namely, S_w, C_w, and λ_c, in the clay-rich and pyrite-bearing depth of X066 m and in the clay-rich depth of X151 m

Model parameters	Depth X066 95% HPD	Depth X066 medians	Depth X151 95% HPD	Depth X151 medians
S_w	(0.201,0.204)	0.203	(1,1)	1
C_w (S/m)	(0.249,0.259)	0.253	(0.408,0.412)	0.410
λ_c (S)	$(1.671 \times 10^{-7}, 1.691 \times 10^{-7})$	1.681×10^{-7}	$(5.888 \times 10^{-8}, 5.953 \times 10^{-8})$	5.924×10^{-8}

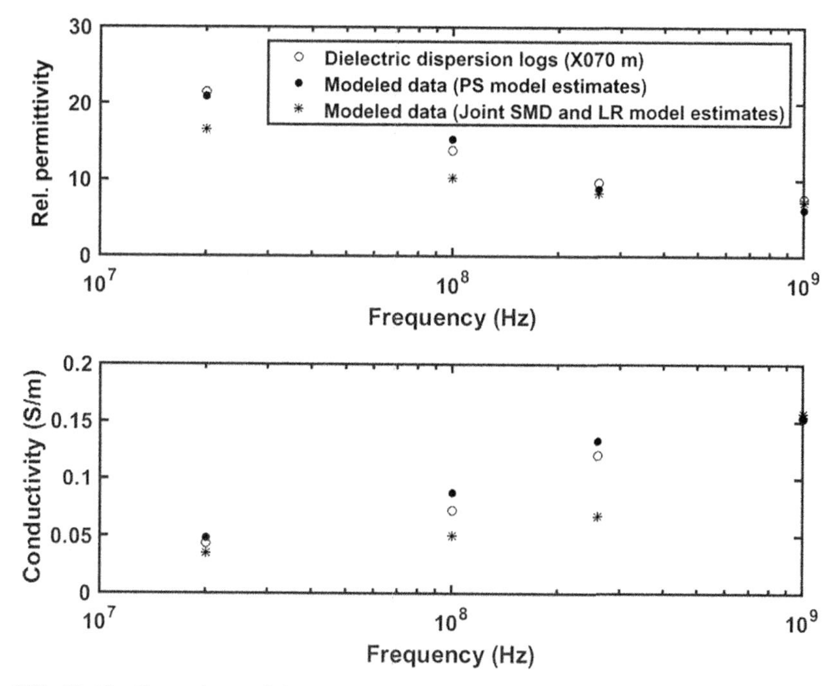

FIG. 13.10 Comparisons of the measured and modeled relative permittivity and conductivity based on the stochastic inversion coupled with clay-pyrite interfacial-polarization model-based inversion-derived estimates with those based on the deterministic inversion coupled with joint SMD and LR model-based inversion-derived estimates.

intervals of interest in the shale gas formation. The proposed method assumes the formation to be homogeneous, isotropic, and rich in clay and conductive minerals. The EM logs being jointly processed in the proposed methodology are acquired at different frequencies and are sensitive to different formation volumes due to varying depths of investigation. Figs. 13.11 and 13.12 show the logs and inversion-derived estimates for the Formation Zone 1 (X058 m–X073 m) and Zone 2 (X545 m–X555 m), respectively. The layout of logs is the same for Figs. 13.11 and 13.12. The high gamma-ray values in both Zone 1 and Zone 2 are due to the high volumetric content of clay and presence of kerogen, which are around 100–120 gAPI. The two intervals in shale gas formation exhibit high resistivity indicated by the laterologs (>100 Ω m). Negligible fluid invasion is confirmed by the multiple DOI laterolog resistivity measurements (RLA1–RLA5) presented in Track 3. Caliper measurement in Track 1 indicates that the borehole is regular along depth in the two zones of interest. Therefore, we can assume that the formation volumes sensed by EM induction, EM propagation, and dielectric dispersion logs have similar properties in the absence of bed boundaries or thin bed layers. EM induction senses the largest formation volume. With the increase in frequency and the reduction in

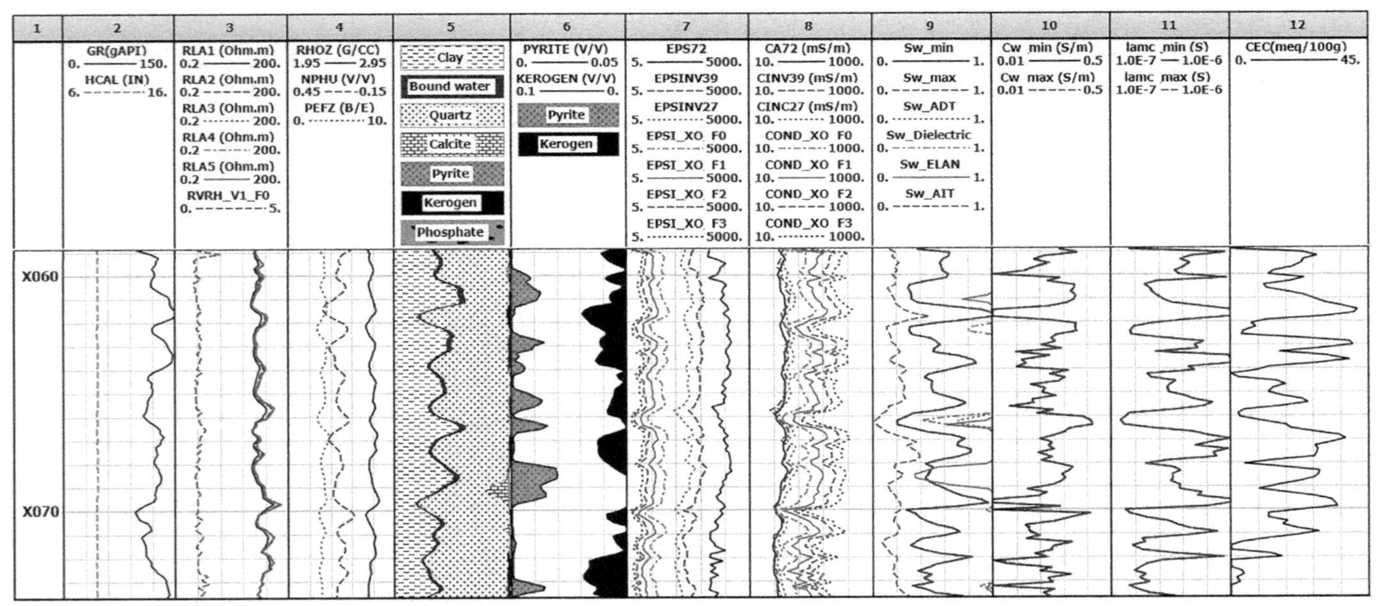

FIG. 13.11 Logs for Zone 1 (X058 m–X073 m). Track 1 is depth, Track 2 contains gamma-ray (GR) and caliper (HCAL), Track 3 contains laterolog resistivity (RLA), Track 4 contains neutron porosity (NPHU), bulk density (RHOZ), and photoelectric factor (PEFZ), Track 5 contains volumetric fractions of various minerals: quartz, clay, kerogen, pyrite, calcite, and phosphate, Track 6 contains volumetric fractions of pyrite and kerogen, Track 7 contains relative permittivity logs at 7 frequencies, Track 8 contains conductivity logs at 7 frequencies, Track 9 contains water saturation (S_w) estimated using various methods, Track 10 contains inversion-derived 95% HPD of water conductivity (C_w) estimates, Track 11 contains inversion-derived 95% HPD of surface conductance of clay (λ_c) estimates, and Track 12 contains formation CEC values interpreted by service company.

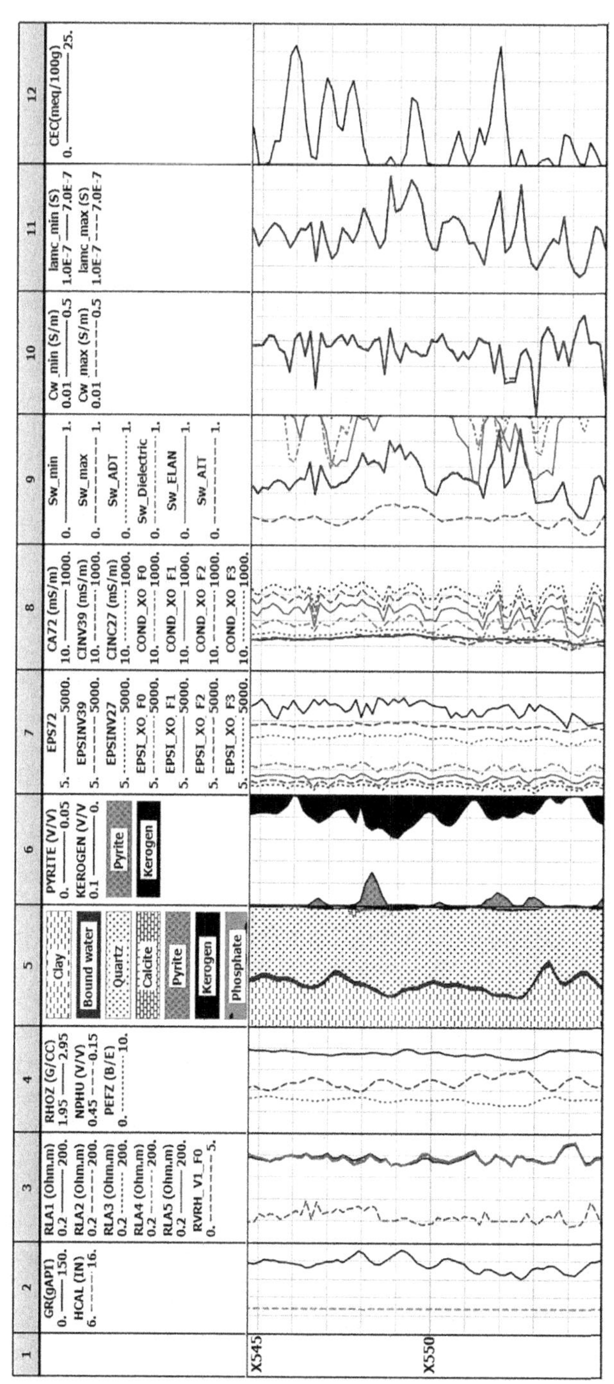

FIG. 13.12 Logs for the Zone 2 (depth X545 m–X555 m), which is a clay-rich and pyrite-bearing interval. The logs and layout presented in this figure are similar to those in Fig. 13.11.

transmitter-receiver spacing, the formation volume sensed by the EM log decreases; consequently, dielectric log at 1 GHz is the shallowest sensing log. The bulk density in those two intervals is around 2.67 g/cc, and the photoelectric factor is around 3, which indicates shale formation (Track 4). ELAN multimineral analysis was performed to continuously estimate total porosity and volumetric fraction of each mineral. The ELAN multimineral analysis is a joint inversion method combining deep and shallow resistivity, gamma-ray, photoelectric factor, bulk density, neutron porosity, sonic, and elemental spectroscopy (ECS) logs. The TOC estimates obtained from Rock-Eval pyrolysis and weight fractions of minerals at some discrete depths obtained from X-ray powder diffraction (XRD) measurements were also included as additional constraints in the ELAN multimineral analysis. The volumetric fractions of various minerals including quartz, clay, kerogen, pyrite, calcite, and phosphate are illustrated in Track 5 indicating that the formation predominantly comprises clay and quartz. The total porosity in the two zones estimated by ELAN multimineral analysis is around 3%–7%, which is not included in the logs in Figs. 13.11 and 13.12. The volumetric fractions of special minerals including pyrite and kerogen are separately illustrated in Track 6. The broadband relative permittivity and conductivity logs at seven log-acquisition frequencies computed from EM measurements [2] are presented in Track 7 and Track 8, which varies from 5 to 1800 and from 0.02 to 0.5 S/m, respectively. The ELAN multimineral analysis results indicate that the lithology is similar within the two zones. The volumetric fraction of clay in both zones is around 30%–50%. At depth X069m–X070m, there is 10%–20% volumetric fraction of calcite. The volumetric fractions of conductive pyrite is around 0%–2% and kerogen is around 1%–4%, which indicates a possibility of hydrocarbon in the two zones. The ratio of the horizontal resistivity (R_H) and vertical resistivity (R_V), which is approximate to 1 in both intervals (Track 3) indicates that the two zones are isotropic.

Tracks 9, 10, and 11 illustrate the inversion-derived estimates. Track 9 includes water saturation estimated using various methods, Tracks 10 and 11 contain inversion-derived estimates of the 95% HPD of connate water conductivity (S_w) and surface conductance of clay (lamc or λ_c), respectively. The ranges of estimated 95% HPD of all three parameters (e.g., S_w_min to S_w_max) are very narrow, as exhibited by the overlap of black continuous and blue dashed curves. The inversion-derived estimates of S_w, C_w, and λ_c in the two zones of interest in the shale gas formation are in the range of 0.2–0.9, 0.1–0.25 S/m, and $1 \times 10^{-7} - 1 \times 10^{-6}$ S, respectively. S_w estimates in the two zones computed using the stochastic inversion of broadband relative permittivity and conductivity logs is between the S_w estimates derived from EM induction resistivity log interpretation and those derived from dielectric dispersion logs interpretation and multimineral analysis solver. The S_w estimates derived from the dielectric dispersion logs interpretation using deterministic inversion of the logs coupled with joint SMD and LR model (S_w_ADT) [3] are close to those obtained by O&G service company's inversion of dielectric dispersion logs (S_w_Dielectric). S_w_ADT and S_w_Dielectric are all almost close to 1 and are typically 0.3–0.5 saturation unit higher than those obtained by the proposed stochastic inversion method, namely S_w_min and S_w_max.

We consider that the S_w_min and S_w_max obtained using the newly developed stochastic inversion is more accurate than the high values of S_w_ADT and S_w_Dielectric because the two zones contain 1%–4% volumetric fraction of kerogen, which tend to be oil-wet. The multimineral analysis solver also obtains higher S_w estimates (S_w_ELAN) than those obtained by the proposed method except for some intervals around X068 m–X069 m (in Zone 1) and X552 m–X553 m (in Zone 2) in which the two methods obtain similar values. The EM induction resistivity log was processed using the popular Simandoux model and the computed S_w_AIT is 0.2–0.4 saturation unit lower than those obtained using the proposed method in both zones. Frequency-independent Simandoux model accounts for the additional conductivity due to clay minerals but ignores the IP effects of clays and pyrite. We compare water saturation estimates by using various methods at two kerogen-rich (around 4%) depths: X072.8 m (in Zone 1) and X549 m (in Zone 2). The S_w estimates derived from EM induction resistivity log, the proposed stochastic inversion of broadband EM dispersion logs, and dielectric dispersion logs at depth X072.8 m is 0.2, 0.47, and 1, and at depth X549 m is 0.31, 0.77, and 1, respectively. The stochastic inversion-derived λ_c in Track 11 exhibits similar trend as the formation CEC values in Track 12 that are obtained by service company dielectric dispersion logs interpretation in both zones, especially in Zone 1. The variation of λ_c is between $1 \times 10^{-7} - 1 \times 10^{-6}$ S in both zones, and this variation can be explained by the variation of the mineral content (illite, chlorite, and kaolinite), which can be confirmed and validated by the ELAN multimineral analysis and XRD data at some discrete depths.

4 Conclusions

Downhole broadband EM dispersion logs were processed for the first time using a MCMC-sampling-based stochastic inversion method coupled with a clay-pyrite interfacial-polarization model to estimate hydrocarbon/water saturation in organic-rich shale gas formations. The newly developed stochastic inversion method simultaneously processes the EM broadband dispersion logs acquired by EM induction tool (26 kHz), EM propagation tool (1 and 2 MHz) and dielectric dispersion logs (20 MHz–1 GHz). The continuous estimations of formation hydrocarbon saturation, connate water conductivity, and surface conductance of clay with ranges of possible values (95% HPD) are obtained for two interested zones in organic-rich shale gas formation. The proposed MCMC-sampling-based stochastic inversion method is robust to noise in the EM logs and suitable for clay and pyrite-rich, low-porosity shale gas formations. The MCMC-based stochastic inversion coupled to clay-pyrite interfacial-polarization model (PS model) performs better in comparison with the Levenberg-Marquardt-based deterministic inversion coupled to the joint SMD and LR model. The hydrocarbon saturation, connate water conductivity, and surface conductance of clay estimates in two clay-rich and pyrite-bearing, homogenous, and isotropic zones in the shale gas formation are in the range of 0.2–0.9, 0.1–0.25 S/m, and $1 \times 10^{-7} - 1 \times 10^{-6}$ S, respectively. The hydrocarbon saturation estimates in the two zones of interest obtained using the proposed stochastic-inversion method are between the estimates derived using conventional interpretation that individually processes the EM induction resistivity log and those derived from deterministic-inversion of dielectric dispersion logs, conventional interpretation of dielectric permittivity log at 1 GHz, and multimineral analysis solver. Hydrocarbon saturation estimates based on the interpretations of dielectric dispersion or dielectric permittivity logs are close to 0 in the two interested zones and are typically 0.3–0.5 saturation unit lower than those obtained by the newly proposed stochastic-inversion of broadband EM logs. The two zones of interest contain 1%–4% (v/v) of kerogen, which tend to be oil wet; consequently, the hydrocarbon saturation estimates derived using the newly proposed stochastic inversion are more physically consistent as compared with those derived using conventional dielectric permittivity/dispersion interpretation and inversion methods.

References

[1] Karcz P, Janas M, Dyrka I. Polish shale gas deposits in relation to selected shale gas prospective areas of Central and Eastern Europe. Prz Geol 2013;61(7):411–23.

[2] Wang H, Poppitt A. The broadband electromagnetic dispersion logging data in a gas shale formation, In: SPWLA 54th annual logging symposiumSociety of Petrophysicists and Well-Log Analysts; 2013.

[3] Han Y, Misra S. Bakken petroleum system characterization using dielectric-dispersion logs. Petrophysics 2018;59(2):201–17.

[4] Tathed P, Han Y, Misra S. Hydrocarbon saturation in Bakken petroleum system based on joint inversion of resistivity and dielectric dispersion logs. Fuel 2018;233:45–55.

[5] Tathed P, Han Y, Misra S. Hydrocarbon saturation in upper Wolfcamp shale formation. Fuel 2018;219:375–88.

[6] Han Y, Misra S. Joint petrophysical inversion of multifrequency conductivity and permittivity logs derived from subsurface galvanic, induction, propagation, and dielectric dispersion measurements. Geophysics 2018;83(3):D97–112.

[7] Archie GE. The electrical resistivity log as an aid in determining some reservoir characteristic. Trans AIME 1942;146(1):54–62.

[8] Das B, Chatterjee R. Well log data analysis for lithology and fluid identification in Krishna-Godavari basin, India. Arab J Geosci 2018;11:231–42.

[9] Gogoi T, Chatterjee R. Estimation of petrophysical parameters using seismic inversion and neural network modeling in upper Assam basin, India. Geosci Front 2018; https://doi.org/10.1016/j.gsf.2018.07.002.

[10] Simandoux P. Dielectric measurements in porous media and application to shaly formation. Rev Inst Fr Pétrol 1963;18:193–215.

[11] Asquith GB. Log analysis by microcomputer. 1st ed. Tulsa: Petroleum Pub. Co; 1980.
[12] Sabouroux P, Ba D. Epsimu, a tool for dielectric properties measurement of porous media: application in wet granular materials characterization. Prog Electromagn Res B 2011;29:191–207.
[13] Brovelli A, Cassiani G. Effective permittivity of porous media: a critical analysis of the complex refractive index model. Geophys Prospect 2008;56(5):715–27.
[14] Stroud D, Milton GW, De BR. Analytical model for the dielectric response of brine-saturated rocks. Phys Rev B 1986;34:5145–53.
[15] Han Y, Misra S, Simpson G. Dielectric dispersion log interpretation in Bakken petroleum system. In: SPWLA 58th Annual Logging Symposium. Society of Petrophysicists and Well-Log Analysts; 2017.
[16] Pirie I, Horkowitz J, Simpson G, Hohman J. Advanced methods for the evaluation of a hybrid-type unconventional play: the Bakken petroleum system. Interpretation 2016;4(2):SF93–111.
[17] Donadille JM, Leech R, Pirie I. Water salinity determination over an extended salinity range using a joint interpretation of dielectric and neutron cross-section measurements, In: SPE annual technical conference and exhibitionSociety of Petroleum Engineers; 2016.
[18] Sanchez-Ramirez JA, Torres-Verdín C, Wolf D, Wang GL, Mendoza A, Liu Z, Schell G. Field examples of the combined petrophysical inversion of gamma-ray, density, and resistivity logs acquired in thinly-bedded clastic rock formations. Petrophysics 2010;51(4):247–63.
[19] Chen H, Heidari Z. Assessment of hydrocarbon saturation in organic-rich source rocks using combined interpretation of dielectric and electrical resistivity measurements, In: SPE annual technical conference and exhibitionSociety of Petroleum Engineers; 2014.
[20] Das PS, Chatterjee R, Dasgupta S, Das R, Bakshi D, Gupta M. Quantification and spatial distribution of pore-filling materials through constrained rock physics template and fluid response modelling in Paleogene clastic reservoir from Cauvery basin, India. Geophys Prospect 2018; https://doi.org/10.1111/1365-2478.12715.
[21] Misra S, Lüling MG, Rasmus J, Homan DM, Barber TD. Dielectric effects in pyrite-rich clays on multifrequency induction logs and equivalent laboratory core measurements, In: SPWLA 57th annual logging symposiumSociety of Petrophysicists and Well-Log Analysts; 2016.
[22] Misra S, Torres-Verdín C, Revil A, Rasmus J, Homan D. Interfacial polarization of disseminated conductive minerals in absence of redox-active species—part 1: mechanistic model and validation. Geophysics 2016;81(2):E139–57.
[23] Chen J, Kemna A, Hubbard SS. A comparison between Gauss-Newton and Markov-chain Monte Carlo-based methods for inverting spectral induced-polarization data for Cole-Cole parameters. Geophysics 2008;73(6):F247–59.
[24] Misra S, Torres-Verdín C, Revil A, Rasmus J, Homan D. Interfacial polarization of disseminated conductive minerals in absence of redox-active species—part 2: effective electrical conductivity and dielectric permittivity. Geophysics 2016;81(2):E159–76.
[25] Wu J, Mukerji T, Journel AG. Improving water saturation prediction with 4D seismic, In: SPE annual technical conference and exhibitionSociety of Petroleum Engineers; 2005.
[26] Minsley BJ. A trans-dimensional Bayesian Markov chain Monte Carlo algorithm for model assessment using frequency-domain electromagnetic data. Geophys J Int 2011;187(1):252–72.
[27] Mosegaard K, Tarantola A. Monte Carlo sampling of solutions to inverse problems. J Geophys Res 1995;100(B7):12431–47.
[28] Brookst SP. Markov chain Monte Carlo method and its application. J R Stat Soc 1998;47:69–100.
[29] Metropolis N, Rosenbluth AW, Rosenbluth MN, Teller AH, Teller E. Equation of state calculations by fast computing machines. J Chem Phys 1953;21(6):1087–92.

[30] Hastings WK. Monte Carlo sampling methods using Markov chains and their applications. Biometrika 1970;57(1):97–109.

[31] Malinverno A. Parsimonious Bayesian Markov chain Monte Carlo inversion in a nonlinear geophysical problem. Geophys J Int 2002;151(3):675–88.

[32] Aster RC, Borchers B, Thurber CH. Parameter estimation and inverse problems. 1st ed. Amsterdam: Elsevier Academic Press; 2005.

[33] Bérubé CL, Chouteau M, Shamsipour P, Enkin RJ, Olivo GR. Bayesian inference of spectral induced polarization parameters for laboratory complex resistivity measurements of rocks and soils. Comput Geosci 2017;105:51–64.

[34] Chib S, Greenberg E. Understanding the Metropolis-Hastings algorithm. Am Stat 1995;49 (4):327–35.

[35] Gelman A, Rubin DB. Inference from iterative simulation using multiple sequences. Stat Sci 1992;7(4):457–72.

[36] Anderson B, Barber T, Luling M. Observations of large dielectric effects on induction logs, or can source rocks be detected with induction measurements, In: SPWLA 47th annual logging symposiumSociety of Petrophysicists and Well-Log Analysts; 2006.

[37] Corley B, Garcia A, Maurer HM, Rabinovich MB, Zhou Z, DuBois P, Shaw N. Study of unusual responses from multiple resistivity tools in the Bossier formation of the Haynesville shale play, In: SPE annual technical conference and exhibitionSociety of Petroleum Engineers; 2010.

Index

Note: Page numbers followed by *f* indicate figures and *t* indicate tables.

D

N

O

P

R

T

U

V